Functional Methods in Quantum Field Theory and Statistical Physics

Functional Methods in Quantum Field Theory and Statistical Physics

A.N. Vasiliev
Department of Theoretical Physics
St Petersburg State University, Russia

Translated from the Russian by Patricia A. Millard

CRC Press
Taylor & Francis Group
Boca Raton London New York

CRC Press is an imprint of the
Taylor & Francis Group, an **informa** business

CRC Press
Taylor & Francis Group
6000 Broken Sound Parkway NW, Suite 300
Boca Raton, FL 3487-2742

First issued in paperback 2020

ISBN-13: 978-0-367-57926-5 (pbk)
ISBN-13: 978-90-5699-035-0 (hbk)

Visit the Taylor & Francis Web site at
http://www.taylorandfrancis.com

and the CRC Press Web site at
http://www.crcpress.com

CONTENTS

Foreword

I am pleased to present to English-speaking physicists this classic of the Russian physics literature, published more than 20 years ago in the USSR. The original Russian edition was very popular throughout the former USSR and Eastern Europe: the initial printing of 3000 copies sold out in one month and then the book became rare. However, it has remained almost completely unknown in the West until a few years ago, when it arrived with the recent wave of Russian physicists. Indeed, it is thanks to one of these physicists, Dr Alexander Bochkarev, that it was first brought to my attention.

Functional Methods in Quantum Field Theory and Statistical Physics is primarily a textbook designed for training young physicists for research in theoretical physics, or for theorists desiring to broaden their technical skills. The techniques it describes remain as important now as they were 20 years ago and continue to be extended into various subfields of, for example, condensed-matter physics. Indeed, this text contains the most complete exposition of the functional variational technique and Legendre transforms still to be found anywhere.

Finally, in this English edition the author has corrected some misprints, and has made a few changes and additions in order to reflect the recent progress in the field.

Patricia A. Millard
University of Pittsburgh, USA

Foreword

[illegible]

Preface

This book is based on a revised and expanded series of lectures given by the author over several years to fifth-year students specializing in theoretical physics at the Physics Faculty of St Petersburg State University (formerly Leningrad State University). It is intended for a well prepared reader familiar with the basics of quantum field theory and statistical physics.

The book is devoted not to physics as such, but to certain techniques used in physics. The term "functional methods" in the title refers to a set of technical methods which allow the quantum-mechanical operator formalism to be interpreted into the language of classical objects: nonlinear functionals. This language is very well suited to describing the expressions which typically arise in field theory, for example, Wick's theorem. In addition, it is universal and can be used in widely differing areas of theoretical physics: quantum field theory, Euclidean theory, the quantum statistics of field and spin systems, and the statistics of the classical nonideal gas.

In Chapter 1 we systematically discuss the mathematical apparatus of ordinary (pseudo-Euclidean) quantum field theory, and in the following chapters this formalism is made specific and generalized to the case of Euclidean theories and statistical physics. The short Chapter 3 is devoted to the theory of the massless Yang–Mills field, and the final Chapter 6, which makes up nearly a third of the book, is devoted to variational methods and functional Legendre transformations.

The material of Chapters 1, 2, 4, and 5, except for that in certain sections, is not new, but it cannot all be found in a single place. The material presented in these chapters would most accurately be thought of as the "folklore" of modern quantum field theory. Of course, this folklore did not appear spontaneously; it was created by the labor of many people, and I have not attempted to give any detailed list of references.

This book consists of six chapters divided into sections, most (but not all) of which are in turn divided into subsections. Throughout the book, parts of the text are referred to by giving the chapter, section, and, if applicable, subsection number. The displayed equations are numbered sequentially within each chapter by the chapter and equation number. The system of units with $\hbar = c = 1$ is used everywhere.

It is my pleasant task to thank everyone who helped me with this book: L.D. Faddeev and V.A. Franke, who read the manuscript and made a number of useful comments which were taken into account in the final version of the text; and A.G. Basuev, N.M. Bogolyubov, A.K. Kazanskii, A.V. Kuz'menko, Yu.M. Pis'mak, and R.A. Radzhabov, who provided essential assistance in the preparation of the original text. I am also grateful to all my colleagues and the graduate students in theoretical nuclear and elementary particle physics at the Physics Faculty of St Petersburg State University for many useful discussions about the topics dealt with here. But above all, the publication of the manuscript can be attributed to my wife, who created the necessary working conditions. It is to her that I dedicate this book.

I am pleased that this English translation is now available to a wider community of readers, and hope that it will prove useful to physicists studying quantum field technique.

Chapter 1

THE BASIC FORMALISM OF FIELD THEORY

1.1 Fields and Products

1.1.1 Canonical quantization

In classical physics the dynamics of a system with a finite number of degrees of freedom are determined by the Lagrangian $\mathcal{L}$, which is given as a function of the generalized coordinates $q = \{q_1 \ldots q_n\}$ and velocities $\dot{q}_i \equiv \partial_t q_i$. The classical equations of motion are obtained by requiring that the action functional $S = \int dt\, \mathcal{L}(t)$ be stationary.

The classical system is usually quantized using the canonical formalism, i.e., in the language of coordinates and momenta. The momenta p_i conjugate to the coordinates q_i are given by $p_i = \partial\mathcal{L}/\partial\dot{q}_i$. If these relations can be solved uniquely for the velocities, it is said that the Lagrangian is nondegenerate. In this case the classical Hamiltonian $\mathcal{H} = \Sigma p_i \dot{q}_i - \mathcal{L}$ can be viewed as a known function of the coordinates and momenta.

In going to the quantum theory the coordinates and momenta are realized as linear Hermitian operators $\hat{q}_i$ and $\hat{p}_i$ acting in some Hilbert space and satisfying the canonical commutation relations $\hat{q}_s\hat{p}_m - \hat{p}_m\hat{q}_s = i\delta_{sm}$. The quantum analog of a classical observable $F(p, q)$ is the operator $F(\hat{p}, \hat{q})$; in particular, the quantum energy operator, the Hamiltonian **H**, is by definition $\mathcal{H}(\hat{p}, \hat{q})$. These rules for going from the classical theory to the quantum theory are the well known recipe for canonical quantization.

A function of noncommuting operators $\mathcal{H}(\hat{p}, \hat{q})$ is not defined uniquely even when the natural additional requirement of Hermiticity is imposed. First of all, there is an arbitrariness in the ordering of the factors $\hat{p}$ and $\hat{q}$, i.e., an arbitrariness in the choice of how the classical Hamiltonian $\mathcal{H}(p, q)$ is written. For example, the identical classical functions $p^2 + q^2$ and $(p + iq)(p - iq)$ correspond to different operators when quantized. This implies that, in general, a given classical system corresponds to many different quantum systems, and the choice of any particular one is always made on the basis of an auxiliary condition.

In addition to this "purely algebraic" ambiguity, there is also an ambiguity in the choice of realization of the operators $\hat{p}$ and $\hat{q}$: the commutation relations do not determine it uniquely, but up to a unitary equivalence (the well known uniqueness theorem of von Neumann pertains only to the Weyl form of the commutation relations). The realization must be chosen on the basis of additional considerations regarding the nature of the canonical variables p and q: whether they are variables like Cartesian coordinates and momenta or variables like angles and angular momenta and so on. For the actual systems usually considered there is some natural realization which is always what is chosen. In the case of Cartesian coordinates and momenta this is the well known representation in terms of multiplication and differentiation operators.

The recipe for canonical quantization, with all these caveats, is easily generalized to systems with an infinite number of degrees of freedom, the generalized coordinates of which, as a rule, are described as a "field" $q(\mathbf{x})$ or a finite set of such fields. The argument $\mathbf{x}$ plays the role of a continuous index numbering the coordinates. The Lagrangian will now be not a function, but a functional of the coordinates $q(t, \mathbf{x})$ and velocities $\dot{q}(t, \mathbf{x})$ on the surface $t = \text{const}$. In the definition of the momentum the ordinary derivative is replaced by the variational derivative, and the δ in the commutation relations is replaced by the δ function [1]:

$$\hat{q}(\mathbf{x})\hat{p}(\mathbf{x}') - \hat{p}(\mathbf{x}')\hat{q}(\mathbf{x}) = i\delta(\mathbf{x}-\mathbf{x}'). \tag{1.1}$$

In this book the term "field" will be used to refer to the notation for the generalized coordinates of any system, independently of whether we are discussing a true field theory or a system with a finite number of degrees of freedom. In particular, the particle coordinate in mechanics is treated as a field depending only on time. To economize on notation and make the formulas more uniform we shall often use *universal notation*, in which the entire set of generalized coordinates is described by a unified field $\varphi(x) \equiv \varphi(t, \mathbf{x})$. Here the argument $\mathbf{x}$ denotes the set of continuous and discrete variables (indices), except for time, on which the field depends. For complex quantities, φ and φ^+ are assumed to be different components of a single field ($\varphi \equiv \varphi_1$, $\varphi^+ \equiv \varphi_2$), and the index distinguishing these components is included in the argument x. The symbol $\int d\mathbf{x}\ldots$ will denote integration over all continuous components and summation over all discrete components of $\mathbf{x}$, and the symbol $\delta(\mathbf{x}-\mathbf{x}')$ will denote the product of δ functions for all continuous components and Kronecker δ symbols for all discrete components of $\mathbf{x}$. Finally, $\delta(x-x') \equiv \delta(t-t')\delta(\mathbf{x}-\mathbf{x}')$ and $\int dx\ldots \equiv \int dt \int d\mathbf{x}\ldots$, where, unless stated otherwise, the integration over time is always understood to run from $-\infty$ to $+\infty$.

In this notation the commutation relation (1.1) holds for all systems.

1.1.2 The classical free theory

In this book the term "theory" is used in various ways: in a broad sense (quantum field theory, relativistic theory) and in a narrow sense, to refer to a particular system or class of systems with certain dynamics [the free theory, the $\lambda\varphi^4$ theory, the theory with action (...), and so on]. For example, the Ising model is a particular theory in this sense. (This terminology has become common in the specialized literature, since it makes it possible to avoid repetitions of constructions like "the system with interaction Lagrangian $\lambda\varphi^4$" and so on.)

Systems for which the Lagrangian is a quadratic form of the field or its derivatives of finite order will be referred to as *free*. For such Lagrangians the classical equation of motion is linear and can be written as

$$K\varphi = 0, \tag{1.2}$$

where K is some linear operator on a set of fields φ, the explicit form of which is determined by the Lagrangian. It is often convenient to take K to be a linear integral operator

$$[K\varphi]\,(x) \;=\; \int dx' K(x,x')\,\varphi(x') \tag{1.3}$$

with suitably chosen kernel $K(x, x')$.

We shall term a *linear operator K* as *t-local* if it is an ordinary differential operator of finite order in the variable t. We shall term a *functional* $F(t, \varphi)$ as *t-local* if it depends only on the field and its derivatives of finite order at a fixed instant of time t. Any Lagrangian is an example of a t-local functional, and the t-locality of the linear operator K in (1.2) follows from the t-locality of the Lagrangian. For a t-local operator the kernel $K(x, x')$ contains $\delta(t-t')$ and finite-order derivatives of it, so the integration over the time t' in (1.3) is removed.

The free action corresponding to (1.3) can formally be written as a quadratic form:

$$S_0'(\varphi) \;=\; \frac{1}{2}\varphi K\varphi \equiv \frac{1}{2}\iint dx dx' \varphi(x)\,K(x,x')\,\varphi(x')\,, \tag{1.4}$$

but actually to go from the usual expression $S_0 = \int dt \mathscr{L}_0(t)$ to the form (1.4) it is necessary to integrate by parts and discard the surface terms. For example, for a free particle in one-dimensional space $\varphi \equiv q(t)$, $\mathscr{L}_0 = m\dot{q}^2/2$, $K = -m\partial_t^2$, and $K(t, t') = -m\,\partial_t^2\delta(t-t')$. The expressions $2S_0 = m\int dt\dot{q}^2$ and

$2S_0' = -m\int dt\, q\ddot{q}$ obviously differ by a surface term. In a true field theory, in going to the form (1.4) it is usually necessary to integrate by parts not only with respect to time, but also with respect to the spatial arguments $\mathbf{x}$.

Equation (1.4) can be identified as the free action only when the latter is treated as a functional on a space of fields which fall off sufficiently rapidly, and the kernel K in this case can without loss of generality be assumed to be symmetric: $K = K^{\mathrm{T}}$, where K^{T} is the transposed operator, the kernel of which is obtained by interchanging the arguments of the original kernel $[K^{\mathrm{T}}(x, x') = K(x', x)]$. This definition of transposition is also applicable to differential operators, since they can be represented by integral operators with δ-like kernels, and it is equivalent to the following simple rule for transposition of the first derivative: $\partial^{\mathrm{T}} = -\partial$.

The equation $K = K^{\mathrm{T}}$ expresses the following: the operator K in (1.2) obtained from the requirement that the free action be stationary will always be a symmetric linear operator on a space of fields which fall off sufficiently rapidly.

Let us be more precise about what we mean by symmetry of the kernel for a complex field φ, φ^+. The form $\varphi^+ L\varphi$ with arbitrary kernel L can always be written as the half-sum $[\varphi^+ L\varphi + \varphi L^{\mathrm{T}}\varphi^+]/2$, which in universal notation $\varphi \equiv \Phi_1$, $\varphi^+ \equiv \Phi_2$ takes the form

$$\varphi^+ L\varphi = \frac{1}{2}\begin{pmatrix}\varphi \\ \varphi^+\end{pmatrix}\begin{pmatrix}0 & L^{\mathrm{T}} \\ L & 0\end{pmatrix}\begin{pmatrix}\varphi \\ \varphi^+\end{pmatrix} \equiv \frac{1}{2}\Phi K\Phi. \tag{1.5}$$

By definition, the transposition of a block operator is the transposition of the corresponding matrix accompanied by the transposition of each of its blocks. Clearly, the matrix kernel K in (1.5) is symmetric in this sense, independently of the properties of the kernel L.

1.1.3 Anticommuting fields

In quantum theory there are also systems whose operators corresponding to canonical variables satisfy not commutation relations (1.1), but anti-commutation relations:

$$\hat{q}(\mathbf{x})\hat{p}(\mathbf{x}') + \hat{p}(\mathbf{x}')\hat{q}(\mathbf{x}) = i\delta(\mathbf{x}-\mathbf{x}').$$

Such relations arise in the theory of identical particles obeying Fermi statistics. Historically, they were arrived at via the well known procedure of second quantization, but they can also be viewed as the natural recipe for the ordinary "first" quantization of not completely ordinary classical systems, the generalized coordinates of which are not simple functions but rather anticommuting functions. Here we shall give only the small amount of information on anticommuting quantities that we shall need in this text (more details can be found in [2]).

In the language of mathematics, a set of a finite number of pairwise anticommuting objects* $\varphi = \varphi_1 \ldots \varphi_n$ forms a set of generators of a finite-dimensional Grassmann algebra, and the algebra itself is defined as the set of all polynomials constructed from $\varphi_1 \ldots \varphi_n$. The number of independent monomials $\varphi_{i_1} \ldots \varphi_{i_k}$ is finite, because, first of all, any two monomials differing only by interchange of the factors coincide up to a sign and, second, no monomial can contain any of the generators φ_i twice: by permutations the two identical factors φ_i could be put next to each other and then it would be possible to use the equation $\varphi_i^2 = 0$ following from the anticommutativity of the generators. From this it follows that the monomial of highest degree is the product of all the generators φ_i. The total number of all independent monomials, including unity, is easily found to be 2^n, i.e., the Grassmann algebra with n generators is a finite-dimensional space of dimension 2^n and its general element can be written as a linear combination

$$f(\varphi) = f_0 + \sum_i f_1(i)\,\varphi_i + \sum_{i<k} f_2(i,k)\,\varphi_i\varphi_k + \ldots \tag{1.6}$$

with arbitrary numerical coefficients $f_k(i_1 \ldots i_k)$.

Ordinary functions of the type $\exp\varphi_i$ are understood as series which, as a rule, are truncated (for example, $\exp\varphi_i = 1 + \varphi_i$).

All odd monomials anticommute with each other, and every even monomial commutes with any other one.

For a Grassmann algebra we introduce the concept of derivatives with respect to φ_i, with a distinction made between "left" derivatives $(\overrightarrow{\partial}/\partial\varphi_i)$ and "right" derivatives $(\overleftarrow{\partial}/\partial\varphi_i)$. The action of the left derivative $\overrightarrow{\partial}/\partial\varphi_i$ on an arbitrary monomial can be determined as follows. If the given monomial does not contain the factor φ_i, the result is zero, while if it does contain φ_i (only once) this factor must be moved to the position farthest to the left by means of permutations and then deleted, or (equivalently) simply deleted followed by multiplication by the sign factor ± 1 according to whether the number of

---

* In this book we shall often encounter functions and functionals of many variables, and for compactness of notation we shall not write out the ellipsis when listing arguments [in, for example, "functional of the variables $\varphi_1 \ldots \varphi_n$" or $W_n(x_1 \ldots x_n)$]. In those relatively rare cases where we are speaking of a product of quantities rather than a list this will be clear from the context or will be indicated by the term "monomial" or "product."

permutations needed to move φ_i to the extreme left position is even or odd. The right derivative is defined analogously, except that here φ_i must be moved to the position farthest to the right. The derivatives are different, because the parities of the permutations to the left and right are in general different. It is easily checked that the left and right derivatives commute with each other, and that identical derivatives anticommute.

When this language is generalized to the case of a field $\varphi(x)$ the argument x is treated as a continuous index. The field φ represents a set of anticommuting generators of the Grassmann algebra [i.e., $\varphi(x)\varphi(x') = -\varphi(x')\varphi(x)$], and functionals $F(\varphi)$ are elements of this algebra. We introduce the left and right variational derivatives with respect to $\varphi(x)$, each of which acts on an individual factor $\varphi(x')$ according to the usual rule:

$$\vec{\delta}\varphi(x')/\delta\varphi(x) = \overleftarrow{\delta}\varphi(x')/\delta\varphi(x) = \delta(x-x') . \tag{1.7}$$

If the differentiated field is separated from the symbol $\delta/\delta\varphi$ by other factors, it must first be moved by permutations to one of the extreme positions (depending on which derivative occurs), and then differentiated according to the rule (1.7). Variational derivatives possess the same commutation properties as ordinary derivatives.

Let us list some useful formulas which we shall need later; their proof is left to the reader.

For any functional $F(\varphi)$ and any even n

$$\frac{\vec{\delta}}{\delta\varphi(x_1)}\cdots\frac{\vec{\delta}}{\delta\varphi(x_n)}F(\varphi) = \frac{\overleftarrow{\delta}}{\delta\varphi(x_n)}\cdots\frac{\overleftarrow{\delta}}{\delta\varphi(x_1)}F(\varphi) . \tag{1.8}$$

From this, taking into account the anticommutativity of like derivatives, we obtain

$$\vec{\delta}/\delta\varphi(x)\cdot\vec{\delta}/\delta\varphi(x') = -\overleftarrow{\delta}/\delta\varphi(x)\cdot\overleftarrow{\delta}/\delta\varphi(x') . \tag{1.9}$$

Now let $\varphi(x)$ and $A(x)$ be a pair of anticommuting (including with each other) fields. Then

$$\exp(-\varphi A)\,F(\vec{\delta}/\delta\varphi)\exp(\varphi A) = F(A+\vec{\delta}/\delta\varphi) , \tag{1.10}$$

$$\exp(A\vec{\delta}/\delta\varphi)\,F(\varphi) = F(\varphi+A) . \tag{1.11}$$

Here and below we use the abbreviated notation

$$\varphi A \equiv \int dx \varphi(x) A(x), \qquad A\vec{\delta}/\delta\varphi \equiv \int dx A(x) \vec{\delta}/\delta\varphi(x).$$

We note that the operation $\delta/\delta\varphi$ anticommutes with the factor A owing to the anticommutativity of A and φ. From (1.10) we have

$$F(\vec{\delta}/\delta\varphi) \exp\varphi A = F(A) \exp\varphi A. \qquad (1.12)$$

Equations (1.10)–(1.12) are direct generalizations of the usual rules for taking variational derivatives. The fact that these generalizations involve the left derivative is a consequence of the notation we have used: all operations are placed to the left of the functional and act sequentially—first the nearest one acts, then the next nearest, and so on.

In what follows systems with ordinary fields will be termed *bosonic* and those with anticommuting fields will be termed *fermionic*. The free equation of motion for fermionic fields also has the form (1.2), but the operator K, the kernel of the form (1.4), will now be antisymmetric rather than symmetric owing to the anticommutativity of the factors $\varphi(x)\varphi(x')$. For brevity we shall term the kernel K symmetric in both cases and write $K = \varkappa K^{T}$, always having in mind the symmetry consistent with the statistics. *Here and below throughout the book* $\varkappa = 1$ *for bosons and* $\varkappa = -1$ *for fermions*. For a complex fermionic field (and in real models the fermionic fields are always complex), the symmetry of the kernel is manifested, as in the bosonic case, only when universal notation is used:

$$\varphi^{+} L\varphi = \frac{1}{2}[\varphi^{+} L\varphi - \varphi L^{T}\varphi^{+}] = \frac{1}{2}\begin{pmatrix}\varphi \\ \varphi^{+}\end{pmatrix}\begin{pmatrix}0 & -L^{T} \\ L & 0\end{pmatrix}\begin{pmatrix}\varphi \\ \varphi^{+}\end{pmatrix}.$$

Integrals on the Grassmann algebra will be studied in Sec. 1.6.2.

1.1.4 The normal-ordered product of free-field operators

The following discussion pertains equally to relativistic quantum field theory, to nonrelativistic quantum mechanics in the second-quantized representation, and, finally, to the ordinary quantum mechanics of systems with a finite number of degrees of freedom. In all these cases the procedure for canonical quantization of the free theory is well known, and we shall not dwell on it. Additional information about various specific systems can be found in Chap. 2, and here we shall restrict ouselves to the following remarks.

The time dependence of the operators of a quantized free field in the interaction representation, which will henceforth be referred to simply as *free-field operators*, is determined by the quantum-mechanical evolution law $\hat{\varphi}(x) \equiv \hat{\varphi}(t, \mathbf{x}) = \exp(i\mathbf{H}_0 t)\,\hat{\varphi}(0, \mathbf{x})\exp(-i\mathbf{H}_0 t)$, where $\hat{\varphi}(0, \mathbf{x})$ is the field operator in the Schrödinger picture and $\mathbf{H}_0$ is the free Hamiltonian. At the same time the operator $\hat{\varphi}(x)$ always satisfies the classical equation of motion (1.2). The field operators at different points (anti)commute with each other to give a c-number.

Let us now turn to the definition of the concept of the *normal-ordered product*. We assume that the free-field operator $\hat{\varphi}(x)$ can be written as the sum of two terms $\hat{a}(x) + \hat{b}(x)$ such that $\hat{a}(x)\hat{a}(x') = \varkappa\hat{a}(x')\hat{a}(x)$, $\hat{b}(x)\hat{b}(x') = \varkappa\hat{b}(x')\hat{b}(x)$, and

$$\hat{a}(x)\hat{b}(x') - \varkappa\hat{b}(x')\hat{a}(x) = n(x, x'), \tag{1.13}$$

where $\varkappa = \pm 1$ depending on the statistics and $n(x, x')$ is some c-number function, henceforth referred to as the *simple contraction* of the field φ.

We shall say that the product of several factors $\hat{a}(x_i)$ and $\hat{b}(y_k)$ is in normal-ordered form if in this product any of the factors $\hat{b}$ is located to the left of any of the factors $\hat{a}$. The normal-ordered product of the factors $\hat{a}(x_1)\ldots\hat{a}(x_n)$, $\hat{b}(y_1)\ldots\hat{b}(y_m)$ arranged in any order is defined as the expression

$$\varepsilon_P \hat{b}(y_1)\ldots\hat{b}(y_m)\hat{a}(x_1)\ldots\hat{a}(x_n), \tag{1.14}$$

in which ε_P is a sign factor, always equal to 1 for a bosonic field. For a fermionic field $\varepsilon_P = 1$ if the original product is reduced to the product in (1.14) by an even number of permutations of the $n + m$ factors $\hat{a}$ and $\hat{b}$, and $\varepsilon_P = -1$ if the number of needed permutations is odd.

Denoting the normal-ordered product as N, we can write

$$N\{P[\hat{b}(y_1)\ldots\hat{b}(y_m)\hat{a}(x_1)\ldots\hat{a}(x_n)]\} =$$

$$= \varepsilon_P \hat{b}(y_1)\ldots\hat{b}(y_m)\hat{a}(x_1)\ldots\hat{a}(x_n),$$

where P denotes an arbitrary permutation of the factors $\hat{a}$ and $\hat{b}$.

This definition can be extended by linearity to any polynomial forms constructed from the operators $\hat{a}$ and $\hat{b}$, i.e.,

$$N\{\sum\prod\dots\} = \sum N\{\prod\dots\},N\{\alpha\prod\dots\} = \alpha N\{\prod\dots\},$$
$$N\{\alpha\} = \alpha, \tag{1.15}$$

where $\Pi\dots$ denotes any product of the factors $\hat{a}$ and $\hat{b}$ and α is an arbitrary constant (a c-number).

The rules (1.15) define the symbol N as a linear operator on a set of polynomial forms constructed from $\hat{a}$ and $\hat{b}$; each such form $\hat{F}$ is uniquely associated with a form $\hat{F}' = N\hat{F}$. However, it does not follow from this that N can be understood as a linear operation on a set of operators. There exist polynomials which are equal as operators, and from the operator equation $\hat{F}_1 = \hat{F}_2$ it does not follow that $N\hat{F}_1 = N\hat{F}_2$. A counter-example is the operator equation (1.13), for which

$$N[\hat{a}(x)\hat{b}(x') - \varkappa\hat{b}(x')\hat{a}(x)] = 0 \neq n(x,x') = N\{n(x,x')\}.$$

This shows that in the symbol $N\hat{F}$ it is in general not possible to replace the operator $\hat{F}$ by the operator $\hat{F}'$ equal to it.

Representing each factor of the product $\hat{\varphi}(x_1)\dots\hat{\varphi}(x_n)$ as a sum $\hat{a}+\hat{b}$ and using the rules (1.15), we thereby define the symbol $N[\hat{\varphi}(x_1)\dots\hat{\varphi}(x_n)]$. In particular, in this manner we obtain

$$N[\hat{\varphi}(x)\hat{\varphi}(x')] = \hat{\varphi}(x)\hat{\varphi}(x') - n(x,x'), \tag{1.16}$$

where n is the same function as in (1.13).

From the definition it is clear that the N-product is symmetric, i.e., the fields inside the N-product behave as classical objects—they commute in the case of bosons and anticommute in the case of fermions: $N\{P[\hat{\varphi}(x_1)\dots\hat{\varphi}(x_n)]\} = = \varepsilon_P N[\hat{\varphi}(x_1)\dots\hat{\varphi}(x_n)]$. Here P is an arbitrary permutation of the factors $\hat{\varphi}(x_i)$ and ε_P has the usual meaning.

1.2 Functional Formulations of Wick's Theorem

1.2.1 Wick's theorem for a simple product

The well known (see, for example, [1]) Wick's theorem specifies the rule for reducing the product $\hat{\varphi}(x_1)\dots\hat{\varphi}(x_n)$ to normal-ordered form. For $n = 1, 2$ we have

$$\hat{\varphi}(x) = N\hat{\varphi}(x), \qquad \hat{\varphi}(x)\hat{\varphi}(x') = N[\hat{\varphi}(x)\hat{\varphi}(x') + n(x,x')]. \quad (1.17)$$

The first of these equations is obvious, and the second is the same as (1.16).

We shall give a proof by induction of the validity of the following rule for reduction to the N-form:

$$\hat{\varphi}(x_1)\ldots\hat{\varphi}(x_n) = N\{\prod_{i<k}\left(1+\frac{\delta}{\delta\varphi_i}n\frac{\delta}{\delta\varphi_k}\right)\varphi_1(x_1)\ldots\varphi_n(x_n)\}\Big|_{\ldots}, \quad (1.18)$$

where the symbol $|_{\ldots}$, as below, stands for $\varphi_1 = \varphi_2 = \ldots = \varphi_n = \hat{\varphi}$. In this and the following expressions $\varphi(x)$ stands for the classical analog of $\hat{\varphi}(x)$, i.e., the ordinary classical field for bosons and the anticommuting field for fermions. The fields $\varphi_1 \ldots \varphi_n$ on the right-hand side of (1.18) are treated as independent functional arguments, and the differential quadratic forms are defined as

$$\frac{\delta}{\delta\varphi_i}n\frac{\delta}{\delta\varphi_k} \equiv \iint dx dx' \frac{\delta}{\delta\varphi_i(x)}n(x,x')\frac{\delta}{\delta\varphi_k(x')}.$$

For $i<k$ and $s<m$ we have

$$\frac{\delta}{\delta\varphi_i}n\frac{\delta}{\delta\varphi_k}\varphi_s(x)\varphi_m(x') = \delta_{is}\delta_{km}n(x,x'). \quad (1.19)$$

For anticommuting fields all the variational derivatives in (1.18) will be assumed to be right ones. Equation (1.19) is then valid for both types of statistics.

The objects inside the curly brackets in (1.18) are classical: they are ordinary functionals and their variational derivatives. As explicitly indicated in (1.18), all the classical fields φ_i remaining after differentiation must be replaced by the quantum operator $\hat{\varphi}$.

Turning now to the proof of the reduction formula (1.18), we note that it coincides with (1.17) for $n=1,2$. We shall assume that it holds for any number $k \le n$ of factors $\hat{\varphi}$, and shall prove that it is also valid when the number of factors is $n+1$. Using the more compact notation $\hat{\varphi}(x_i) \equiv \hat{\varphi}_i$, we write the last factor $\hat{\varphi}_{n+1}$ as the sum $\hat{a}_{n+1} + \hat{b}_{n+1}$. Using the commutation relations for the operators $\hat{a}$ and $\hat{b}$, we obtain

$$\hat{\varphi}_1 \ldots \hat{\varphi}_{n+1} = \hat{\varphi}_1 \ldots \hat{\varphi}_n \hat{a}_{n+1} + \varepsilon \hat{b}_{n+1} \hat{\varphi}_1 \ldots \hat{\varphi}_n +$$

$$+ \sum_{k=1}^{n} \varepsilon_k n (x_k, x_{n+1}) [\hat{\varphi}_1 \ldots \hat{\varphi}_n]_k, \tag{1.20}$$

where $[\hat{\varphi}_1 \ldots \hat{\varphi}_n]_k$ denotes the product $\hat{\varphi}_1 \ldots \hat{\varphi}_n$ without the factor $\hat{\varphi}_k$, and ε and ε_k are sign factors. For bosons they are always equal to 1, and for fermions $\varepsilon = (-1)^n$ and $\varepsilon_k = (-1)^{n-k}$.

The factors $\hat{a}$ and $\hat{b}$ in the first two terms on the right-hand side of (1.20) are arranged in the correct ("normal") order and, using the rule (1.18) for the n factors $\hat{\varphi}$, the sum of these two terms can be written as

$$N\{\mathcal{P}_n \varphi_1 \ldots \varphi_n|_{\ldots} \hat{a}_{n+1} + \varepsilon \hat{b}_{n+1} \mathcal{P}_n \varphi_1 \ldots \varphi_n|_{\ldots}\}, \tag{1.21}$$

where $\mathcal{P}_n$ is the *reduction operator* for n factors:

$$\mathcal{P}_n \equiv \prod_{i<k=1}^{n} \left(1 + \frac{\delta}{\delta\varphi_i} n \frac{\delta}{\delta\varphi_k}\right), \tag{1.22}$$

and the symbol $|_{\ldots}$ has the same meaning as in (1.18).

The expression in the curly brackets in (1.21) is an operator inside the N-product. Using the symmetry of the latter, the factor $\hat{b}_{n+1}$ can be moved to the extreme right-hand position. Here it is important that all the monomials in the operator polynomial $\mathcal{P}_n \varphi_1 \ldots \varphi_n|_{\ldots}$ have the same parity, since the fields are contracted in pairs. From this it follows that the permutation of this polynomial with $\hat{b}_{n+1}$ gives rise to an additional (for fermions) sign factor ε which is exactly the same as in (1.21), so that the operators $\hat{a}$ and $\hat{b}$ are grouped together in the extreme right-hand position to form the field operator $\hat{\varphi}_{n+1}$. This proves that Eq. (1.21) can be written as $N\{\mathcal{P}_n \varphi_1 \ldots \varphi_{n+1}\}|_{\ldots}$. We recall that the reduction operator $\mathcal{P}_n$ does not contain a derivative with respect to φ_{n+1}.

Let us now consider the sum over k on the right-hand side of (1.20). Using Eq. (1.19) and the reduction rule (1.18) for the product $[\hat{\varphi}_1 \ldots \hat{\varphi}_n]_k$ containing $n-1$ factors, we write this sum as

$$N \sum_{k=1}^{n} \varepsilon_k \mathcal{P}_n \left(\frac{\delta}{\delta\varphi_k} n \frac{\delta}{\delta\varphi_{n+1}} \varphi_k \varphi_{n+1}\right) [\varphi_1 \ldots \varphi_n]_k|_{\ldots} \,. \tag{1.23}$$

We have written the reduction operator for $[\hat{\varphi}_1 ... \hat{\varphi}_n]_k$ in the same form (1.22) as for the operator with n factors, because the extra terms containing derivatives with respect to φ_k do not contribute, as the field φ_k does not appear in the differentiated expression.

It is now necessary to return the factors φ_k and φ_{n+1} to their natural location. In the fermionic case, taking both derivatives to be right ones, we have

$$\frac{\delta}{\delta\varphi_k} n \frac{\delta}{\delta\varphi_{n+1}} \varphi_1 ... \varphi_{n+1} = (-1)^{n-k} \left(\frac{\delta}{\delta\varphi_k} n \frac{\delta}{\delta\varphi_{n+1}} \varphi_k \varphi_{n+1} \right) [\varphi_1 ... \varphi_n]_k .$$

The differentiation with respect to φ_{n+1} from the right does not introduce a sign factor, while the next differentiation with respect to φ_k from the right gives a factor $(-1)^{n-k}$ equal to the parity of the permutation of φ_k to the right. The sign factor ε_k in (1.23) is also equal to $(-1)^{n-k}$, which allows us to write (1.23) as

$$N \sum_{k=1}^{n} \mathscr{P}_n \left[\frac{\delta}{\delta\varphi_k} n \frac{\delta}{\delta\varphi_{n+1}} \right] \varphi_1 ... \varphi_{n+1} |_{...} .$$

Adding to this the expression obtained earlier for the sum of the first two terms on the right-hand side of (1.20), we conclude that the complete reduction operator for $n + 1$ factors is written as

$$\mathscr{P}_n \cdot \left[1 + \sum_{k=1}^{n} \frac{\delta}{\delta\varphi_k} n \frac{\delta}{\delta\varphi_{n+1}} \right] \cong \mathscr{P}_n \prod_{k=1}^{n} \left(1 + \frac{\delta}{\delta\varphi_k} n \frac{\delta}{\delta\varphi_{n+1}} \right)$$

(the equality $\cong$ is valid because the differentiated expression is linear in φ_{n+1}). Comparing this with the definition of the operator $\mathscr{P}_n$ in (1.22), we find that the operator given above is $\mathscr{P}_{n+1}$, and thus Eq. (1.18) is proved by induction.

We again note that in the fermionic case all variational derivatives are assumed to be right ones.

1.2.2 *The Sym-product and the T-product*

There are also other "products" in which the field operators behave like classical objects. First there is the ordinary *symmetrized product* of field operators:

$$\text{Sym}\,[\hat{\varphi}(x_1)\ldots\hat{\varphi}(x_n)] = \frac{1}{n!}\sum_P \varepsilon_P P\,[\hat{\varphi}(x_1)\ldots\hat{\varphi}(x_n)]\,. \tag{1.24}$$

The summation runs over all $n!$ permutations P of the operators $\hat{\varphi}$, and ε_P has the usual meaning. From this definition and from (1.17) we find

$$\text{Sym}\,[\hat{\varphi}(x)\,\hat{\varphi}(x')] = N[\hat{\varphi}(x)\,\hat{\varphi}(x')] + n_s(x,x')\,,$$

where n_s is the *symmetric part of the simple contraction* n:

$$n_s \equiv \frac{1}{2}(n + \varkappa n^{\mathrm{T}})\,. \tag{1.25}$$

The *time-ordered product* or *T-product* plays a very important role in field theory:

$$T[\hat{\varphi}(x_1)\ldots\hat{\varphi}(x_n)] = \sum_P \varepsilon_P P\,\{\theta(t_1\ldots t_n)\,\hat{\varphi}(x_1)\ldots\hat{\varphi}(x_n)\}\,. \tag{1.26}$$

Here and throughout this book we define $\theta_{tt'} \equiv \theta(t-t')$ and

$$\theta(t_1\ldots t_n) \equiv \prod_{k=1}^{n-1} \theta(t_k - t_{k+1})\,. \tag{1.27}$$

The summation in (1.26) runs over all simultaneous permutations of the factors $\hat{\varphi}(x_i)$ and times t_i in the θ function.

Equation (1.26) is uniquely (and symmetrically) defined only when all the times $t_1\ldots t_n$ are different, and when some or all of the arguments t_i are the same an auxiliary condition is needed. For this we require that the Sym-*product and the T-product be equal on the surface* $t = \text{const.}$* This is equivalent to supplementing the definition of the θ function (1.27) on its surfaces of discontinuity (the surfaces on which some subgroups of the arguments t_i coincide) by means of complete symmetrization relative to permutations of coincident arguments. In particular, for $t = t'$ we take $\theta_{tt'} = \theta_{t't} = 1/2$. With this auxiliary condition the T-product remains symmetric also when any subgroups of time arguments coincide.

---

* Sometimes the N-product rather than the Sym-product is used here; see, for example, [2].

For the field to the "zeroth" power the symbols T and Sym are defined using the usual auxiliary condition Sym 1 = T 1 = 1. We note that Sym and T, like N, are not actually linear operators on an operator space: neither $T\hat{F}_1 = T\hat{F}_2$ nor Sym $\hat{F}_1$ = Sym $\hat{F}_2$ follows from the operator equation $\hat{F}_1 = \hat{F}_2$.

The *time-ordered contraction* or *propagator* of the field $\hat{\varphi}$ is the c-number function $\Delta(x, x')$ defined by the relation

$$T[\hat{\varphi}(x)\hat{\varphi}(x')] = N[\hat{\varphi}(x)\hat{\varphi}(x')] + \Delta(x, x'). \tag{1.28}$$

From the definition (1.26) and Eq. (1.17) we obtain

$$\Delta(x, x') = \theta_{tt'} n(x, x') + \varkappa\theta_{t't} n(x', x). \tag{1.29}$$

We can write this compactly as $\Delta = \theta_{tt'} n + \varkappa\theta_{t't} n^{\mathrm{T}}$. By convention, for coincident times the contraction Δ is defined by the auxiliary condition

$$\Delta(x, x')|_{t=t'} = \frac{1}{2}[n(x, x') + \varkappa n(x', x)]|_{t=t'} = n_s(x, x')|_{t=t'}. \tag{1.30}$$

In contrast to n, the contraction Δ is always symmetric, i.e., $\Delta = \varkappa\Delta^{\mathrm{T}}$. The contractions have another important property: the simple contraction n satisfies the free equation (1.2) in each of its arguments, while the time-ordered contraction (1.29) is always the Green function (up to a factor i) of this equation*:

$$Kn = Kn^T = 0, \quad K\Delta = i. \tag{1.31}$$

The left-hand sides are understood as products of linear operators, and the i on the right-hand side is a multiple of the unit operator [the kernel of the latter is $\delta(x - x')$]. We shall not present the general proof of Eqs. (1.31), since they are easily verified for all the specific systems considered in the next chapter. A special case is that of a massless vector field, which is studied separately in Chap. 3.

---

* Often $i\Delta$ instead of Δ is written in (1.28), and then the equation $K\Delta = i$ takes the form $K\Delta = 1$. In this notation $\Delta = K^{-1}$ is the true Green function (without the i). We prefer the notation (1.28), because then the contraction $\Delta = iK^{-1}$ enters as a whole into all the expressions, but for brevity we shall refer to it simply as the Green function, understanding that this means "up to a factor of i".

1.2.3 Wick's theorem for symmetric products

The statement of Wick's theorem is simplified considerably if the symmetrized product of field operators rather than the simple product appears on the left-hand side.

Let us first consider the Sym-product. Reducing each term of the sum (1.24) to normal form using the rule (1.18), we obtain

$$\mathrm{Sym}\,[\hat{\varphi}(x_1)\ldots\hat{\varphi}(x_n)] = \\ = N\{\prod_{i<k}\left(1+\frac{\delta}{\delta\varphi_i}n\frac{\delta}{\delta\varphi_k}\right)\mathrm{Sym}\,[\varphi_1(x_1)\ldots\varphi_n(x_n)]\}\Big|_{\ldots} . \tag{1.32}$$

The symmetrization on the right-hand side is done with respect to all permutations of $x_1 \ldots x_n$ while preserving the order of the functional arguments $\varphi_1 \ldots \varphi_n$. For example,

$$\mathrm{Sym}\,[\varphi_1(x_1)\varphi_2(x_2)] = \frac{1}{2}[\varphi_1(x_1)\varphi_2(x_2)+\varkappa\varphi_1(x_2)\varphi_2(x_1)]\,. \tag{1.33}$$

The fields φ_i in (1.32) are classical objects, and by permutations of the factors φ_i we can return the arguments x_i to the original order $x_1 \ldots x_n$ in each term of the sum on the right-hand side of (1.32). The extra sign factor arising in the φ_i permutations obviously coincides with the factor ε_P in the definition (1.24) of the symmetrization operation, and these two factors cancel each other out. From this we see that the symmetrization on the right-hand side of (1.32) can also be understood as symmetrization of the bosonic type (all terms with plus sign) relative to permutations of the functional arguments $\varphi_1 \ldots \varphi_n$ while preserving the order $x_1 \ldots x_n$. In the example of (1.33),

$$\varphi_1(x_1)\varphi_2(x_2)+\varkappa\varphi_1(x_2)\varphi_2(x_1) = \varphi_1(x_1)\varphi_2(x_2)+\varphi_2(x_1)\varphi_1(x_2)\,.$$

We stress the fact that the sign on the right-hand side of this equation is independent of the statistics.

Let us now turn to the reduction operator and expand each of the differential quadratic forms into parts which are symmetric and antisymmetric under permutation of the indices:

$$\frac{\delta}{\delta\varphi_i}n\frac{\delta}{\delta\varphi_k} = \frac{1}{2}\left[\frac{\delta}{\delta\varphi_i}n\frac{\delta}{\delta\varphi_k}+\frac{\delta}{\delta\varphi_k}n\frac{\delta}{\delta\varphi_i}\right]+ \\ +\frac{1}{2}\left[\frac{\delta}{\delta\varphi_i}n\frac{\delta}{\delta\varphi_k}-\frac{\delta}{\delta\varphi_k}n\frac{\delta}{\delta\varphi_i}\right]. \tag{1.34}$$

Using the facts that (*i*) the differentiated functional is proved to be even under any permutation of the functional arguments φ_i, (*ii*) it is linear in each of the arguments φ_i, and (*iii*) after differentiation all the φ_i are assumed to be equal to the same operator $\hat{\varphi}$, we conclude that the odd parts of any of the forms (1.34) do not contribute to the final expression and can be discarded.

Therefore, each factor of the reduction operator can be written as

$$1+\frac{1}{2}\left[\frac{\delta}{\delta\varphi_i}n\frac{\delta}{\delta\varphi_k}+\frac{\delta}{\delta\varphi_k}n\frac{\delta}{\delta\varphi_i}\right]\cong\exp\frac{1}{2}\left[\frac{\delta}{\delta\varphi_i}n\frac{\delta}{\delta\varphi_k}+\frac{\delta}{\delta\varphi_k}n\frac{\delta}{\delta\varphi_i}\right]. \quad (1.35)$$

The replacement of this by an exponential is justified because the extra higher powers of derivatives do not contribute since the differentiated functional is linear in each of the arguments φ_i. The full reduction operator in (1.32) is the product of $i<k$ operators (1.35):

$$\mathscr{P} = \exp\left(\frac{1}{2}\sum_{i\neq k}\frac{\delta}{\delta\varphi_i}n\frac{\delta}{\delta\varphi_k}\right) . \quad (1.36)$$

Again using the linearity of the differentiated functional in φ_i, we can add diagonal terms with $i = k$ to the argument of the exponential in (1.36), after which the exponent takes the form of a "perfect square":

$$\sum_{ik}\frac{\delta}{\delta\varphi_i}n\frac{\delta}{\delta\varphi_k} = \left(\sum_i\frac{\delta}{\delta\varphi_i}\right)n\left(\sum_k\frac{\delta}{\delta\varphi_k}\right). \quad (1.37)$$

If we change from $\varphi_1 \ldots \varphi_n$ to new functional arguments, the average $\varphi \equiv (\varphi_1 + \ldots + \varphi_n)/n$ and the differences $\varphi_i - \varphi_{i+1}$, the form (1.37) reduces to the second derivative with respect to φ:

$$\frac{\delta}{\delta\varphi} = \sum_i\frac{\delta\varphi_i}{\delta\varphi}\cdot\frac{\delta}{\delta\varphi_i} = \sum_i\frac{\delta}{\delta\varphi_i}\ ;\quad \sum_{ik}\frac{\delta}{\delta\varphi_i}n\frac{\delta}{\delta\varphi_k} = \frac{\delta}{\delta\varphi}n\frac{\delta}{\delta\varphi}. \quad (1.38)$$

After performing the differentiation we must replace all the φ_i by $\hat{\varphi}$. Clearly, it is possible to set $\varphi_1 = \varphi_2 = \ldots = \varphi_n \equiv \varphi$ even before the differentiation, because the form (1.38) does not contain derivatives with respect to the differences $\varphi_i - \varphi_{i+1}$. This means that Wick's theorem (1.32) can be rewritten as follows:

$$\mathrm{Sym}\,[\hat{\varphi}(x_1)\ldots\hat{\varphi}(x_n)] = N\{\exp(\frac{1}{2}\frac{\delta}{\delta\varphi}n\frac{\delta}{\delta\varphi})\,\varphi(x_1)\ldots\varphi(x_n)\}\Big|_{\varphi=\hat{\varphi}}. \tag{1.39}$$

The symbol Sym on the right-hand side has been omitted because the product $\varphi(x_1)\ldots\ \varphi(x_n)$ is automatically symmetric. We also note that the contraction n in (1.39) can be replaced by its symmetric part (1.25), because the kernel of the differential quadratic form is automatically symmetrized owing to the commutativity of the derivatives in the bosonic case and the anti-commutativity of like derivatives (in this case, right ones) in the fermionic case:

$$\frac{\delta}{\delta\varphi_i}n\frac{\delta}{\delta\varphi_k} = \frac{\delta}{\delta\varphi_k}\varkappa n^{\mathrm{T}}\frac{\delta}{\delta\varphi_i}\,;\quad \frac{\delta}{\delta\varphi}n\frac{\delta}{\delta\varphi} = \frac{\delta}{\delta\varphi}n_s\frac{\delta}{\delta\varphi}. \tag{1.40}$$

Equation (1.39) is better than (1.18) in the sense that the reduction operator in (1.39) is universal, i.e., it is independent of the type of expression reduced.

Let us now turn to the T product. From the definition (1.26) and Eq. (1.18) we have

$$T[\hat{\varphi}(x_1)\ldots\hat{\varphi}(x_n)] = \tag{1.41}$$

$$= N\{\prod_{i<k}\left(1+\frac{\delta}{\delta\varphi_i}n\frac{\delta}{\delta\varphi_k}\right)\sum_P \varepsilon_P P[\theta(t_1\ldots t_n)\,\varphi_1(x_1)\ldots\varphi_n(x_n)]\}\Big|_{\ldots}.$$

The summation on the right-hand side runs over all simultaneous permutations of the arguments $x_1\ldots x_n$ and times $t_1\ldots t_n$ in the θ function with the order of the functional arguments $\varphi_1\ldots\ \varphi_n$ preserved. It is important to note that in each term of the sum over permutations the field with the higher index always has the smaller time, i.e., from $i<k$ it follows that the time argument t_i of the field φ_i is larger than that of the field φ_k. This follows directly from the definition of the T-product and (1.18): the time arguments of the fields decrease in going from left to right in all the terms of the sum (1.26), and the indices of the fields in (1.18) increase in going from left to right.

Let us first assume that among the times $t_1\ldots t_n$ there are none which coincide. It is then clear from the above discussion that in all the differential forms $(\delta/\delta\varphi_i)n(\delta/\delta\varphi_k)$ of the reduction operator in (1.41) it is possible to replace the contraction n by Δ, using the fact that for $t > t'$ these two contractions coincide. Then, using the symmetry of the contraction Δ, the form $(\delta/\delta\varphi_i)\Delta(\delta/\delta\varphi_k)$ can be symmetrized relative to interchange of the indices:

$$\frac{\delta}{\delta\varphi_i}\Delta\frac{\delta}{\delta\varphi_k} \to \frac{1}{2}\left[\frac{\delta}{\delta\varphi_i}\Delta\frac{\delta}{\delta\varphi_k} + \frac{\delta}{\delta\varphi_k}\Delta\frac{\delta}{\delta\varphi_i}\right].$$

Arguing as in the proof of the preceding theorem, the reduction operator can be represented as the exponential of a "perfect square," after which all the fields φ_i are set equal even before differentiation. Then the right-hand side of (1.41) takes the form

$$N\exp\left[\frac{1}{2}\cdot\frac{\delta}{\delta\varphi}\Delta\frac{\delta}{\delta\varphi}\right]\sum_P \varepsilon_P P\left[\theta(t_1 \ldots t_n)\,\varphi(x_1)\ldots\varphi(x_n)\right]\Bigg|_{\ldots} . \tag{1.42}$$

It should be noted that the product of classical fields is symmetric, i.e., $\varepsilon_P P[\varphi(x_1)\ldots\varphi(x_n)] = \varphi(x_1)\ldots\varphi(x_n)$, and so the sum over P in (1.42) is

$$\varphi(x_1)\ldots\varphi(x_n)\sum_P P\theta(t_1\ldots t_n) = \varphi(x_1)\ldots\varphi(x_n),$$

since the sum of all permutations of the θ function is unity. Therefore,

$$T[\hat{\varphi}(x_1)\ldots\hat{\varphi}(x_n)] =$$
$$=N\left\{\exp\left[\frac{1}{2}\cdot\frac{\delta}{\delta\varphi}\Delta\frac{\delta}{\delta\varphi}\right]\varphi(x_1)\ldots\varphi(x_n)\right\}\Bigg|_{\varphi=\hat{\varphi}} . \tag{1.43}$$

We have derived this expression assuming that all of the times $t_1\ldots t_n$ are different, but it can easily be checked that it is true also in the general case: if the contraction Δ in (1.43) for coincident times is given by (1.30), the right-hand side of (1.43) defines the T-product when arbitrary groups of time arguments coincide, just as in the preceding section.

Equation (1.43) is the functional statement of the Wick time-ordering theorem. It was obtained in this form by Hori [3].

1.2.4 Reduction formulas for operator functionals

Expressions of the form

$$F(\hat{\varphi}) = \sum_{n=0}^{\infty} \int \ldots \int dx_1 \ldots dx_n \; F_n(x_1 \ldots x_n)\, \hat{\varphi}(x_1) \ldots \hat{\varphi}(x_n)\,, \qquad (1.44)$$

which are specified by a set of their, possibly generalized, coefficient functions $F_n(x_1 \ldots x_n)$, will be called operator functionals. We shall term a functional symmetric if all its coefficient functions are symmetric. As always, here we mean the symmetry consistent with the statistics. A symmetric operator functional can be associated with a classical functional, replacing the free-field operator $\hat{\varphi}$ in (1.44) by the classical argument φ. The symmetric functions F_n are uniquely determined by the functional $F(\varphi)$. We also note that for a symmetric functional $F(\hat{\varphi}) = \text{Sym}\, F(\hat{\varphi})$.

To avoid misunderstandings, we again note that an operator functional (1.44) is specified by its coefficient functions, and not by the operator $F(\hat{\varphi})$. The functions F_n uniquely determine the operator $F(\hat{\varphi})$, but the converse is not true, because the free field $\hat{\varphi}$ is not a completely arbitrary argument, but one of the solutions of (1.2). The set of symmetric functions F_n is in one-to-one correspondence with the classical functional $F(\varphi)$, and to uniquely determine the operator $F(\hat{\varphi})$ it is sufficient to know the functional $F(\varphi)$ on the *mass shell*, i.e., on the set of all solutions of the free equation (1.2). The concepts "operator" and "operator functional" are not identical.

Equations (1.39) and (1.43) of the preceding section also can be generalized directly to operator functionals owing to the universality of the corresponding reduction operators. This is the real virtue of universality. The same is true for *t*-local functionals (see the definition in Sec. 1.1.2), if they depend only on the field on the surface $t = \text{const}$, and not on its time derivatives. The point is that we have taken special care to supplement the definition of the *T*-product of fields for coincident times, but we have not yet done this for the analogous expressions involving time derivatives of the fields (see the next section). Excluding such functionals from consideration for now, we rewrite the reduction formulas (1.39) and (1.43) as

$$\text{Sym}F(\hat{\varphi}) = N\exp\left[\frac{1}{2} \cdot \frac{\delta}{\delta\varphi} n \frac{\delta}{\delta\varphi}\right] F(\varphi)\Big|_{\varphi = \hat{\varphi}}, \qquad (1.45)$$

$$TF(\hat{\varphi}) = N\exp\left[\frac{1}{2} \cdot \frac{\delta}{\delta\varphi} \Delta \frac{\delta}{\delta\varphi}\right] F(\varphi)\Big|_{\varphi = \hat{\varphi}}. \qquad (1.46)$$

From this we easily find the expressions for the inverse transformations and all possible combined transformations. For example,

$$NF(\hat{\varphi}) = \text{Sym}\exp\left[-\frac{1}{2}\cdot\frac{\delta}{\delta\varphi}n\frac{\delta}{\delta\varphi}\right]F(\varphi)\bigg|_{\varphi=\hat{\varphi}}, \tag{1.47}$$

$$TF(\hat{\varphi}) = \text{Sym}\exp\left[\frac{1}{2}\cdot\frac{\delta}{\delta\varphi}(\Delta-n)\frac{\delta}{\delta\varphi}\right]F(\varphi)\bigg|_{\varphi=\hat{\varphi}}.$$

Starting from the general statement of Wick's theorem (1.18) and arguing exactly as in the derivation of (1.39), we obtain the following expression generalizing (1.45) to the case of the product of several symmetric operator functionals:

$$\prod_{i=1}^{n}[\text{Sym}F_i(\hat{\varphi})] = N\{\exp\left[\frac{1}{2}\sum_i\frac{\delta}{\delta\varphi_i}n\frac{\delta}{\delta\varphi_i}+\right.$$

$$\left.+\sum_{i<k}\frac{\delta}{\delta\varphi_i}n\frac{\delta}{\delta\varphi_k}\right]\prod_{i=1}^{n}F_i(\varphi_i)\}\bigg|_{\varphi_1=\ldots=\varphi_n=\hat{\varphi}}. \tag{1.48}$$

Here and below the noncommuting factors on the left-hand side are assumed to be arranged in increasing order:

$$\prod_{i=1}^{n}[\text{Sym}\ F_i(\hat{\varphi})] \equiv \text{Sym}\ F_1(\hat{\varphi})\cdot\ldots\cdot\text{Sym}\ F_n(\hat{\varphi}).$$

The diagonal terms of the quadratic form of the derivatives in (1.48) generate contractions inside the cofactors F_i which reduce them to normal-ordered form, and the nondiagonal terms give contractions between different cofactors.

Equation (1.48) has an obvious generalization to the case where several of the factors on the left-hand side are written in N-ordered, rather than Sym-ordered, form:

$$\prod_{i=1}^{n}[\mathscr{A}F_i(\hat{\varphi})] =$$

$$= N\{\exp\left[\frac{1}{2}\sum_{\text{Sym}}\frac{\delta}{\delta\varphi_i}n\frac{\delta}{\delta\varphi_i}+\sum_{i<k}\frac{\delta}{\delta\varphi_i}n\frac{\delta}{\delta\varphi_k}\right]\prod_{i=1}^{n}F_i(\varphi_i)\}\bigg|_{\ldots}. \tag{1.49}$$

Here $\mathcal{A}$ stands for Sym or N, and the summation in the diagonal terms of the quadratic form runs over only those fields which correspond to factors in the Sym-ordered form. If all the factors are written in N-ordered form, there are no diagonal terms.

If a product of operator functionals appears inside some kind of symmetric product it is automatically symmetrized, and the usual expressions (1.45) and (1.46) can be used. For example,

$$T\{\prod_{i=1}^{n} F_i(\hat{\varphi})\} = N\{\exp(\frac{1}{2}\cdot\frac{\delta}{\delta\varphi}\Delta\frac{\delta}{\delta\varphi})\prod_{i=1}^{n} F_i(\varphi)\}\Big|_{\varphi=\hat{\varphi}} . \quad (1.50)$$

However, it is sometimes convenient to explicitly split the contractions between different cofactors. This can be done by rewriting (1.50) as follows:

$$T\{\prod_{i=1}^{n} F_i(\hat{\varphi})\} =$$

$$= N\{\exp\left[\frac{1}{2}\sum_i \frac{\delta}{\delta\varphi_i}\Delta\frac{\delta}{\delta\varphi_i} + \sum_{i<k}\frac{\delta}{\delta\varphi_i}\Delta\frac{\delta}{\delta\varphi_k}\right]\prod_{i=1}^{n} F_i(\varphi_i)\}\Big|_{\ldots} . \quad (1.51)$$

The transformation from this representation to the previous one is accomplished using the usual trick based on (1.38).

By means of the simple Wick theorem (1.18), any (not necessarily symmetric) operator functional $F(\hat{\varphi})$ can be reduced to the form $NF_{(N)}(\hat{\varphi})$, and then (1.45) and (1.46) can be used to make the replacements $N\to\text{Sym}\to T$. The equations $F(\hat{\varphi}) = NF_{(N)}(\hat{\varphi}) = \text{Sym}\, F_{(\text{Sym})}(\hat{\varphi}) = TF_{(T)}(\hat{\varphi})$ define, in terms of the original $F(\hat{\varphi})$, the symmetric (by condition) functionals $F_{(N)}(\varphi)$, $F_{(\text{Sym})}(\varphi)$, and $F_{(T)}(\varphi)$, which we shall respectively refer to below as the N-form, the Sym-form, and the T-form of the original $F(\hat{\varphi})$. These three functionals are related to each other by expressions following from (1.45)–(1.47). We note that for any $F(\hat{\varphi})$, by definition $F(\hat{\varphi}) = \text{Sym}\, F_{(\text{Sym})}(\hat{\varphi}) = F_{(\text{Sym})}(\hat{\varphi})$ owing to the assumed symmetry of $F_{(\text{Sym})}(\hat{\varphi})$, but the equation $F(\varphi) = F_{(\text{Sym})}(\varphi)$ holds only for symmetric $F(\hat{\varphi})$.

1.2.5 The Wick and Dyson T-products

In this section we shall discuss the question of defining the T-product in those cases where the expression contains derivatives of the field with respect to the argument x, in particular, derivatives with respect to time. We define

$$T[\mathcal{D}_1(x_1)\dots\mathcal{D}_n(x_n)\hat{\varphi}(x_1)\dots\hat{\varphi}(x_n)]=$$
$$=\mathcal{D}_1(x_1)\dots\mathcal{D}_n(x_n)T[\hat{\varphi}(x_1)\dots\hat{\varphi}(x_n)]\,, \tag{1.52}$$

where $\mathcal{D}_i(x_i)$ are arbitrary differential operators acting on the arguments x_i. The T-product defined with this auxiliary condition is called the *Wick product*, distinguished from the Dyson T-product, which will be discussed below.

The expression inside the T-product on the left-hand side of (1.52) can be formally represented as an operator functional (1.44) with a singular kernel:

$$\int\dots\int dy_1\dots dy_n[\mathcal{D}_1^{\mathrm{T}}(y_1)\dots\mathcal{D}_n^{\mathrm{T}}(y_n)\delta(x_1-y_1)\dots\delta(x_n-y_n)]\times$$
$$\times\hat{\varphi}(y_1)\dots\hat{\varphi}(y_n)\,. \tag{1.53}$$

We recall that the transpose of the differential operator $\mathcal{D}(x)$ is given by the rule $\partial^{\mathrm{T}}=-\partial$. If we assume that (1.53) is an ordinary operator functional and write

$$T\int\dots\int dy_1\dots dy_n F(y_1\dots y_n)\hat{\varphi}(y_1)\dots\hat{\varphi}(y_n)=$$
$$=\int\dots\int dy_1\dots dy_n F(y_1\dots y_n)T[\hat{\varphi}(y_1)\dots\hat{\varphi}(y_n)]\,,$$

the right-hand side of this equation for the singular kernel (1.53) is exactly the right-hand side of (1.52). In other words, the rule (1.52) is equivalent to extending the reduction formula (1.46) to the case of functionals with any singular kernel, including to t-local functionals with time derivatives of the field. We note that an equation analogous to (1.52) is always satisfied for the N- and Sym-products.

Let us now define the *Dyson T-product*. Let $\hat{F}_i(t_i)$ be arbitrary time-dependent operators (other arguments, if they exist, are treated as fixed parameters), and each of the $\hat{F}_i$ be either a bosonic or a fermionic operator, but not a mixture of the two types. The Dyson T-product of these operators is defined as

$$T_{\mathrm{D}}[\hat{F}_1(t_1)\dots\hat{F}_n(t_n)]=\sum_P\varepsilon_P P[\theta(t_1\dots t_n)\hat{F}_1(t_1)\dots\hat{F}_n(t_n)]\,, \tag{1.54}$$

in which the summation runs over all simultaneous permutations of the times t_i in the θ function and the operators $\hat{F}_i(t_i)$ as a whole, and the sign factor ε_P

is defined, as usual, as the parity of the permutation of fermion factors. In the special case where each of the $\hat{F}_i$ is a free-field operator $\hat{\varphi}(x_i)$, the definition (1.54) coincides with the usual Wick T-product (1.26). However, in those cases where the $\hat{F}_i$ contain time derivatives of the field, the Dyson T-product (1.54) will not coincide with the Wick T-product defined by (1.52).

For example, let $\hat{F}_1(t) = \hat{\varphi}(x)$ and $\hat{F}_2(t') = \partial_{t'}\hat{\varphi}(x')$. Then

$$T_D[\hat{\varphi}(x)\partial_{t'}\hat{\varphi}(x')] = \theta_{tt'}\hat{\varphi}(x)\partial_{t'}\hat{\varphi}(x') + \varkappa\theta_{t't}\partial_{t'}\hat{\varphi}(x')\hat{\varphi}(x),$$

while

$$T[\hat{\varphi}(x)\partial_{t'}\hat{\varphi}(x')] = \partial_{t'}T[\hat{\varphi}(x)\varphi(x')] =$$

$$= \partial_{t'}[\theta_{tt'}\hat{\varphi}(x)\hat{\varphi}(x') + \varkappa\theta_{t't}\hat{\varphi}(x')\hat{\varphi}(x)].$$

These two expressions clearly differ, because the differentiation with respect to t' in the second produces additional terms containing derivatives of θ functions.

1.3 The *S* Matrix and Green Functions

1.3.1 Definitions

Let us now consider interacting quantum systems, for which $\mathbf{H} = \mathbf{H}_0 + \mathbf{V}$, where $\mathbf{H}$ and $\mathbf{H}_0$ are respectively the full and the free Hamiltonians and $\mathbf{V}$ is the interaction Hamiltonian.

The dynamics of the quantum system can be described by a unitary evolution operator $\mathbf{U}(\tau_1, \tau_2)$ in the interaction picture. This operator is the solution of the equation $i\partial\mathbf{U}(\tau_1, \tau_2)/\partial\tau_1 = \mathbf{V}(\tau_1)\mathbf{U}(\tau_1,\tau_2)$, in which $\mathbf{V}(\tau)$ is the interaction Hamiltonian in the interaction picture: $\mathbf{V}(\tau) = \exp(i\mathbf{H}_0\tau) \times \mathbf{V}\exp(-i\mathbf{H}_0\tau)$. The solution of the equation of motion for $\mathbf{U}$ with the initial condition $\mathbf{U}(\tau, \tau) = 1$ is written as

$$\mathbf{U}(\tau_1, \tau_2) = e^{i\mathbf{H}_0\tau_1}e^{i\mathbf{H}(\tau_2-\tau_1)}e^{-i\mathbf{H}_0\tau_2} =$$

$$= \sum_{n=0}^{\infty}(-i)^n\int\ldots\int dt_1\ldots dt_n\theta(t_1\ldots t_n)\mathbf{V}(t_1)\ldots\mathbf{V}(t_n). \tag{1.55}$$

The integration over each time t_i runs from τ_2 to τ_1.

By definition, the limit of the evolution operator $\mathbf{U}(\tau_1,\tau_2)$ for $\tau_1 \to \infty$ and $\tau_2 \to -\infty$, $\mathbf{U} = \mathbf{U}(+\infty, -\infty)$, is called the *S*-matrix operator $\mathbf{U}$.

Mathematically, $\mathbf{U}(\tau_1,\tau_2)$ is a well defined unitary operator when $\mathbf{H}_0$ and $\mathbf{H}$ are Hermitian (not simply symmetric) operators. For translationally invariant systems, in particular, relativistic ones, as a rule the operator $\mathbf{H}$ does not have this property. However, even if the operator $\mathbf{U}(\tau_1,\tau_2)$ is well defined, the limit $\mathbf{U} \equiv \mathbf{U}(\infty, -\infty)$ may not exist in a mathematically rigorous sense. The necessary condition for this limit to exist is that the spectra of the operators $\mathbf{H}_0$ and $\mathbf{H}$ coincide, and as a rule this condition is not satisfied. Therefore, in most cases of practical importance the operators that we consider are only certain formal constructions from which it is nevertheless possible to extract all the needed physical information about the system.

Let us now return to Eq. (1.55). After symmetrizing the general term of the series with respect to all permutations of the times $t_1 \ldots t_n$, we express it in terms of the Dyson *T*-product of the operators $\mathbf{V}(t_i)$:

$$\int \ldots \int dt_1 \ldots dt_n \theta(t_1 \ldots t_n) \mathbf{V}(t_1) \ldots \mathbf{V}(t_n) =$$

$$= \frac{1}{n!} \int \ldots \int dt_1 \ldots dt_n T_{\mathrm{D}} [\mathbf{V}(t_1) \ldots \mathbf{V}(t_n)] \, , \tag{1.56}$$

and the entire series (1.55) can then be written as a Dyson *T*-ordered exponential:

$$\mathbf{U}(\tau_1, \tau_2) = T_{\mathrm{D}} \exp \left[-i \int_{\tau_2}^{\tau_1} dt \mathbf{V}(t) \right] . \tag{1.57}$$

Equation (1.56) is true only when $\mathbf{V}(t)$ is a bosonic operator, which is what we shall always assume, although in practice one encounters cases where this is not so. For example, the interaction operator of a two-level system with quantized electromagnetic field is written in the form $a^+b + b^+a$, where b^+ and b are the fermionic raising and lowering operators of the two-level system and a^+ and a are bosonic operators. The interaction in this case is a fermionic operator and the Volterra series (1.55) does not reduce to the Dyson *T*-ordered exponential.

An important object of study in field theories are the Green functions of a field, which are defined as

$$H_n(x_1 \ldots x_n) = (0| T_{\mathrm{D}} [\hat{\varphi}_{\mathrm{H}}(x_1) \ldots \hat{\varphi}_{\mathrm{H}}(x_n)] |0) \, . \tag{1.58}$$

Here $|0)$ denotes the ground state of the full Hamiltonian, the "physical vacuum," which is assumed to be nondegenerate and normalized to unity; $\hat{\varphi}_H(x)$ is the field operator in the Heisenberg picture $\hat{\varphi}_H(t, \mathbf{x}) = \exp(i\mathbf{H}t)\,\hat{\varphi}(0, \mathbf{x})\exp(-i\mathbf{H}t)$. The symbol T_D is defined in the preceding section.

1.3.2 Transformation to the interaction picture in the evolution operator

In field theories the operator $\mathbf{V}(t)$ is given explicitly in the form of a certain t-local operator functional. In this and in the following sections it is assumed that the operator functional $\mathbf{V}(t)$ does not contain time derivatives of the field. We shall refer to such functionals as *simple*.

Any operator functional can be written in N-ordered form, after which it is possible, if desired, to transform to the Sym-ordered form using Eq. (1.47). In the discussion which follows, a special role will be played by the symmetric functional $\mathcal{V}(t, \hat{\varphi})$, which is the Sym-form of the interaction operator: $\mathbf{V}(t) = \mathcal{V}(t, \hat{\varphi}) = \text{Sym}\ \mathcal{V}(t, \hat{\varphi})$. By assumption, $\mathcal{V}(t, \varphi)$ is a simple t-local functional.*
The quantity

$$S_v(\tau_1, \tau_2; \varphi) \equiv -\int_{\tau_2}^{\tau_1} dt\,\mathcal{V}(t, \varphi) \tag{1.59}$$

will be called the *interaction functional* in the interval $\tau_2 \le t \le \tau_1$ for a given quantum interaction $\mathbf{V}$. The interaction functional on the entire time axis will be denoted $S_v(\varphi)$.

Let us immediately emphasize the fact that S_v is treated simply as some classical image of the given quantum interaction, and not as a characteristic of the classical system whose canonical quantization leads to the quantum theory in question. The functional S_v determines the operator $\mathbf{V}$ uniquely, while the canonical quantization in general is not unique.

We can transform to the interaction picture in (1.57) on the basis of the following statement: for any set $\hat{F}_1(t_1)\ldots\hat{F}_n(t_n)$ of simple t-local operator functionals,

---

*As noted in Sec. 1.2.4, a classical functional corresponds to a quantum operator only up to an arbitrary addition which vanishes on the mass shell, i.e., on the set of solutions of the equation $K\varphi = 0$. However, for ordinary K such an addition must necessarily contain time derivatives of the field, so in the class of simple functionals the Sym-form of $\mathbf{V}$ is determined uniquely.

$$T_D[\hat{F}_1(t_1)\ldots\hat{F}_n(t_n)] = T[\hat{F}_1(t_1)\ldots\hat{F}_n(t_n)]. \tag{1.60}$$

The symbol T on the right-hand side denotes the Wick T-product, in which all operators $\hat{F}_i$ are assumed to be represented by their Sym-forms: $\hat{F}_i(t_i) = \mathrm{Sym}\,\mathcal{F}_i(\hat{\varphi}) = \mathcal{F}_i(\hat{\varphi})$. The need for this representation arises from our choice of definition of the T-product for coincident times (Sec. 1.2.2).

To prove (1.60), we reduce both sides of this equation to N-form and verify that the resulting expressions are identical.

The right-hand side of (1.60) is reduced to N-form using the rule (1.50):

$$T[\hat{F}_1(t_1)\ldots\hat{F}_n(t_n)] =$$
$$= N\{\exp\left[\frac{1}{2}\cdot\frac{\delta}{\delta\varphi}\Delta\frac{\delta}{\delta\varphi}\right]\mathcal{F}_1(\varphi)\ldots\mathcal{F}_n(\varphi)\}\Big|_{\ldots}. \tag{1.61}$$

In order to reduce the left-hand side of (1.60) to N-form, we use the definition (1.54), and by means of the rule (1.48) we reduce each term in the sum over permutations to N-form. As a result, we arrive at the expression

$$N\{\exp\left[\frac{1}{2}\sum_i\frac{\delta}{\delta\varphi_i}n\frac{\delta}{\delta\varphi_i} + \sum_{i<k}\frac{\delta}{\delta\varphi_i}n\frac{\delta}{\delta\varphi_k}\right]\times$$
$$\times\sum_P\varepsilon_P P[\theta(t_1\ldots t_n)\mathcal{F}_1(\varphi_1)\ldots\mathcal{F}_n(\varphi_n)]\}\Big|_{\ldots}, \tag{1.62}$$

in which the summation runs over all simultaneous permutations of the times t_i in the θ function and the subscripts of the functionals $\mathcal{F}_i$ while preserving the order of the functional arguments $\varphi_1\ldots\varphi_n$; ε_P is determined, as usual, by the parity of the permutation of the fermionic factors.

The further arguments are nearly identical to the proof of the Wick time-ordering theorem (1.43). The main consideration is that in all the terms of the sum over permutations in (1.62) the arguments φ_i are ordered with index increasing from left to right, and this ordering exactly coincides with the ordering of decreasing times in the T-product. Therefore, in the off-diagonal terms of the quadratic form of the derivatives the contraction n can be replaced by Δ, followed by symmetrization in $i \rightleftarrows k$ exactly as was done in the proof of Eq. (1.43):

$$\frac{\delta}{\delta\varphi_i}n\frac{\delta}{\delta\varphi_k} \to \frac{\delta}{\delta\varphi_i}\Delta\frac{\delta}{\delta\varphi_k} = \frac{1}{2}\left[\frac{\delta}{\delta\varphi_i}\Delta\frac{\delta}{\delta\varphi_k} + \frac{\delta}{\delta\varphi_k}\Delta\frac{\delta}{\delta\varphi_i}\right]. \tag{1.63}$$

The diagonal terms of the quadratic form of the derivatives were unimportant in the proof of (1.43) because they did not contribute. Now they are important, but in these terms it is also possible to replace n by Δ using, first, the fact that the kernel of the form $(\delta/\delta\varphi_i)n(\delta/\delta\varphi_i)$ is automatically symmetrized and, second, the fact that these operators act on simple t-local functionals and therefore the results involve only the values of n_s on the surface $t = t'$, where, according to (1.30), n_s and Δ coincide. This proves that the quadratic form of the derivatives in (1.62) can be written as a perfect square like (1.38); then, repeating word for word the corresponding part of the proof of Wick's theorem (1.43) we obtain the desired result.

Using Eq. (1.60) for the case where each of the $\hat{F}_i(t_i)$ is the interaction operator $\mathbf{V}(t_i)$, we represent the evolution operator and the S matrix as a Wick T-ordered exponential, which can then be reduced to normal-ordered form using the rule (1.46):

$$\mathbf{U}(\tau_1, \tau_2) = T \exp iS_v(\tau_1, \tau_2; \hat{\varphi}) = \\ = N \exp \left(\frac{1}{2}\frac{\delta}{\delta\varphi}\Delta\frac{\delta}{\delta\varphi}\right) \exp iS_v(\tau_1, \tau_2; \varphi)\Big|_{\varphi = \hat{\varphi}}. \tag{1.64}$$

In these equations $S_v(\tau_1, \tau_2; \varphi)$ is the classical interaction functional (1.59), the Sym-form of the quantum interaction.

1.3.3 Transformation to the interaction picture for the Green functions

Let us consider the transformation from the Dyson T-product to the Wick T-product in the Green functions (1.58), assuming as before that the interaction operator does not contain time derivatives of the field.

Let $\hat{F}_i(t_i)$ be certain simple t-local operator functionals and $\hat{F}_{i\mathrm{H}}(t_i)$ be the same operators in the Heisenberg picture:

$$\hat{F}_{i\mathrm{H}}(t) = e^{i\mathbf{H}t} e^{-i\mathbf{H}_0 t} \hat{F}_i(t) e^{i\mathbf{H}_0 t} e^{-i\mathbf{H}t} = \mathbf{U}(0, t)\hat{F}_i(t)\mathbf{U}(t, 0). \tag{1.65}$$

Here $\hat{F}_i(t)$ is the operator in the interaction picture and $\mathbf{U}$ is the evolution operator (1.55). By definition,

$$T_{\mathrm{D}}[\hat{F}_{1\mathrm{H}}(t_1)\ldots\hat{F}_{n\mathrm{H}}(t_n)] = \sum_P \varepsilon_P P[\theta(t_1\ldots t_n)\hat{F}_{1\mathrm{H}}(t_1)\ldots\hat{F}_{n\mathrm{H}}(t_n)].$$

Using Eq. (1.65) and the group property $\mathbf{U}(t_1, t_2)\mathbf{U}(t_2, t_3) = \mathbf{U}(t_1, t_3)$ for the evolution operator, we obtain

$$\hat{F}_{1\mathrm{H}}(t_1)\ldots\hat{F}_{n\mathrm{H}}(t_n) =$$

$$= \mathbf{U}(0,t_1)\hat{F}_1(t_1)\mathbf{U}(t_1,t_2)\hat{F}_2(t_2)\ldots\hat{F}_n(t_n)\mathbf{U}(t_n,0)\,. \qquad (1.66)$$

Let τ_1 and τ_2 be arbitrary numbers such that $\tau_2 \le t_i \le \tau_1$ for any of the times t_i in the product (1.66). Using the equations $\mathbf{U}(0, t_1) = \mathbf{U}(0, \tau_1)\mathbf{U}(\tau_1, t_1)$ and $\mathbf{U}(t_n, 0) = \mathbf{U}(t_n, \tau_2)\mathbf{U}(\tau_2, 0)$, we obtain

$$T_{\mathrm{D}}[\hat{F}_{1\mathrm{H}}(t_1)\ldots\hat{F}_{n\mathrm{H}}(t_n)] = \mathbf{U}(0,\tau_1)M(t_1\ldots t_n;\tau_1,\tau_2)\mathbf{U}(\tau_2,0)\,,$$

where

$$M(t_1\ldots t_n;\tau_1,\tau_2) \equiv \sum_P \varepsilon_P P[\theta(t_1\ldots t_n)\mathbf{U}(\tau_1,t_1)\times$$

$$\times\hat{F}_1(t_1)\mathbf{U}(t_1,t_2)\ldots\hat{F}_n(t_n)\mathbf{U}(t_n,\tau_2)\,]\,. \qquad (1.67)$$

The summation in the last expression runs over all simultaneous permutations of the operators $\hat{F}_i(t_i)$ as a whole and the times t_i in the θ function and in the evolution operators $\mathbf{U}$.

We shall prove that the right-hand side of (1.67) is

$$T[\mathcal{F}_1(\hat{\varphi})\ldots\mathcal{F}_n(\hat{\varphi})\ \exp\ iS_{\mathrm{v}}(\tau_1,\tau_2;\hat{\varphi})\,]\,, \qquad (1.68)$$

where T denotes the Wick T-product, S_{v} is the interaction functional (1.59), and $\mathcal{F}_i(\varphi)$ are symmetric functionals corresponding to the Sym-form of the operators $\hat{F}_i(t_i)$ in the interaction picture: $\hat{F}_i(t_i) = \mathrm{Sym}\,\mathcal{F}_i(\hat{\varphi}) = \mathcal{F}_i(\hat{\varphi})$.

The equality of (1.68) and (1.67) is proved just as for the analogous equation (1.60), namely, the right-hand sides of (1.67) and (1.68) are reduced to N-form and it is checked that the results coincide.

The general term in (1.67) is the product of n operators $\hat{F}_i(t_i) = \mathcal{F}_i(\hat{\varphi})$ and $n+1$ evolution operators $\mathbf{U}$, each of which can be written in N-form using (1.64). Each term of the sum over permutations (1.67) can be reduced to N-form using (1.49), introducing $2n+1$ independent arguments φ_i which will be arranged in the reduction formula with the index increasing from left to right. The argument of the exponential of the complete reduction operator will look like:

$$\frac{1}{2}\sum_i{}' \frac{\delta}{\delta\varphi_i}\Delta\frac{\delta}{\delta\varphi_i}+\frac{1}{2}\sum_i{}'' \frac{\delta}{\delta\varphi_i}n\frac{\delta}{\delta\varphi_i}+\sum_{i<k}\frac{\delta}{\delta\varphi_i}n\frac{\delta}{\delta\varphi_k}, \tag{1.69}$$

and in the reduced functional each of the operators $\hat{F}_i$ must be replaced by $\mathcal{F}_i$ and each of the operators U by exp iS_v for the corresponding time interval. In the form (1.69) the first sum (with the prime) runs over the $n+1$ fields φ_i which are the arguments of the classical interaction functionals S_v, and the factor exp (1/2 Σ'...) is the contribution of all the reduction operators (1.64) for the evolution operators **U**. The second group of diagonal terms (the sum with two primes) is the contribution of the reduction operators to the N-form of all the symmetric functionals $\mathcal{F}_i$; the t-locality of these functionals together with Eq. (1.30) allows n to be replaced by Δ in these terms, exactly as in the proof of (1.60). As far as the off-diagonal terms in (1.69) are concerned, the usual arguments according to which the field with highest index has the smallest time remain valid also for the right-hand side of (1.67), because the time arguments of the fields in the functional $\mathbf{U}(t_1, t_2)$ are located between t_1 and t_2 (here the assumption $\tau_2 \le t_i \le \tau_1$ is important). This allows us to substitute (1.63) in the off-diagonal terms of Eq. (1.69), so that the latter becomes a perfect square and all the fields φ_i can, as usual, be set equal even before differentiation. The product of the factors exp iS_v for all the evolution operators in (1.67) is then combined to form the complete exponential for the interval $\tau_2 \le t \le \tau_1$; the sum over permutations, as in the proof of (1.43), leads to the substitution $\theta \to 1$, and in the end we arrive at the right-hand side of (1.68).

The statement we have proved can be written as follows:

$$T_{\rm D}\,[\hat{F}_{1\rm H}(t_1)\ldots\hat{F}_{n\rm H}(t_n)\,] =$$

$$= \mathbf{U}(0,\tau_1)\,T[\hat{F}_1(t_1)\ldots\hat{F}_n(t_n)\ \exp\ iS_v(\tau_1,\tau_2;\hat{\varphi})\,]\,\mathbf{U}(\tau_2,0)\ . \tag{1.70}$$

We note that in deriving (1.45), (1.46), (1.60), and (1.70) we have essentially used only the fact that for $t > t'$ the contractions $\Delta(x, x')$ and $n(x, x')$ coincide, and for $t = t'$ they are related by (1.30). In the chapters which follow we shall study Euclidean field theory and quantum statistics. In these theories the contractions Δ and n will be different, but the functional form of the reduction operators and the equalities $\Delta = n$ for $t > t'$ and $\Delta = n_s$ for $t = t'$ will be preserved. Therefore, the proof of Eqs. (1.45), (1.46), (1.60), and (1.70) remains valid also for these theories.

Let us now consider the expectation value of the operator (1.70) in the true ground state $|0)$. We assume that the following asymptotic expressions of nonstationary perturbation theory for a discrete nondegenerate level are well known: for $\tau_1 \to \infty$ and $\tau_2 \to -\infty$

$$\mathbf{U}(\tau_2, 0)\,|0) \cong \alpha(\tau_2)\,|0\rangle;\ \mathbf{U}(\tau_1, 0)\,|0) \cong \beta(\tau_1)\,|0\rangle. \tag{1.71}$$

Here $|0\rangle$ is the ground state of the free Hamiltonian, which is assumed to be nondegenerate, and α and β are phase factors. The proof of Eq. (1.71) and the generalization to the case of a degenerate level are given in Appendix 1.

For $\tau_1 \to \infty$ and $\tau_2 \to -\infty$ we have

$$(0|\,\mathbf{U}(0, \tau_1)\ldots\mathbf{U}(\tau_2, 0)\,|0) \cong \beta^*(\tau_1)\alpha(\tau_2)\langle 0|\ldots|0\rangle. \tag{1.72}$$

The meaning of the phase factor $\beta^*(\tau_1)\alpha(\tau_2)$ is easily found from (1.71): $\beta(\tau_1)\alpha^*(\tau_2) \cong \langle 0|\,\mathbf{U}(\tau_1, \tau_2)\,|0\rangle$. Therefore, for $\tau_1 \to \infty$, $\tau_2 \to -\infty$

$$0\,|\,T_D[\hat{F}_{1H}(t_1)\ldots\hat{F}_{nH}(t_n)]\,|0) =$$

$$= \frac{\langle 0|\,T\{\hat{F}_1(t_1)\ldots\hat{F}_n(t_n)\ \exp iS_v(\hat{\varphi})\}\,|0\rangle}{\langle 0|\,T\exp iS_v(\hat{\varphi})\,|0\rangle}, \tag{1.73}$$

where S_v is the interaction functional (1.59) on the entire time axis.*

The denominator on the right-hand side of (1.73) is the vacuum expectation value of the S matrix $\mathbf{U} \equiv \mathbf{U}(\infty, -\infty)$. This quantity determines ΔE_0, the energy shift of the ground state when the interaction is switched on, because for any nondegenerate discrete level for $\tau_1 - \tau_2 \to \infty$

$$\ln\langle 0|\,\mathbf{U}(\tau_1, \tau_2)\,|0\rangle = -i\Delta E_0(\tau_1 - \tau_2) + 0(1). \tag{1.74}$$

The proof of this asymptotic expression is given in Appendix 1. The statement (1.74) as applied to the S matrix will be written as $\langle 0|\mathbf{U}|0\rangle = \exp[-i\Delta E_0 \int dt]$.

The numerator on the right-hand side of (1.73) is called the *full Green function* of the operators $\hat{F}_{iH}$, while the quotient (1.73) is called the *Green*

---

* The S-matrix operator is sometimes written in this expression instead of $\exp iS_v(\hat{\varphi})$. This notation is careless: strictly speaking, "T-operator" has no meaning at all, in contrast to "T-operator functional" (cf. the remark in Sec. 1.2.1).

function without vacuum loops. The meaning of this terminology will become clear later. The quantities

$$G_n(x_1...x_n) \equiv \langle 0|\,T\{\hat{\varphi}(x_1)\ldots\hat{\varphi}(x_n)\exp iS_v(\hat{\varphi})\}|0\rangle \qquad (1.75)$$

are by definition the full Green functions of the field $\hat{\varphi}_{\mathrm{H}}$. The "zeroth-order function" $G_0 \equiv \langle 0|\,T\exp iS_v(\hat{\varphi})\,|0\rangle$ is a number, the vacuum expectation value of the S matrix. The functions

$$H_n(x_1...x_n) = (0\,|T_{\mathrm{D}}[\hat{\varphi}_{\mathrm{H}}(x_1)\ldots\hat{\varphi}_{\mathrm{H}}(x_n)]\,|0) = G_0^{-1}G_n(x_1...x_n) \qquad (1.76)$$

defined by (1.58) can now be refered to as the Green functions without vacuum loops of the field $\hat{\varphi}_{\mathrm{H}}$.

1.3.4 Interactions containing time derivatives of the field

In obtaining Eqs. (1.60) and (1.70) reducing Dyson T-products to Wick products, we have assumed that there are no time derivatives of the field in the interaction operator $\mathbf{V}(t)$. Now we consider the general case where $\mathbf{V}(t)$ is a t-local functional depending not only on the field $\hat{\varphi}(x)$ itself, but also on derivatives $\hat{\varphi}_n \equiv \partial_t^n\hat{\varphi}$ of finite order. The general case reduces immediately to the one we have studied if each of the derivatives $\hat{\varphi}_n$ is assumed to be an independent field and we define the matrix of contractions of the set of fields $\hat{\varphi}_n$ in terms of the Dyson T-product:

$$\Delta_{nm}(x,x') = T_{\mathrm{D}}[\hat{\varphi}_n(x)\hat{\varphi}_m(x')] - N[\hat{\varphi}_n(x)\hat{\varphi}_m(x')]\,. \qquad (1.77)$$

The symbol T in this expression can also be understood as the ordinary Wick T-product of the independent fields $\hat{\varphi}_n$, thereby treating (1.77) as a special case of the general definition (1.28). This means that in the language of a system of independent fields $\hat{\varphi}_n$, all the standard reduction formulas remain valid (the index n, if desired, can be included in the argument x). In particular, it is possible to use Eqs. (1.60) and (1.70), since in the language of the fields φ_n the interaction functional $S_v(\varphi_0, \varphi_1, \ldots)$, which is the Sym-form of the quantum interaction, does not contain time derivatives of the fields. Therefore,

$$\mathbf{U} = T\exp\, iS_v(\hat{\varphi}_0, \hat{\varphi}_1, \ldots) =$$

$$=N\exp\left(\frac{1}{2}\sum_{nm}\frac{\delta}{\delta\varphi_n}\Delta_{nm}\frac{\delta}{\delta\varphi_m}\right)\exp iS_v(\varphi_0,\varphi_1,\dots)\Big|_{\dots}. \tag{1.78}$$

The symbol $|_{\dots}$ indicates that after the differentiation each of the classical fields φ_i must be replaced by the quantum field $\hat{\varphi}_i$.

The resulting theory can also be formulated in the language of a single field $\varphi \equiv \varphi_0$ by changing the interaction functional in a certain way. To do this, we isolate from the contraction matrix Δ_{nm} the "Wick part" Δ'_{nm}:

$$\Delta'_{nm}(x,x') \equiv \partial_t^n\partial_{t'}^m\Delta(x,x')\,; \quad \Delta_{nm} = \Delta'_{nm} + a_{nm}. \tag{1.79}$$

The extra piece a_{nm} is t-local, i.e., it contains the factor $\delta(t - t')$ and finite-order derivatives of it. It is also clear that $a_{00} = 0$.

Let us now define the functional $S'_v(\varphi_0, \varphi_1, \dots)$ as

$$\exp iS'_v(\varphi_0,\varphi_1,\dots) =$$

$$= \exp\left(\frac{1}{2}\sum_{nm}\frac{\delta}{\delta\varphi_n}a_{nm}\frac{\delta}{\delta\varphi_m}\right)\exp iS_v(\varphi_0,\varphi_1,\dots). \tag{1.80}$$

The S-matrix functional in N-form (1.78) is expressed in terms of S'_v as

$$\mathbf{U} = N\exp\left(\frac{1}{2}\sum_{nm}\frac{\delta}{\delta\varphi_n}\Delta'_{nm}\frac{\delta}{\delta\varphi_m}\right)\exp iS'_v(\varphi_0,\varphi_1,\dots)\Bigg|_{\varphi_i=\hat{\varphi}_i}. \tag{1.81}$$

The final thing to note is that this expression can be viewed as the reduction formula of the ordinary Wick T-exponential

$$\mathbf{U} = N\exp\left(\frac{1}{2}\cdot\frac{\delta}{\delta\varphi}\Delta\frac{\delta}{\delta\varphi}\right)\exp iS_v^{\text{eff}}(\varphi)\Big|_{\varphi=\hat{\varphi}} \tag{1.82}$$

for the interaction

$$S_v^{\text{eff}}(\varphi) = S'_v(\varphi_0,\varphi_1,\dots)\Big|_{\varphi_k=\partial_t^k\varphi}. \tag{1.83}$$

In fact, let $\hat{\varphi}_n = K_n\hat{\varphi}$ and $\Delta'_{nm} = K_nK_m\Delta = K_n\Delta K_m^{\mathrm{T}}$ where K_n are arbitrary linear operators (for us $K_n = \partial_t^n$). We have

$$\mathcal{D} \equiv \sum_{nm} \frac{\delta}{\delta\varphi_n} \Delta'_{nm} \frac{\delta}{\delta\varphi_m} = \sum_{nm} \frac{\delta}{\delta\varphi_n} K_n \Delta K^{\mathrm{T}}_m \frac{\delta}{\delta\varphi_m} =$$

$$= \sum_{nm} \left(K^{\mathrm{T}}_n \frac{\delta}{\delta\varphi_n} \right) \Delta \left(K^{\mathrm{T}}_m \frac{\delta}{\delta\varphi_m} \right),$$

i.e., the differential form reduces to a perfect square. From this we find

$$\mathcal{D}F(\varphi_0, \varphi_1, \ldots)\Big|_{\varphi_n = K_n\varphi} = \frac{\delta}{\delta\varphi} \Delta \frac{\delta}{\delta\varphi} F(K_0\varphi, K_1\varphi, \ldots),$$

since

$$\frac{\delta}{\delta\varphi} = \sum_n \frac{\delta\varphi_n}{\delta\varphi} \cdot \frac{\delta}{\delta\varphi_n} = \sum_n K^{\mathrm{T}}_n \frac{\delta}{\delta\varphi_n}.$$

The quantity $\delta/\delta\varphi_n$ on the right-hand side is understood as the partial derivative with respect to $\varphi_n = K_n\varphi$, and $\delta/\delta\varphi$ on the left-hand side is the "total derivative." These arguments prove Eq. (1.82).

A closed expression for the *effective interaction* (1.83) can be obtained when the original interaction functional $S_v(\varphi_0, \varphi_1, \ldots)$ has sufficiently simple (linear or quadratic) dependence on the derivatives φ_n, $n \geq 1$. We note that the effective interaction defined by (1.80) and (1.83) will always have Lagrangian form, i.e., it will be an integral over time of some t-local functional, a Lagrangian. But this Lagrangian is, in general, complex rather than real, as in the case usually encountered (see Sec. 1.5.2 regarding the unitarity of the S-matrix).

It should also be noted that for most cases of practical importance the scheme we have described of constructing the effective interaction is too general. Actually, the free-field operator from which the interaction is constructed satisfies a t-local equation of motion (1.2). For ordinary Lagrangians this equation is either first- or second-order in the time derivative (see Chap. 2 for more details). In the first-order equation the derivative $\hat{\varphi}_1$ is expressed in terms of the field $\hat{\varphi}$ itself or its derivatives with respect to other arguments, and in this case it can be assumed without loss of generality that the interaction does not contain any time derivatives of the field at all. In the second-order equation the second derivative of the field with respect to time can be eliminated in exactly the same way, and it can be assumed that any interaction is expressed only in terms of the field operator and its first derivative with respect to time.

Up to now we have considered only the quantum theory, assuming **V** to be some given operator functional whose origin we are not interested in. Now let us say a few words about the quantization of classical systems whose interaction Lagrangian contains time derivatives of a field. Without loss of generality, it can be stated that the Lagrangian depends only on first derivatives with respect to time, i.e., velocities. If the Lagrangian is nondegenerate (the special features of the quantization of systems with degenerate Lagrangians are described in, for example, [4]), it can be used to uniquely determine the classical interaction Hamiltonian $\mathscr{V}$ as a function of coordinates and momenta. In quantization in the Schrödinger picture, q and p are replaced by operators, with different arrangements of noncommuting factors corresponding to different operators $\mathbf{V} = \mathscr{V}(\hat{q}, \hat{p})$, i.e., different "quantization recipes" (see the example in Sec. 2.6). The operator realization of the coordinates and momenta in the Schrödinger picture can always be taken to be the same as for the free theory. Then in going to the interaction picture the time dependence of these operators will be determined by the equations of the free theory, so that the momentum operators in the interaction picture can be expressed in terms of time derivatives of the coordinate operators (of the free field). As a result, we obtain the explicit representation of the interaction operator $\mathbf{V}(t)$ in the form of a t-local operator functional of the free field, and the further study of the quantum theory proceeds as described above. Different versions of the quantization corresponding to different operators $\mathbf{V}(t)$ will be associated with different functionals S_v^{eff}, since this functional determines $\mathbf{V}(t)$ uniquely. In particular cases the choice of quantization recipe is limited by various additional considerations, for example, the requirement of relativistic invariance in relativistic theories.

1.3.5 Generating functionals for the S*-matrix and Green functions*

The *functional*

$$R(\varphi) = \exp\left(\frac{1}{2}\cdot\frac{\delta}{\delta\varphi}\Delta\frac{\delta}{\delta\varphi}\right)\exp iS_v(\varphi) \tag{1.84}$$

representating the S*-matrix operator in normal-ordered form* will be referred to as the S-matrix generating functional. For an interaction involving time derivatives of the fields, S_v in this expression should be understood as the effective interaction studied in the preceding section. By definition, the S-matrix operator is $NR(\hat{\varphi})$. The field $\hat{\varphi}$ satisfies the free equation (1.2), from which we see that to define the S matrix as an operator it is sufficient to know the functional $R(\varphi)$ restricted to the set of φ satisfying (1.2). This restricted functional will be referred to as the *on-shell S-matrix functional*, in contrast to the functional (1.84), which provides the *off-shell* representation of the S matrix.

Let us also define the generating functional for the full Green functions (1.75):

$$G(A) = \sum_{n=0}^{\infty} \frac{1}{n!} G_n (iA)^n \equiv$$

$$\equiv \sum_{n=0}^{\infty} \frac{i^n}{n!} \int \dots \int dx_1 \dots dx_n G_n (x_1 \dots x_n) A(x_n) \dots A(x_1) . \qquad (1.85)$$

The argument of the functional $A(x)$ is a classical object of the same nature as the field $\varphi(x)$, i.e., an ordinary function for bosons and an anticommuting function for fermions. In the latter case only functions G_n with even n are nonzero.

Using Eqs. (1.75) and (1.76), we obtain

$$G(A) = G_0 (0| T_{\mathrm{D}} \exp i\hat{\varphi}_{\mathrm{H}} A| 0) = \langle 0| T \exp i [S_v(\hat{\varphi}) + \hat{\varphi} A] |0\rangle . \qquad (1.86)$$

All the notation is standard and needs no explanation.

Usually the N-product is defined such that $\langle 0| N [\hat{\varphi}(x_1) \dots \hat{\varphi}(x_n)] |0\rangle = 0$ for any $n \geq 1$. Then for an arbitrary functional

$$\langle 0| NF(\hat{\varphi}) |0\rangle = F(0) . \qquad (1.87)$$

Assuming this to be true, from (1.86) and (1.46) we obtain

$$G(A) = \exp \left[\frac{1}{2} \cdot \frac{\delta}{\delta \varphi} \Delta \frac{\delta}{\delta \varphi} \right] \exp i [S_v(\varphi) + \varphi A] \bigg|_{\varphi = 0} . \qquad (1.88)$$

By convention, in the fermionic case the derivatives in the differential reduction operators are assumed to be right ones. This is convenient for writing out the reduction formulas (1.45) and (1.46), which turn out to be identical for the two statistics. But here we are planning to use Eqs. (1.10)–(1.12), which requires transforming to left derivatives; we see from (1.9) that this leads only to a change of sign in the exponent of the reduction operator. After this replacement, using (1.12) we find the functional (1.88) for the free theory:

$$G^{(0)}(A) = \exp\left[\frac{\varkappa}{2}\cdot\frac{\vec{\delta}}{\delta\varphi}\Delta\frac{\vec{\delta}}{\delta\varphi}\right]\exp i\varphi A\bigg|_{\varphi=0} = \exp\left[-\frac{\varkappa}{2}A\Delta A\right]. \quad (1.89)$$

Similarly,

$$\exp i\,[S_v(\varphi) + \varphi A] = \exp iS_v(\varkappa\vec{\delta}/\delta iA)\ \exp i\varphi A. \quad (1.90)$$

Combining these expressions, we obtain

$$G(A) = \exp iS_v(\varkappa\,\vec{\delta}/\delta iA)\ \exp\,[-\varkappa A\Delta A/2]\,. \quad (1.91)$$

We note that all these equations remain valid also for an interaction involving time derivatives if $S_v(\varphi)$ is understood to be the effective interaction. Let us also write down the useful relation obtained by comparing (1.88) and (1.91) for $A = 0$:

$$\exp\left(\frac{\varkappa}{2}\cdot\frac{\vec{\delta}}{\delta\varphi}\Delta\frac{\vec{\delta}}{\delta\varphi}\right)F(\varphi)\bigg|_{\varphi=0} = F\left(\varkappa\frac{\vec{\delta}}{\delta\varphi}\right)\exp\left(\frac{\varkappa}{2}\varphi\Delta\varphi\right)\bigg|_{\varphi=0}. \quad (1.92)$$

There is a simple relation between the functionals R and G. Using Eq. (1.10), we obtain

$$G(A) = \exp\left[\frac{\varkappa}{2}\left(iA + \frac{\vec{\delta}}{\delta\varphi}\right)\Delta\left(iA + \frac{\vec{\delta}}{\delta\varphi}\right)\right]\exp iS_v(\varphi)\bigg|_{\varphi=0},$$

from which, taking into account the symmetry of Δ and the definition (1.84), we find

$$G(A) = \exp\left(-\frac{\varkappa}{2}A\Delta A\right)\exp\left[(i\Delta A)\frac{\vec{\delta}}{\delta\varphi}\right]R(\varphi)\bigg|_{\varphi=0}.$$

According to (1.11), the remaining differential operator gives a shift:

$$G(A) = \exp\left(-\frac{\varkappa}{2}A\Delta A\right)R(i\Delta A) = G^{(0)}(A)\,R(i\Delta A)\,. \quad (1.93)$$

This is the desired expression showing that knowledge of the off-shell S matrix is equivalent to knowledge of all the Green functions. Making the replacement $A \to -i\Delta^{-1}A$ in (1.93), we obtain the inverse relation:

$$R(A) = \exp\left(-\frac{1}{2}A\Delta^{-1}A\right) G(-i\Delta^{-1}A). \tag{1.94}$$

The absence of the factor $\varkappa$ in the Gaussian exponential should be noted.

In conclusion, we note that for a complex field ψ, ψ^{+} the Green-function generating functional is usually defined as $G(a^{+}, a) = \langle 0|\; T\exp i[S_v(\hat{\psi}^{+}, \hat{\psi}) + \hat{\psi}^{+}a + a^{+}\hat{\psi}]\;|0\rangle$. In universal notation

$$A = \begin{pmatrix} a \\ a^{+} \end{pmatrix}, \varphi = \begin{pmatrix} \psi \\ \psi^{+} \end{pmatrix}, \psi^{+}a + a^{+}\psi = \begin{pmatrix} \psi \\ \psi^{+} \end{pmatrix}\begin{pmatrix} 0 & \varkappa \\ 1 & 0 \end{pmatrix}\begin{pmatrix} a \\ a^{+} \end{pmatrix} \equiv \varphi g A, \tag{1.95}$$

i.e., it is necessary to make the replacement $A \to gA$ in the functional (1.86); so $A\Delta A \to Ag^{\mathrm{T}}\Delta gA$.

1.4 Graphs

1.4.1 *Perturbation theory*

From the definition (1.84) we obtain the perturbation series for the S matrix:

$$R(\varphi) = \sum_{N=0}^{\infty} \frac{1}{N!} \exp\left[\frac{1}{2} \cdot \frac{\delta}{\delta\varphi}\Delta\frac{\delta}{\delta\varphi}\right]\mathcal{M}^{N}(\varphi), \tag{1.96}$$

where $\mathcal{M}(\varphi) \equiv iS_v(\varphi)$ is the argument of the differentiated exponential in (1.84).

The effect of the reduction operation on $\mathcal{M}^{N}(\varphi)$ is conveniently described in diagrammatic language. The factor $\mathcal{M}(\varphi)$ is depicted graphically as a point, and a power $\mathcal{M}^{N}(\varphi)$ is depicted as a graph consisting of N isolated points.

The operation $(\delta/\delta\varphi)\Delta(\delta/\delta\varphi)$ graphically corresponds to the addition of a line Δ connecting a pair of points. The line is connected in all possible ways, because each derivative $\delta/\delta\varphi$ can act on any factor of the product $\mathcal{M}^{N}(\varphi) = \mathcal{M}(\varphi)\ldots\mathcal{M}(\varphi)$. In particular, it is possible to have the case where the two derivatives of the quadratic form act on the same $\mathcal{M}(\varphi)$. In diagrammatic language this means that a line leaving a point returns to the same point. We shall refer to such lines as *self-contracted*.

The result of the complete reduction operation on $\mathcal{M}^N(\varphi)$ is represented as the sum of graphs consisting of N points with any number of additional lines. The points are referred to as the *vertices of the graph*. A vertex to which n lines are attached corresponds to the factor

$$\mathcal{M}_n(x_1...x_n;\varphi) \equiv \delta^n \mathcal{M}(\varphi)/\delta\varphi(x_1)...\delta\varphi(x_n), \tag{1.97}$$

since the addition of a line is accompanied by differentiation of the vertex factor with respect to φ. The arguments x of the factors in (1.97) are contracted with the corresponding arguments of the lines Δ.

The factor $\mathcal{M}(\varphi) \equiv \mathcal{M}_0(\varphi)$ will henceforth be referred to as the *generating vertex* of a graph. If the interaction is a polynomial in the field, only the first few vertex factors in (1.97) will be nonzero. In general, the number of lines converging to a vertex is not bounded. Among nonpolynomial interactions, a particularly important one is the exponential interaction, which will be studied separately in Sec. 1.5.

Using Eq. (1.51), we can write down the following representation for the general term of the series (1.96):

$$\frac{1}{N!}\exp\left[\frac{1}{2}\sum_{ik}\frac{\delta}{\delta\varphi_i}\Delta\frac{\delta}{\delta\varphi_k}\right]\mathcal{M}(\varphi_1)...\mathcal{M}(\varphi_N)\Bigg|_{\varphi_1=...=\varphi_N=\varphi} . \tag{1.98}$$

In this expression the various vertices $\mathcal{M}(\varphi)$ are numbered by the arguments $\varphi_1...\varphi_N$, and in the reduction operator we have explicitly separated the contributions of operators adding lines between different pairs of vertices. The diagonal terms of the quadratic form of the derivatives are responsible for the addition of self-contracted lines. They can be taken into account exactly by introducing the *reduced vertex*

$$\mathcal{M}_{\text{red}}(\varphi) \equiv \exp\left(\frac{1}{2}\cdot\frac{\delta}{\delta\varphi}\Delta\frac{\delta}{\delta\varphi}\right)\mathcal{M}(\varphi) \tag{1.99}$$

and by writing, instead of (1.98),

$$\frac{1}{N!}\exp\left[\frac{1}{2}\sum_{i\neq k}\frac{\delta}{\delta\varphi_i}\Delta\frac{\delta}{\delta\varphi_k}\right]\mathcal{M}_{\text{red}}(\varphi_1)...\mathcal{M}_{\text{red}}(\varphi_N)\Bigg|_{...} . \tag{1.100}$$

The other terms of the reduction operator add lines only between pairs of different vertices.

It is clear from the definition (1.99), the reduction formula (1.45), and Eq. (1.30) that the reduced vertex (1.99) represents the N-form of the quantum interaction, while $\mathcal{M}(\varphi)$ represents its Sym-form.

For a more precise description of the diagrammatic series of perturbation theory it is convenient to use concepts from graph theory (see, for example, [5]), which will be introduced in the following section.

1.4.2 Some concepts from graph theory

By definition, a *graph* (diagram) is a set of points: vertices and lines connecting them. In a *labeled graph* the vertices are labeled 1 ... N, and a graph with unlabeled vertices is referred to as a *free graph*.

An adjacency matrix $\pi(d)$ is associated with a labeled graph d. By definition, the matrix element π_{ik} is equal to the number of lines (directly) connecting the vertices i and k. It is clear from the definition that the adjacency matrix is always symmetric. If there are no self-contracted lines in a graph, the adjacency matrix has zeros along its diagonal.

Two labeled graphs are termed equal if their adjacency matrices are equal.

By changing the labeling of the vertices in a given graph d, we obtain a new graph d', the adjacency matrix of which is related to the original one by a similarity transformation $\pi' = P\pi P^{\mathrm{T}}$, where P is the permutation matrix: $P_{ik} = 1$ if k is transformed into i in the permutation, and $P_{ik} = 0$ otherwise. Labeled graphs differing only by permutation of the vertex labels are termed *equivalent*. Equal graphs are equivalent, but the converse is not true in general. Equivalent labeled graphs correspond to the same free graph.

The *symmetry group* of a given labeled graph is the subgroup of permutations taking the graph into itself (i.e., into a graph equal to it). A given permutation belongs to the symmetry group of the graph when and only when the permutation matrix corresponding to it commutes with the adjacency matrix of the graph in question.

The *symmetry number* of a graph is the order of its symmetry group. Equivalent graphs have identical symmetry numbers, since their symmetry groups are isomorphic. Therefore, the symmetry number s can be viewed as a characteristic of a free graph.

The total number of ways of labeling the vertices of a free graph is $N!$. The resulting equivalent labeled graphs are divided into $N!/s$ classes, each containing s identical labeled graphs. Therefore, the total number of different labeled graphs corresponding to a given free graph is equal to $N!/s$.

1.4.3 Symmetry coefficients

Let us return to the expression (1.98), representing the general term of the perturbation series, and change from summation over i, k to summation over pairs $\alpha \equiv (i, k) = (k, i)$:

$$\frac{1}{2}\sum_{ik} \frac{\delta}{\delta\varphi_i} \Delta \frac{\delta}{\delta\varphi_k} = \sum_{\alpha} \varepsilon_\alpha \Delta_\alpha \mathscr{D}_\alpha.$$

Here $\mathscr{D}_\alpha \equiv \delta^2/\delta\varphi_i\delta\varphi_k$, Δ_α is the line connecting the pair α, and the numerical coefficient $\varepsilon_\alpha = 1/2$ for pairs (i, i) and $\varepsilon_\alpha = 1$ for pairs with $i \neq k$. Writing the exponential of the sum as a product of exponentials, we obtain

$$\prod_\alpha \exp\,[\varepsilon_\alpha \Delta_\alpha \mathscr{D}_\alpha] = \sum_{m_1 \ldots m_\alpha \ldots} \prod_\alpha \left[\frac{1}{(m_\alpha)!}(\varepsilon_\alpha \Delta_\alpha \mathscr{D}_\alpha)^{m_\alpha}\right]. \qquad (1.101)$$

The action of an individual term of this sum on the product of vertices $\mathscr{M}(\varphi_1) \ldots \mathscr{M}(\varphi_N)$ generates a labeled graph whose adjacency matrix is uniquely determined by the set of numbers m_α: $\pi_{ik} = m_\alpha$, where α is the number of the pair (i, k). The following two observations are crucial to the rest of the discussion: first, there are no identical sets m_α in the sum (1.101), so that different labeled graphs correspond to different terms of this sum. Second, it is clear that the series (1.101) generates all possible labeled graphs. This means that the expression (1.98) is written as the sum of all possible different labeled graphs with N vertices, where each of these graphs enters with the coefficient

$$C(d) = \left[N!\,2^r \prod_i \pi_{ii}! \prod_{i<k} \pi_{ik}!\right]^{-1}, \qquad (1.102)$$

where π_{ik} are the elements of the adjacency matrix and $r = \Sigma_i \pi_{ii}$ is the total number of self-contracted lines in the graph. The coefficients (1.102) are identical for equivalent labeled graphs, because only invariants under the permutation group enter into (1.102).

The differential operators $\mathscr{D}_\alpha$ entering into (1.101) transform the original vertices $\mathscr{M}$ into the vertices (1.97) without introducing additional coefficients. In the bosonic case all the graphs enter with plus sign, and in the fermionic case the sign of a graph is determined by the parity of the permutation of the anticommuting objects when the differentiation is performed. The sign rule for a Yukawa-type interaction will be obtained in Sec. 1.6.

Setting $\varphi_1 = \varphi_2 = \ldots = \varphi_N \equiv \varphi$ in (1.98), we change to the language of free graphs. The coefficient of a free graph D is

$$C(D) = \left[s \cdot 2^r \cdot \prod_i \pi_{ii}! \prod_{i<k} \pi_{ik}! \right]^{-1}, \tag{1.103}$$

since (1.98) contains all possible labeled graphs, among which there are no identical ones, and the total number of different labeled graphs corresponding to a given free graph is equal to $N!/s$.

A graph in which any two vertices are connected by no more than one line and in which there are no self-contracted lines is referred to as a *Mayer* graph. From (1.103) we see that the coefficient of a free Mayer graph is simply the inverse symmetry number.

1.4.4 *Recursion relation for the symmetry coefficients*

Differentiating the functional (1.84) with respect to the parameter Δ, we obtain the equation [6]

$$\frac{\delta}{\delta\Delta(x, x')} R(\varphi) = \frac{1}{2} \cdot \frac{\delta}{\delta\varphi(x)} \cdot \frac{\delta}{\delta\varphi(x')} R(\varphi), \tag{1.104}$$

which can be used to derive the recursion relation between the symmetry coefficients for the graphs $R(\varphi)$.

Let D_1 and D_2 be free graphs such that D_2 is obtained by connecting a single line to D_1. Differentiation with respect to Δ corresponds graphically to removing a line from the graph in all possible ways, so that the derivative with respect to Δ of the graph D_2 contains, among others, the graph D_1 with two specific vertices, which correspond to changed (compared to D_1) vertex factors (1.97). This graph enters into the left-hand side of (1.104) with coefficient $C_2 N(D_2 \to D_1)$, where C_2 is the coefficient of D_2 in $R(\varphi)$ and $N(D_2 \to D_1)$ is the number of ways of transforming D_2 into D_1 by removing a single line. On the right-hand side of (1.104) the same graph enters with the coefficient $\varepsilon C_1 N(D_1 \to D_2)$, where C_1 is the coefficient of D_1 in $R(\varphi)$ and $N(D_1 \to D_2)$ is the number of ways of transforming D_1 into D_2 by adding a single line, i.e., the number of different pairs of vertices, the joining of which by a line transforms D_1 into D_2. The additional factor ε is equal to unity if the line is connected to different vertices; in this case taking the second derivative with respect to φ for a given pair of vertices leads to an additional factor of 2 which cancels the coefficient 1/2 on the right-hand side of (1.104). If the connected line is a self-contracted one, $\varepsilon = 1/2$.

Equating the coefficients of D_1 on both sides of Eq. (1.104), we obtain the desired recursion relation:

$$C_2 = C_1 \varepsilon N(D_1 \to D_2)/N(D_2 \to D_1), \tag{1.105}$$

which is convenient for the practical calculation of the symmetry coefficients.

Let us give some examples.

1. Let D_1 be a square and D_2 be a square one of whose sides is doubled. In this case $N(D_2 \to D_1) = 2$, because one of the double lines can be broken, and $N(D_1 \to D_2) = 4$, because D_2 is obtained from D_1 by doubling any of the four sides of the square.
2. Let D_1 be a square with a single diagonal and D_2 be a square with two diagonals. In this case $N(D_1 \to D_2) = 1$, while $N(D_2 \to D_1) = 6$, since a square with two diagonals is a tetahedron for which all six lines are equivalent.

Equation (1.105) can be used to calculate the coefficient of any graph starting from the simplest one, which consists of N isolated vertices with coefficient $1/N!$.

1.4.5 *Transformation to Mayer graphs for the exponential interaction*

The generating vertex $\mathcal{M}(\varphi) = i\rho\int dx \exp \lambda\varphi(x)$ corresponds to the exponential interaction operator $\mathbf{V} = -\rho\int d\mathbf{x} \exp [\lambda\hat{\varphi}(0, \mathbf{x})]$ without normal-ordering. In these expressions φ is a bosonic field and ρ and λ are arbitrary numbers. The general term of the series (1.96) then has the form

$$\frac{(i\rho)^N}{N!}\int \ldots \int dx_1 \ldots dx_N \exp \left(\frac{1}{2} \cdot \frac{\delta}{\delta\varphi}\Delta\frac{\delta}{\delta\varphi}\right) \exp \sum_i \lambda\varphi(x_i). \tag{1.106}$$

The sum of all the $\varphi(x_i)$ can be represented as the scalar product of the field φ and the function $h(x) = \Sigma_i \delta(x - x_i)$. Then Eq. (1.12) can be used to bring (1.106) to the form

$$\frac{(i\rho)^N}{N!}\int \ldots \int dx_1 \ldots dx_N \exp \left\{\lambda\sum_i \varphi(x_i) + \frac{\lambda^2}{2}\sum_{ik}\Delta(x_i, x_k)\right\}. \tag{1.107}$$

The diagonal elements of the quadratic form redefine the vertices:

$$i\rho \exp \lambda\varphi(x) \to i\rho \exp \left[\lambda\varphi(x) + \frac{\lambda^2}{2}\Delta(x, x)\right] \equiv m(x),$$

and in the series expansion of the exponential the nondiagonal elements generate all possible graphs with lines Δ connecting different vertices.

Let us introduce the *Mayer line* or the *superpropagator*

$$g(x, x') = -1 + \exp[\lambda^2 \Delta(x, x')] = \lambda^2 \Delta(x, x') + \lambda^4 \Delta^2(x, x')/2 + \ldots.$$

Equation (1.107) can then be rewritten as

$$\frac{1}{N!}\int\ldots\int dx_1\ldots dx_N \prod_i m(x_i) \prod_{i<k} (1 + g(x_i, x_k))$$

and represented as a finite sum of graphs with vertices $m(x)$ and lines $g(x, x')$. All these graphs are Mayer graphs, i.e., they do not contain any self-contracted lines and any two vertices are connected by no more than one line; the coefficient of the free graph is defined according to the usual rule (the inverse symmetry number).

The transformation to Mayer lines is equivalent to a certain partial summation of the graphs of ordinary perturbation theory with the line Δ and is possible because for the exponential interaction the vertex factors (1.97) have a power-law dependence on the number of lines converging at a vertex. A similar summation can be performed also for a nonexponential interaction if it is first expanded into a Fourier or Laplace integral. We note that the graphical technique for the exponential interaction was first used in the equilibrium statistics of a nonideal classical gas (see Chap. 5 for more detail). We also note that if the original operator interaction is written in normal-ordered form, the diagonal terms of the quadratic form vanish in (1.107).

1.4.6 Graphs for a Yukawa-type interaction

In universal notation the representation (1.84) and the diagrammatic technique generated by it are identical for all theories. However, in the practical calculation of the contributions of various graphs it is in the end still necessary to transform from universal to ordinary notation; in addition, the universal notation masks the particular features of different theories. We shall therefore briefly discuss the diagrammatic technique in ordinary language for the important special case of a Yukawa-type interaction.

This interaction involves a complex field ψ, ψ^+, which can be either a bosonic or a fermionic field, and a real bosonic field φ. The interaction is written in cubic form

$$iS_v(\psi^+, \psi, \varphi) = \iiint dx dx' dy \psi^+(x) \Gamma(x, x', y) \psi(x') \varphi(y), \quad (1.108)$$

linear in each of the fields ψ^+, ψ, and φ. For an ordinary t-local interaction the kernel Γ contains the δ function of the three time arguments, and otherwise it is completely arbitrary [the interaction of quantum electrodynamics is of the type (1.108)].

The free action S_0 of the field ψ, ψ^+ is usually a bilinear form $\psi^+ L\psi$ with some kernel L. In universal notation (see Sec. 1.1)

$$\psi^+ L\psi = \frac{1}{2}[\psi^+ L\psi + \varkappa\psi L^{\mathrm{T}}\psi^+] = \frac{1}{2}\begin{pmatrix}\psi \\ \psi^+\end{pmatrix}\begin{pmatrix}0 & \varkappa L^{\mathrm{T}} \\ L & 0\end{pmatrix}\begin{pmatrix}\psi \\ \psi^+\end{pmatrix}.$$

Equation (1.31), which relates the propagator to the kernel of the free action, takes the following form in this case:

$$\Delta \equiv \begin{pmatrix}\Delta_{11} & \Delta_{12} \\ \Delta_{21} & \Delta_{22}\end{pmatrix} = i\begin{pmatrix}0 & \varkappa L^{\mathrm{T}} \\ L & 0\end{pmatrix}^{-1} = i\begin{pmatrix}0 & L^{-1} \\ \varkappa L^{-1\mathrm{T}} & 0\end{pmatrix}, \qquad (1.109)$$

and determines the matrix of contractions of the field $\psi \equiv \psi_1$, $\psi^+ \equiv \psi_2$: the contraction of ψ with ψ^+ is $\Delta_{12} = iL^{-1}$, the contraction of ψ^+ with ψ is $\Delta_{21} = \varkappa\Delta^{\mathrm{T}}_{12}$, and the contractions of ψ with ψ and ψ^+ with ψ^+ are zero.

The universal quadratic form of the derivatives in the reduction operator for the field ψ, ψ^+ is now written as

$$\frac{1}{2}\begin{pmatrix}\delta/\delta\psi \\ \delta/\delta\psi^+\end{pmatrix}\begin{pmatrix}0 & \Delta_{12} \\ \Delta_{21} & 0\end{pmatrix}\begin{pmatrix}\delta/\delta\psi \\ \delta/\delta\psi^+\end{pmatrix} = \frac{\delta}{\delta\psi}\Delta_{12}\frac{\delta}{\delta\psi^+}$$

[we have used Eq. (1.40) and the equation $\Delta_{21} = \varkappa\Delta^{\mathrm{T}}_{12}$], and the representation (1.84) for the S-matrix functional takes the form

$$R(\psi^+, \psi, \varphi) = \exp\left[\frac{\delta}{\delta\psi}\Delta_{12}\frac{\delta}{\delta\psi^+} + \frac{1}{2}\frac{\delta}{\delta\varphi}\Delta'\frac{\delta}{\delta\varphi}\right]\exp iS_v(\psi^+, \psi, \varphi), \qquad (1.110)$$

where Δ' is the propagator of the field φ.

The graphs corresponding to (1.110) contain two types of line: the ordinary lines of the field φ and the *directed* or *oriented* lines denoting contractions of the field ψ with ψ^+ (the lines leave the vertex from which the field ψ is taken and go to the vertex from which ψ^+ is taken). The complete graph is the superposition of the ordinary graph depicting the lines of the field φ and the *directed graph* depicting the lines of the field ψ, ψ^+. Any labeled graph can be uniquely specified by the pair of matrices $\bar{\pi}$, π, where π is the ordinary adjacency matrix of the graph of the φ lines and $\bar{\pi}$ is the adjacency matrix of the directed graph of the ψ, ψ^+ lines. By definition, the matrix element $\bar{\pi}_{ik}$ is equal to the number of directed lines leaving the vertex i and arriving at the vertex k. In contrast to π, the matrix $\bar{\pi}$ is not necessarily symmetric.

The concepts of equality, equivalence, and symmetry group of a labeled graph introduced in Sec. 1.4.2 have an obvious generalization if instead of the matrix π we always take the pair of matrices π, $\bar{\pi}$. Graphs are equal if the pair π, $\bar{\pi}$ is equal to the pair π', $\bar{\pi}'$, i.e., $\pi = \pi'$ and $\bar{\pi} = \bar{\pi}'$. In a permutation of vertices $(\pi, \bar{\pi}) \to (P\pi P^{T}, P\bar{\pi}P^{T})$, where P is the permutation matrix; the symmetry group of a graph is obviously the intersection of the symmetry groups of the ordinary graph described by the matrix π and the directed graph described by the matrix $\bar{\pi}$.

As before, the expansion (1.110) contains all possible labeled graphs among which there are no ordinary graphs; repeating the arguments of Sec. 1.4.3, we easily obtain the following expression for the symmetry coefficient of a free graph:

$$C(D) = \left[s \cdot 2^{r} \prod_{i} \pi_{ii}! \prod_{i<k} \pi_{ik}! \prod_{i,k} \bar{\pi}_{ik}!\right]^{-1}. \tag{1.111}$$

Here s is the symmetry number of the graph, i.e., the order of its symmetry group, and $r = \Sigma_i \pi_{ii}$ is the total number of self-contracted φ lines in the graph.

Equation (1.111) greatly simplifies for the interaction (1.108), since in this case there are no self-contracted φ lines at all, and the matrix elements π_{ik}, $\bar{\pi}_{ik}$ are either zero or unity. Therefore, the coefficient of the free graph for the Yukawa interaction is simply the inverse symmetry number.

In some cases the specific form of the interaction (1.108) considerably simplifies the calculation of the symmetry numbers themselves. It will be useful in what follows to introduce a few more concepts from graph theory.

The *degree of a vertex* is the set of the following three numbers: the number of φ-lines attached to it, the number of incoming ψ lines, and the number of outgoing ψ lines. For the interaction (1.108) each of these numbers is either zero or unity. We shall say that a vertex is *unique* if its degree differs from the degree of any other vertex of the graph.

The vertices i and k of a labeled graph are termed *similar* if the symmetry group contains a permutation taking i into k. A vertex similar only to itself is termed *stationary*. Similar vertices necessarily have identical degrees, so that any unique vertex is stationary. If all the vertices of a graph are stationary, its symmetry group contains only the identity permutation and the symmetry number s is equal to unity.

A graph is termed *connected* if in it it is possible to go from any vertex to any other vertex by following lines. The following statement is true for the interaction (1.108): if one of the vertices of a connected graph is stationary, then all the vertices are stationary and $s = 1$.

Let the vertex i be stationary. Since the graph is connected, there is a line attached to this vertex. For definiteness let us assume that this is a line of the field φ connecting the vertex i to some vertex k. Let P be an arbitrary permutation: $(i, k) \to (i', k')$, $\pi \to \pi' = P\pi P^{\mathrm{T}}$. By assumption, $\pi_{ik} = \pi'_{i'k'} = 1$. If P belongs to the symmetry group of the graph and the vertex i is stationary, then $\pi' = \pi$ and $i' = i$, from which $\pi_{ik} = \pi_{ik'} = 1$. It remains to be noted that for the interaction (1.108) any row of the matrix π has no more than a single nonzero matrix element, because no more than one φ-line leaves a vertex. Therefore, from the equation $\pi_{ik} = \pi_{ik'} = 1$ it follows that $k' = k$, and since this is true for any permutation from the symmetry group of the graph we conclude that the vertex k is stationary.

Therefore, any vertex k connected by a φ line to a stationary vertex i is also stationary. Clearly, the argument remains valid also when the vertex k is connected to the vertex i by a directed ψ-field line, since for the interaction (1.108) each row and each column of the matrix $\bar{\pi}$ also contains no more than one nonzero element. Going from one vertex to another in succession, we prove that the stationarity of one implies the stationarity of all.

From this it follows that if at least one vertex of a connected graph is unique, the symmetry number of the graph is equal to unity.

Let us now discuss the question of the signs of the graphs for the case of a fermionic field ψ, ψ^+. The interaction (1.108) is a quadratic form $\psi^+ L\psi$ in this field with kernel $L(x, x') = \int dy\Gamma(x, x', y)\varphi(y)$. The differentiation with respect to ψ and ψ^+ in (1.110) can be performed explicitly using the relation (proved in Sec. 1.6.5)

$$\exp\frac{\delta}{\delta\psi}\Delta_{12}\frac{\delta}{\delta\psi^+}\exp\psi^+ L\psi = \det[1 - L\Delta_{12}]^{-\varkappa}\exp\psi^+[L^{-1} - \Delta_{12}]^{-1}\psi, \tag{1.112}$$

in which $\det M$ denotes the determinant of the linear operator M.

We write $Q \equiv [L^{-1} - \Delta_{12}]^{-1} = L + L\Delta_{12}L + \dots$. The quadratic form $\psi^+ Q\psi$ is represented graphically as the progression

 (1.113)

in which the directed lines correspond to the propagator Δ_{12} and the points correspond to the vertex factors (1.97) for the interaction (1.108).

Using the well known formula $\det M = \exp \operatorname{tr} \ln M$ expressing the determinant of a linear operator in terms of the trace of its logarithm, we write $\det [1 - L\Delta_{12}]^{-\varkappa} = \exp \varkappa \mathcal{P}(\varphi)$. The functional $\mathcal{P}(\varphi) = -\operatorname{tr} \ln (1 - L\Delta_{12})$ is represented graphically as the sum of closed loops:

$\mathcal{P}(\varphi) = \sum_{n=1}^{\infty} \frac{1}{n} tr (L\Delta_{12})^n \equiv$  $+ \frac{1}{2}$ 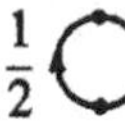 $+ \frac{1}{3}$ 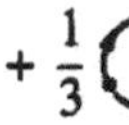 $+ \dots, \qquad (1.114)$

containing only the field φ at the vertices. The representation (1.110) for the S-matrix functional can now be rewritten as

$$R(\psi^+, \psi, \varphi) = \exp \left(\frac{1}{2} \frac{\delta}{\delta\varphi} \Delta' \frac{\delta}{\delta\varphi}\right) \exp \left[\varkappa \mathcal{P}(\varphi) + \psi^+ Q \psi\right] =$$

$$= \sum_{N=0}^{\infty} \frac{1}{N!} \exp \left(\frac{1}{2} \frac{\delta}{\delta\varphi} \Delta' \frac{\delta}{\delta\varphi}\right) [\psi^+ Q \psi]^N \exp [\varkappa \mathcal{P}(\varphi)] . \qquad (1.115)$$

The remaining operation of reduction in φ adds lines Δ' between the vertices of the chains (1.113) and loops (1.114) in all possible ways.

If the field ψ, ψ^+ is a bosonic field, then $\varkappa = 1$ and all the graphs in (1.115) enter with plus sign. If the field ψ, ψ^+ is a fermionic field, then $\varkappa = -1$ and an additional sign factor -1 appears on each closed fermionic loop (1.114). This factor completely determines the sign in front of the graph with the condition that the uncontracted anticommuting fields ψ, ψ^+ remaining in R are grouped, as in (1.115), into blocks $\psi^+ Q \psi$ which commute with each other. Each of these blocks corresponds to a continuous chain of ψ lines passing through the entire graph.

Terms with $N = 1$ in (1.115) have only a single continuous chain of ψ lines, so that the ends of this chain will be unique vertices in the graphs. It follows from the above analysis of the symmetry coefficients that all free graphs with a single continuous chain of ψ lines enter into (1.115) with the same coefficient equal to 1. If the number of continuous chains is greater than one, the symmetry group of the graph can be nontrivial, but it can contain only permutations under which continuous chains of ψ lines are permuted as a whole. These arguments greatly simplify the computation of the symmetry coefficients for an interaction of the Yukawa type.

1.4.7 Graphs for a pair interaction

In nonrelativistic theory one often encounters the pair interaction of a complex field ψ, ψ^+. Its Hamiltonian is given in the next chapter [Eq. (2.29)], and the generating vertex has the following form up to unimportant terms quadratic in the field:

$$iS_v(\psi, \psi^+) = \frac{1}{2}\iint dx dx' \psi^+(x)\psi^+(x')\Gamma(x, x')\psi(x')\psi(x). \quad (1.116)$$

For the Hamiltonian (2.29) $\Gamma(x, x') = -i\delta(t - t')\mathcal{V}(\mathbf{x}, \mathbf{x}')$, where $\mathcal{V}$ is the symmetric potential of the pair interaction. Instead of (1.110) we have

$$R(\psi, \psi^+) = \exp\left[\frac{\delta}{\delta\psi}\Delta_{12}\frac{\delta}{\delta\psi^+}\right] \exp iS_v(\psi, \psi^+). \quad (1.117)$$

Using the standard diagrammatic technique of Sec. 1.4.1, the functional (1.117) can be written as the sum of directed graphs with vertices associated with derivatives of (1.116), the analogs of the vertex factors (1.97).

In practice, however, a different diagrammatic technique is often used. In it the interaction is associated with not one but two vertices of a graph. The starting point is the following representation of the functional (1.117):

$$R(\psi, \psi^+) = \exp\left[\frac{\delta}{\delta\psi}\Delta_{12}\frac{\delta}{\delta\psi^+} + \frac{1}{2}\cdot\frac{\delta}{\delta\varphi}\Gamma\frac{\delta}{\delta\varphi}\right] \exp \rho\varphi\Big|_{\varphi=0}, \quad (1.118)$$

in which $\varphi(x)$ is an extra bosonic field which is introduced, $\rho(x) \equiv \psi^+(x)\psi(x)$, and $\rho\varphi \equiv \int dx \rho(x)\varphi(x)$. According to (1.12), we have

$$\exp\left(\frac{1}{2}\frac{\delta}{\delta\varphi}\Gamma\frac{\delta}{\delta\varphi}\right)\exp \rho\varphi\Big|_{\varphi=0} = \exp\left(\frac{1}{2}\rho\Gamma\rho\right) = \exp iS_v(\psi^+, \psi),$$

which proves the equivalence of the representations (1.117) and (1.118).

The right-hand side of (1.118) has the form of an S-matrix functional for the Yukawa interaction and can be represented by the corresponding graphs. The interaction kernel $\Gamma(x, x')$ plays the role of the propagator of the field φ, and the condition $\varphi = 0$ in (1.118) implies that only graphs in which all the fields φ are contracted are considered. The interaction (1.116) is now represented by not one, but two vertices connected by a line Γ. Here the number of graphs grows, but this is compensated for by the simplicity of the calculation of the symmetry coefficients and, especially, by the presence of the exact sign rule for the case where ψ, ψ^+ is a fermionic field.

1.4.8 *Connectedness of the logarithm of $R(\varphi)$*

We shall prove the following topological statement, first made in the equilibrium statistics of a nonideal classical gas and known there as the *first Mayer theorem* [7, 8]:

$$\ln R(\varphi) = \text{connected part of } R(\varphi). \qquad (1.119)$$

We recall that a graph is termed connected if one can go from any vertex to any other vertex by following lines in the graph. The functional $R(\varphi)$ has the form 1 + graphs, among which there are both connected and unconnected graphs; the contribution of an unconnected graph is equal to the product of the contributions of its connected components. *The connected part of $R(\varphi)$ is the sum of the contributions of all the connected graphs of $R(\varphi)$ with their coefficients and signs.*

To prove the statement (1.119) we shall use universal notation, in which the diagrammatic technique is the same for all theories, and we shall restrict ourselves to the case of a bosonic field in order to avoid the question of the sign of each graph. Let us consider the expression $\exp\Sigma_\alpha C_\alpha d_\alpha$, in which the summation runs over all connected free graphs, d_α is the contribution of such a graph to $R(\varphi)$, and C_α is the coefficient of this graph. We must show that the quantity of interest coincides with the functional $R(\varphi)$ itself:

$$R(\varphi) = \prod_\alpha \exp\,(C_\alpha d_\alpha) = \sum_{\dots m_\alpha \dots} \prod_\alpha \left[\frac{1}{m_\alpha!} C_\alpha^{m_\alpha} d_\alpha^{m_\alpha}\right]. \qquad (1.120)$$

The general term of the series on the right-hand side represents the graph in which the connected graph d_1 is repeated m_1 times, the graph d_2 is repeated m_2 times, and so on. This graph will be connected if one of the numbers m_α is equal to unity and all the others are zero; clearly, the connected graph d_α enters into (1.120) with the correct coefficient C_α. It remains to be shown that for unconnected graphs the coefficients on the right-hand side of (1.120) are exactly the same as in $R(\varphi)$. If this is true, then

$$C(\dots d_\alpha^{m_\alpha} \dots) = \prod_\alpha [C_\alpha^{m_\alpha}/m_\alpha!], \qquad (1.121)$$

where $C(\dots d_\alpha^{m_\alpha}\dots)$ is the coefficient of the graph containing the graph d_1 m_1 times, the graph d_2 m_2 times, and so on. Using (1.103), Eq. (1.121) can be rewritten as a relation between symmetry numbers:

$$s(\ldots d_\alpha^{m_\alpha}\ldots) = \prod_\alpha [m_\alpha! s_\alpha^{m_\alpha}], \tag{1.122}$$

the validity of which is already obvious, because the symmetry group of the unconnected graph in question consists of the symmetry groups of each of its connected components [which gives $\Pi_\alpha(s_\alpha^{m_\alpha})$] and permutations of all the vertices of identical connected components as a whole [which gives $\Pi_\alpha(m_\alpha!)$]. By proving Eq. (1.122) we have proved the statement (1.119).

For a fermionic theory it would still be necessary to show that the sign of an unconnected graph is equal to the product of the signs of all the connected components of this graph. This is obvious for a Yukawa interaction when we recall the sign rule: a factor of minus one for each closed fermionic loop. For the time being we shall limit ourselves to this remark, and in Sec. 1.7.4 we shall give a version of the proof of the connectedness theorem which does not require knowledge of the coefficients and signs of the graphs, so that the generalization to fermionic theories is trivial.

1.4.9 Graphs for the Green functions

Comparison of Eqs. (1.84), (1.87), and (1.88) shows that the generating functional of the Green functions $G(A)$ coincides with the vacuum expectation value of the S matrix for a theory with interaction $S_v(\varphi) + \varphi A$ depending on the functional parameter A. The addition φA is often interpreted as a real interaction with a "source" or an "external field" A, but this is purely a question of terminology.

It is clear from the above discussion that the functional $G(A)$ has exactly the same diagrammatic representation as the S matrix, but instead of (1.97) the vertex factors in the graphs for $G(A)$ will be

$$\mathcal{M}_n' = \mathcal{M}_n + i\delta_{n1}A, \tag{1.123}$$

(the arguments are understood to be $x_1 \ldots x_n$), where δ_{n1} is the Kronecker delta and $\mathcal{M}_n$ are the vertex factors (1.97) for $\varphi = 0$.

Classifying graphs according to the number of vertices $\mathcal{M}_1'$, we write

$$G(A) = \sum_{n=0}^{\infty} \frac{1}{n!}$$  . (1.124)

The crosses denote the vertices $\mathcal{M}_1' = \mathcal{M}_1 + iA$ to which one line Δ is attached, the circle represents the interior of the graph, which does not contain $\mathcal{M}_1'$

vertices, and the factor $1/n!$ is isolated simply for convenience. Each of the free graphs enters into (1.124) with the usual symmetry coefficient (1.103).

According to the definition (1.85), the full Green function G_n is the coefficient of $(iA)^n/n!$ in (1.124). Therefore,

$$G_n(x_1 \ldots x_n) = \sum_{m=0}^{\infty} \frac{1}{m!} \quad (1.125)$$

The general term of this series has $n+m$ external lines Δ. The outermost arguments of n lines [bottom in (1.125)] are free and correspond to the arguments $x_1 \ldots x_n$ of the function G_n; the arguments of the remaining m lines [top in (1.125)] are contracted with the vertex factors $\mathcal{M}_1$ shown by the points in (1.125). We note that for most of the theories commonly studied the interaction functional $S_v(\varphi)$ does not contain terms linear in φ, so that $\mathcal{M}_1 = 0$. Then only the term with $m = 0$ remains in (1.125).

According to the general rules of Sec. 1.4.1, Eq. (1.124) is the sum of unity and all graphs, both connected and unconnected. It follows from the Mayer theorem proved in the preceding section that

$$W(A) \equiv \ln G(A) = \text{connected part of } G(A). \quad (1.126)$$

Connected Green functions $W_n(x_1 \ldots x_n)$ are defined as the coefficients in the expansion of the functional $W(A)$ analogous to (1.85):

$$W(A) = \sum_{n=0}^{\infty} 1/n! \cdot W_n (iA)^n \quad (1.127)$$

[in the abbreviated notation of (1.85)]. Therefore, $W(A)$ is by definition the generating functional of connected Green functions W_n.

It follows from Eq. (1.126) that the W_n are the connected parts of the full Green functions G_n and are obtained from (1.125) by discarding the contributions of all the unconnected graphs (and unity for $n = 0$). To explain this, we recall that $G(A)$ is the sum of unity and graphs, and therefore $W(A) = \ln G(A)$ consists only of graphs. Consequently, all powers $W^2(A)$, $W^3(A)$, etc. do not contain connected graphs at all [here it is crucial that $W(A)$ does not contain any term like unity], so that on the right-hand side of the equation $G(A) = \exp W(A) = 1 + W(A) + W^2(A)/2 + \ldots$ the connected graphs enter only in the term $W(A)$. Taking the coefficient of $(iA)^n/n!$ in this equation, on the left-hand side we obtain by definition the full function G_n, and on the right-hand side we obtain W_n from $W(A)$ plus the contributions of higher powers of $W(A)$. The latter, as noted above, contain only unconnected graphs, so the connected graphs of G_n and W_n coincide. On the other hand, it follows from (1.126) that all the graphs of W_n are connected, so that W_n is simply the sum of the contributions of all the connected graphs of G_n.

We see from (1.89) that for the free theory $W(A) = -\varkappa A\Delta A/2$. Comparing this expression with the quadratic term $W_2(iA)^2/2$ in the expansion (1.127), it should be noted that the ordering of the arguments used in (1.85) does not correspond to the matrix notation of the quadratic form, and an additional sign factor arises in the permutation of the fields A: $W_2A^2 = \varkappa AW_2A$. Taking this into account, we find that for the free theory $W_2 = \Delta$, and the other connected functions W_n are equal to zero.

The graphs of W_0 without external lines Δ with free arguments are referred to as *connected vacuum loops*, and the graphs of W_1 with a single free external line are referred to as tadpoles. We shall not use the last term. We shall refer to the second connected function $W_2(x, x')$ as the *full propagator*, in contrast to the *bare propagator* $\Delta(x, x')$ (the time-ordered contraction). It was shown above that for the free theory the full propagator coincides with the bare one.

Returning to Eq. (1.126), we note that it leads to the equation $G_0 = \exp W_0$ for zeroth-order Green functions. The zeroth-order function G_0 is the vacuum expectation value of the S matrix (see Sec. 1.3.3). The equation obtained above shows that its logarithm is the sum of all the connected vacuum loops.

The constant W_0 enters into (1.127) additively and appears in the form of a factor $G_0 = \exp W_0$ in each full Green function: $G_n = H_n \exp W_0$. Comparison with (1.76) shows that the H_n are the Green functions (1.58). Clearly, the functions H_n are obtained from the G_n by discarding all graphs containing vacuum loops as unconnected subgraphs, which explains the term "Green function without vacuum loops." The corresponding generating functional $H(A)$ is expressed in terms of the connected functions W_n with $n > 0$:

$$H(A) \equiv \sum_{n=0}^{\infty} \frac{1}{n!} H_n (iA)^n = \exp \sum_{n=1}^{\infty} \frac{1}{n!} W_n (iA)^n. \tag{1.128}$$

Equating the coefficients of powers of iA, we obtain

$$H_0 = 1, \quad H_1 = W_1, \quad H_2 = W_2 + W_1 W_1, \tag{1.129}$$

and the general expression is written as

$$H_n(x_1...x_n) = (-i)^n\left[\frac{\vec{\delta}W}{\delta A(x_1)}+\frac{\vec{\delta}}{\delta A(x_1)}\right]...$$

$$...\left[\frac{\vec{\delta}W}{\delta A(x_n)}+\frac{\vec{\delta}}{\delta A(x_n)}\right]\cdot 1\Bigg|_{A=0}. \tag{1.130}$$

Let us also give the definition of the *amputated Green functions* W_n^{am}:

$$W_n(x_1...x_n) = \int...\int dx_1'...dx_n' D(x_1,x_1')...$$

$$D(x_n,x_n')W_n^{\text{am}}(x_1'...x_n'). \tag{1.131}$$

In this equation the kernel $D(x, x')$ denotes the full propagator $W_2(x, x')$. This notation will be used often, and in what follows expressions of the type (1.131) will be written in the abbreviated form $W_n = D^n W_n^{\text{am}}$, $W_n^{\text{am}} = D^{-n} W_n$.

Using the definitions (1.85) and (1.131), it is easily shown that $(D^n W_n^{\text{am}})A^n = W_n A^n = W_n^{\text{am}}(D^{\text{T}}A)^n$, i.e., the transfer of the linear operator from the kernel W_n^{am} to the arguments A is accompanied by its transposition. Taking this into account, we obtain the following relations for the generating functionals:

$$W(A) = W^{\text{am}}(D^{\text{T}}A), \quad W^{\text{am}}(A) = W(D^{-1\text{T}}A). \tag{1.132}$$

Owing to the symmetry of the propagator, D^{T} can be replaced by $\varkappa D$.

In conclusion, we note that the functional $W(A)$ can be related simply to the logarithm of the S matrix using Eq. (1.93).

1.5 Unitarity of the *S* Matrix

1.5.1 The conjugation operation

Let us discuss the features of the conjugation operation, which we shall need below to formulate the unitarity condition.

We shall understand this operation as *ordinary Hermitian conjugation for operators, complex conjugation for classical bosonic fields, and a formally*

defined involution operation on the Grassmann algebra for anticommuting classical fermionic fields. The involution operation on the Grassmann algebra possesses the same properties as Hermitian conjugation of operators, in particular, $(AB)^+ = B^+A^+$.

If all the fields in question are real, the conjugation is the identity transformation. In general, when universal notation is used conjugation is a linear transformation $\varphi \to \varphi^+ = I\varphi$ of a unified field φ, because by convention φ includes both the fields φ_α and φ_α^+ as independent components. The kernel of the linear operator I is a matrix in the discrete indices which, when included in the arguments x, distinguish the components φ_α and φ_α^+. In all the other arguments, including the time, the kernel I is a multiple of the unit operator. We also note the equality $I^*I = 1$ (for linear operators the star denotes complex conjugation), which means that double conjugation is the identity transformation.

The equation $F(\hat\varphi)^+ = F^+(\hat\varphi)$ is the definition of the conjugate operator functional. It is easily verified that the symmetry of the kernels F_n is preserved in the conjugation.

From the definition (1.24) and the equation $\hat\varphi^+ = I\hat\varphi$ it is easy to find

$$\{\mathrm{Sym}[\hat\varphi(x_1)\ldots\hat\varphi(x_n)]\}^+ = \mathrm{Sym}\{[\hat\varphi(x_1)\ldots\hat\varphi(x_n)]^+\}. \quad (1.133)$$

This shows that the conjugation and symmetrization operations commute. An equation analogous to (1.133) will hold for the N-product when the conjugation operation transforms the terms $\hat a$ and $\hat b$ in the decomposition $\hat\varphi = \hat a + \hat b$ (see Sec. 1.1.4) into each other up to a linear transformation. This condition is usually satisfied (the only exception is the N-product, which will be introduced in Sec. 2.1.2), and in what follows we shall assume that

$$\{N[\hat\varphi(x_1)\ldots\hat\varphi(x_n)]\}^+ = N\{[\hat\varphi(x_1)\ldots\hat\varphi(x_n)]^+\}. \quad (1.134)$$

It follows from (1.133) and (1.134) that $[NF(\hat\varphi)]^+ = NF^+(\hat\varphi)$ and analogously for the Sym-product. An equation of this type is not valid for the T-product.

Below we shall also need the rules for how derivatives transform under conjugation. Directly from the definitions we obtain

$$\left[\frac{\overleftarrow{\delta}}{\delta\varphi}F(\varphi)\right]^+ = \frac{\overrightarrow{\delta}}{\delta\varphi^+}F^+(\varphi) = \left[I^+\frac{\overrightarrow{\delta}}{\delta\varphi}\right]F^+(\varphi). \quad (1.135)$$

This expression is written for the more complicated case of a fermionic field: right derivatives transform into left ones and vice versa. The last equation in (1.135) follows from the relation $I^* = I^{-1}$ and the rule $\delta/\delta(M\varphi) = M^{-1\mathrm{T}}\delta/\delta\varphi$, which is valid for any linear change of variable. For a bosonic field the right and left derivatives in (1.135) are not distinguished.

Using Eqs. (1.135) and (1.9), we obtain

$$\left[\frac{\overleftarrow{\delta}}{\delta\varphi}L\frac{\overleftarrow{\delta}}{\delta\varphi}F(\varphi)\right]^{+} = \left[\frac{\overrightarrow{\delta}}{\delta\varphi^{+}}L^{*}\frac{\overrightarrow{\delta}}{\delta\varphi^{+}}\right]F^{+}(\varphi) = \left[\frac{\overleftarrow{\delta}}{\delta\varphi}\bar{L}\frac{\overleftarrow{\delta}}{\delta\varphi}\right]F^{+}(\varphi)\,, \tag{1.136}$$

where we have written $\bar{L} \equiv \varkappa I^{*}L^{*}I^{+}$. Let us write down this expression in abbreviated form for the two statistics, assuming in the fermionic case that all derivatives are right ones:

$$\left[\frac{\delta}{\delta\varphi}L\frac{\delta}{\delta\varphi}\right]^{+} = \frac{\delta}{\delta\varphi}\bar{L}\frac{\delta}{\delta\varphi},\quad \bar{L} \equiv \varkappa I^{*}L^{*}I^{+}. \tag{1.137}$$

We conclude by presenting a useful expression for the simple contraction n. Using the conjugation of Eq. (1.16) taking into account (1.134) and the relation $\hat{\varphi}^{+} = I\hat{\varphi}$, we obtain $n = I^{*}n^{+}I^{+}$. From this it is easily shown that the symmetric part of the simple contraction n_s, defined by Eq. (1.25), is invariant under the operation (1.137), i.e., $n_s = \varkappa I^{*}n_s{}^{*}I^{+} \equiv \bar{n}_s$.

1.5.2 *Formal unitarity of the off-shell* S *matrix*

The S-matrix operator $\mathbf{U}$, being the formal limit of the unitary evolution operator (1.55), must satisfy the unitarity condition $\mathbf{U}^{+}\mathbf{U} = 1 = \mathbf{U}\mathbf{U}^{+}$ in each order of perturbation theory. Of course, in field theory it is impossible to require more, because in a strict sense the desired limit of the evolution operator does not, as a rule, exist.

For an interaction which does not contain time derivatives the S-matrix operator is given by Eq. (1.64), where the interaction functional $S_{\mathrm{v}}(\varphi)$ involved in it is real owing to the Hermiticity of the quantum interaction Hamiltonian. According to the rules of the preceding section we obtain

$$\mathbf{U}^{+} = N\exp\left[\frac{1}{2}\cdot\frac{\delta}{\delta\varphi}\bar{\Delta}\frac{\delta}{\delta\varphi}\right]\exp\left[-iS_{\mathrm{v}}(\varphi)\right]\Big|_{\varphi=\hat{\varphi}}\,, \tag{1.138}$$

where $\bar{\Delta} \equiv \varkappa I^{*}\Delta^{*}I^{+}$. Reducing the product $\mathbf{U}^{+}\mathbf{U}$ to N-form using (1.49), we obtain $\mathbf{U}^{+}\mathbf{U} = NQ(\hat{\varphi})$, where $Q(\varphi)$ has the following form:

$$\exp\left[\frac{1}{2}\cdot\frac{\delta}{\delta\varphi_1}\bar{\Delta}\frac{\delta}{\delta\varphi_1}+\frac{1}{2}\cdot\frac{\delta}{\delta\varphi_2}\Delta\frac{\delta}{\delta\varphi_2}+\frac{\delta}{\delta\varphi_1}n\frac{\delta}{\delta\varphi_2}\right]\times$$
$$\times\ \exp i\left[S_v(\varphi_2)-S_v(\varphi_1)\right]\Big|_{\varphi_1=\varphi_2=\varphi}. \qquad (1.139)$$

For the operator equation $\mathbf{U}^{+}\mathbf{U} = NQ(\hat{\varphi}) = 1$ to be valid it is sufficient that the functional $Q(\varphi)$ become unity on the mass shell, i.e., on the set of solutions of the free equation (1.2). However, it actually turns out that the functional $Q(\varphi)$ is identically unity in each order of perturbation theory. This property, which is stronger than ordinary operator unitarity, is referred to as *off-shell unitarity*.

For the proof, to make the notation more compact we introduce the two-component field $\Phi \equiv (\varphi_1, \varphi_2)$ and symmetrize the quadratic form of the derivatives in (1.139):

$$\frac{1}{2}\cdot\frac{\delta}{\delta\Phi}\begin{pmatrix}\bar{\Delta} & 2n\\ 0 & \Delta\end{pmatrix}\frac{\delta}{\delta\Phi}=\frac{1}{2}\cdot\frac{\delta}{\delta\Phi}\begin{pmatrix}\bar{\Delta} & n\\ \varkappa n^{\mathrm{T}} & \Delta\end{pmatrix}\frac{\delta}{\delta\Phi}\equiv\frac{1}{2}\cdot\frac{\delta}{\delta\Phi}M\frac{\delta}{\delta\Phi}.$$

We expand the functional (1.139) in a series* in powers of $S_v(\varphi_2) - S_v(\varphi_1)$. Since the zeroth-order contribution is obviously equal to unity, we must show that for all $n \geq 1$ and any φ

$$\exp\left[\frac{1}{2}\frac{\delta}{\delta\Phi}M\frac{\delta}{\delta\Phi}\right](S_v(\varphi_2)-S_v(\varphi_1))^n\Big|_{\varphi_1=\varphi_2=\varphi}=0. \qquad (1.140)$$

Let us transform from the variables φ_1, φ_2 to the half-sum $\psi_1 \equiv (\varphi_1 + \varphi_2)/2$ and the difference $\psi_2 \equiv \varphi_1 - \varphi_2$. Writing this substitution in matrix form $\psi = g\Phi$, we obtain

---

* A beautiful trick making the series expansion unnecessary is suggested in [9]. One uses the identity $\exp i[S_v(\varphi_2) - S_v(\varphi_1)] = 1 - i\int dt[\mathcal{V}(\varphi_2, t) - \mathcal{V}(\varphi_1, t)] \exp i[S_v(t, -\infty; \varphi_2) - S_v(t, -\infty; \varphi_1)]$ [in the notation of (1.59)] and formulas of the type (1.51), with the role of one of the factors F played by $\mathcal{V}(\varphi_2, t) - \mathcal{V}(\varphi_1, t)$ and that of the other played by $\exp i[S_v(t, -\infty; \varphi_2) - S_v(t, -\infty; \varphi_1)]$. However, we shall follow the traditional proof here.

$$\frac{\delta}{\delta\Phi} = g^{\mathrm{T}}\frac{\delta}{\delta\psi}, \quad \frac{\delta}{\delta\Phi}M\frac{\delta}{\delta\Phi} = \frac{\delta}{\delta\psi}gMg^{\mathrm{T}}\frac{\delta}{\delta\psi}.$$

From the known matrices M and g we find the matrix $\widetilde{M} \equiv gMg^{\mathrm{T}}$: $\widetilde{M}_{11} = (\bar{\Delta} + \Delta + n + \varkappa n^{\mathrm{T}})/4$, $\widetilde{M}_{22} = \bar{\Delta} + \Delta - n - \varkappa n^{\mathrm{T}}$, $\widetilde{M}_{12} = \varkappa \widetilde{M}_{21}^{\mathrm{T}} = (\bar{\Delta} - \Delta + \varkappa n^{\mathrm{T}} - n)/2$. Using the definition (1.137) for $\bar{\Delta}$, the relation (1.29) between the contractions Δ and n, and the equation $n = I^* n^+ I^+$ given at the end of the preceding section, we obtain the following explicit expressions for the matrix elements of $\widetilde{M}$:

$$\widetilde{M}_{11} = \frac{1}{2}(n + \varkappa n^{\mathrm{T}}), \quad \widetilde{M}_{22} = 0, \quad \widetilde{M}_{12} = \varkappa \widetilde{M}_{21}^{\mathrm{T}} = \theta_{tt'}[\varkappa n^{\mathrm{T}} - n]\,. \quad (1.141)$$

It is clear from the definitions that $\widetilde{M}$ can be interpreted as the matrix of contractions for the field ψ, and for what follows it is important to note that the component ψ_2 has nonzero contraction only with ψ_1, and this contraction is the advanced one: $\widetilde{M}_{21}(x, x') = 0$ for $t > t'$.

Equation (1.140) coincides up to a coefficient with the general term (1.96) of the perturbation series for the S-matrix generating functional of the pair of fields ψ_1, ψ_2 with the contractions (1.141) and interaction $S_v(\varphi_2) - S_v(\varphi_1)$. After performing the differentiation in (1.140) the field $\psi_2 = \varphi_1 - \varphi_2$ is set equal to zero, in other words, only graphs in which all fields ψ_2 are contracted are considered.

Graphs with self-contracted lines are excluded, as usual, by transforming to the reduced vertex (1.99), which in the present case coincides with the left-hand side of (1.140) for $n = 1$ before setting $\varphi_1 = \varphi_2$. The cross terms in the derivative form do not contribute and we obtain

$$\exp\left[\frac{1}{2}\cdot\frac{\delta}{\delta\varphi_2}\bar{\Delta}\frac{\delta}{\delta\varphi_2}\right]S_v(\varphi_2) - \exp\left[\frac{1}{2}\cdot\frac{\delta}{\delta\varphi_1}\Delta\frac{\delta}{\delta\varphi_1}\right]S_v(\varphi_1)\,. \quad (1.142)$$

This reduced vertex, like the original one, is odd under the replacement $\varphi_1 \leftrightarrow \varphi_2$. Actually, owing to the t-locality of the interaction the contraction Δ in (1.142) can be replaced by n_s [see (1.30)] and, analogously, $\bar{\Delta} \to \bar{n}_s$. We still need to use the equation $n_s = \bar{n}_s$ noted at the end of the preceding section.

After transforming to the reduced interaction (1.142), Eq. (1.140) contains only graphs without self-contracted lines. Since the functional (1.142) is odd in the argument ψ_2 and we have the condition $\psi_2 = 0$, at least one directed line representing the contraction of ψ_2 with ψ_1 must leave each vertex of a graph. From this it follows that any graph will necessarily contain a closed loop of directed lines $\widetilde{M}_{21}$, and since the corresponding contraction (1.141) is the advanced one, the contribution of the graph will be zero. This is true for any graph, which proves the statement.

The proof of unitarity given above is based on the t-locality and Hermiticity of the interaction and on the general relation (1.29) between the simple and time-ordered contractions. We recall that we have started from the representation (1.64) for the S-matrix generating functional and have assumed that the interaction functional does not contain time derivatives of the field. The proof generalizes directly to interactions with time derivatives if such a derivative is treated as a set of independent fields $\hat{\varphi}_n \equiv \partial_t^n \hat{\varphi}$ (see Sec. 1.3.4). However, in the language of the effective interaction (1.83) the situation is more complicated.

As shown in Sec. 1.3.4, the effective interaction is always in Lagrangian form, but the corresponding Lagrangian is in general non-Hermitian. This non-Hermiticity must apparently lead to violation of unitarity. However, for an interaction involving time derivatives there is another effect leading to unitarity violation, and these two effects cancel each other out. The second one is the following. We have assumed it obvious that the presence of a closed loop of advanced functions inside a graph implies that the contribution of the graph vanishes. However, this is true only when it is known *a priori* that the expression corresponding to the graph cannot be so singular as to contain δ-like singularities when all or some of the time arguments t_i (assigned to the vertices of the graph) coincide. If there are such singularities and the closed loop of advanced functions is superimposed on one of them, the resulting expression is, strictly speaking, undefined, but after appropriate redefinition it may turn out to be nonzero. This is exactly what happens in those cases where the interaction functional contains time derivatives of the field, and for a Hermitian interaction this effect would lead to unitarity violation. As already stated above, in the case of the effective interaction (1.83) the presence of time derivatives is compensated for by the non-Hermiticity, and the S-matrix functional turns out to be unitary in each order of perturbation theory (see, for example, [10,11,12] for more detail).

1.6 Functional Integrals

1.6.1 Gaussian integrals

In the sections that follow we shall study the representation of various objects by functional (path) integrals. We begin this discussion with the simplest such integral, the Gaussian integral.

Let us first recall the rules for calculating Gaussian integrals on finite-dimensional spaces. Let x and y be real n-dimensional vectors, $xy \equiv \Sigma x_i y_i$ be a real scalar product, $xKx \equiv \Sigma x_i K_{is} x_s$ be a positive-definite quadratic form, and $Dx = dx_1 \ldots dx_n$. Then

$$J(K, y) \equiv \int Dx \exp\left[-\frac{1}{2}xKx + xy\right] = \\ = \det\left[\frac{K}{2\pi}\right]^{-1/2} \exp\left[\frac{1}{2}yK^{-1}y\right], \tag{1.143}$$

where K^{-1} is the inverse of the matrix K. In calculating the integral we make the substitution $x = x' + K^{-1}y$, removing the linear term in the argument of the exponential: $xKx - 2xy = x'Kx' - yK^{-1}y$. The factor $\exp[yK^{-1}y/2]$ is taken out of the integral, and the remaining integral $J(K, 0)$ is calculated using the substitution $x' = K^{-1/2}x''$. The Jacobian $Dx'/Dx'' = \det K^{-1/2}$ appears as a coefficient, and we are left with the integral $J(1, 0)$, which is equal to $(2\pi)^{n/2}$, where n is the dimensionality of the space. We have introduced the factor of 2π inside the determinant using the fact that $\det[\lambda K] = \lambda^n \det K$. In the case of complex variables the expression analogous to (1.143) has the form

$$\int D\mathrm{Re}z \, D\mathrm{Im}z \exp[-z^+Kz + y^+z + z^+y] = \\ = \det[K/\pi]^{-1} \exp[y^+K^{-1}y]. \tag{1.144}$$

In what follows we shall use Eqs. (1.143) and (1.144) not only for positive-definite matrices, but also for any nondegenerate matrix K. This is equivalent to defining improper Gaussian integrals by analytic continuation in the elements of K. Equations (1.143) and (1.144) have an obvious generalization to infinite-dimensional functional spaces:

$$\int D\varphi \exp\left[-\frac{1}{2}\varphi K\varphi + \varphi A\right] = \det[K/2\pi]^{-1/2} \exp\left[\frac{1}{2}AK^{-1}A\right], \tag{1.145}$$

$$\int D\mathrm{Re}\varphi \, D\mathrm{Im}\varphi \exp[-\varphi^+K\varphi + \varphi^+A + A^+\varphi] = \\ = \det[K/\pi]^{-1} \exp[A^+K^{-1}A]. \tag{1.146}$$

The symbol $\int D\varphi \ldots$ is understood as the integral over some linear space of functions, $\varphi K\varphi$ is a nondegenerate quadratic form on this space, K^{-1} is the inverse of K, and $\varphi A \equiv \int dx \varphi(x) A(x)$ is a linear form with the function A playing the role of a parameter.

The determinants of linear operators are usually calculated using the expression $\ln\det K = \mathrm{tr}\ln K$. If $K = 1 + M$, then the logarithm of the operator $1 + M$ can be expanded in a series $M - M^2/2 + M^3/3 - \ldots$. The trace of the general term of the series is easily calculated if M is a linear integral operator with known kernel $M(x, x')$: the product of operators corresponds to convolution of the kernels, and the trace of a linear integral operator with kernel $L(x, x')$ is equal to $\int dx L(x, x)$. A useful formula for calculating determinants is

$$\delta \ln \det K = \delta\, \mathrm{tr} \ln K = \mathrm{tr}\,(K^{-1}\delta K)\,, \qquad (1.147)$$

where δK is the variation of K when any (numerical or functional) parameter is changed. Here it is assumed that a change of a parameter entering into K does not change the linear space on which the operators K and K^{-1} are defined.

Returning to Eqs. (1.145) and (1.146), let us show that in universal notation, where φ and φ^+ are assumed to be independent components of the unified field Φ, Eq. (1.145) can be used also for a complex field. Writing the argument of the integrated exponential in (1.146) as

$$-\frac{1}{2}\begin{pmatrix}\varphi\\ \varphi^+\end{pmatrix}\begin{pmatrix}0 & K^{\mathrm{T}}\\ K & 0\end{pmatrix}\begin{pmatrix}\varphi\\ \varphi^+\end{pmatrix}+\begin{pmatrix}\varphi\\ \varphi^+\end{pmatrix}\begin{pmatrix}0 & 1\\ 1 & 0\end{pmatrix}\begin{pmatrix}A\\ A^+\end{pmatrix}\equiv-\frac{1}{2}\Phi\mathcal{K}\Phi+\Phi g\mathcal{A}\,,$$

we compute the integral (1.146) using the rule (1.145). Here it is necessary to use the fact that in the one-dimensional case the replacement φ, $\varphi^+ \to \mathrm{Re}\varphi$, $\mathrm{Im}\varphi$ corresponds formally to the Jacobian $-2i$, so that $D\mathrm{Re}\varphi\, D\mathrm{Im}\varphi = (-2i)^{-\nu} D\Phi$, where $D\Phi \equiv D\varphi D\varphi^+$ and ν is the "dimensionality" of the space φ (the dimensionality is numerically equal to the trace of the unit operator in the space). Therefore, after calculating the integral, written in universal notation, on the left-hand side of (1.146) using the rule (1.145), we obtain $(-2i)^{-\nu} \det\,[\mathcal{K}/2\pi]^{-1/2} \exp\left[\frac{1}{2}\mathcal{A}g^{\mathrm{T}}\mathcal{K}^{-1}g\mathcal{A}\right]$. We want to show that this expression coincides with the right-hand side of (1.146). The equality of the arguments of the exponentials is easily verified:

$$\frac{1}{2}\mathcal{A}g^{\mathrm{T}}\mathcal{K}^{-1}g\mathcal{A} = \frac{1}{2}\begin{pmatrix}A\\ A^+\end{pmatrix}\begin{pmatrix}0 & 1\\ 1 & 0\end{pmatrix}\begin{pmatrix}0 & \mathrm{K}^{-1}\\ K^{-1\mathrm{T}} & 0\end{pmatrix}\begin{pmatrix}0 & 1\\ 1 & 0\end{pmatrix}\begin{pmatrix}A\\ A^+\end{pmatrix} = A^+K^{-1}A,$$

and the equality of the preexponential factors is easily verified using the equation

$$\det \mathcal{K} = \det\begin{pmatrix} 0 & K^{\mathrm{T}} \\ K & 0 \end{pmatrix} = (-1)^{\nu}\det K^2 \tag{1.148}$$

and taking into account the fact that $\mathcal{K}$ is an operator in a space of dimensionality 2ν and therefore $\det[\mathcal{K}/2\pi] = (2\pi)^{-2\nu}\det \mathcal{K}$.

We have therefore verified the fact that when universal notation is used all Gaussian integrals can be calculated using the rule (1.145).

Equations (1.145) and (1.146) will henceforth be viewed as the definition of the Gaussian functional integral. This definition cannot, of course, claim to be mathematically rigorous, but in some sense it is "less ambiguous" than the Feynman (who first introduced functional integrals into physics) definition in terms of an interpolation procedure [13].* In what follows we shall treat functional integrals just like well convergent ordinary integrals: in particular, we shall make substitutions of the integration variables and differentiate with respect to a parameter inside the integral. We shall not attempt to justify the correctness of such operations, but only note that this is the fastest way of obtaining various closed expressions which in perturbation theory are derived using infinite sums of graphs.

1.6.2 Integrals on a Grassmann algebra

The integral on a finite-dimensional Grassmann algebra (see Sec. 1.1.3) with generators $\varphi \equiv \varphi_1 \dots \varphi_n$ is defined as a linear functional $I(f)$ on the space of functions (1.6), given on the basis monomials $\varphi_{i_1} \dots \varphi_{i_k}$ by the relations [2]

$$I(\varphi_{i_1} \dots \varphi_{i_k}) = 0 \quad \text{for} \quad k < n, \quad I(\varphi_1 \varphi_2 \dots \varphi_n) = (2\pi)^{-n/2}. \tag{1.149}$$

These relations are assumed to be valid for any choice of generators, i.e., the "integral is independent of the notation used for the integration variables." If

* There the functional integration variable φ is approximated by some interpolating function specified by a finite set of numerical parameters (for example, a function which is nearly piecewise-linear). The integral over all φ is replaced by an ordinary multidimensional integral over these parameters, and then the limit is studied as the accuracy of the approximation is improved. If the limit exists, it is taken to define the value of the functional integral. It is well known (see[14,15], for example) that this definition is ambiguous: the result depends explicitly on the choice of interpolation method. We shall therefore not use it, and instead base our discussion on equations of the type (1.145), which ensure uniqueness at least when functional integrals are used to calculate the terms of the perturbation series (see also the comment in Sec. 2.4).

we write the symbol $I(f)$ as $\int D\varphi f(\varphi)$ and understand it as a multiple integral ($D\varphi \equiv D\varphi_n ... D\varphi_1$), the definition (1.149) can be reformulated as the following rule for calculating single integrals for each of the generators φ_i: $\int D\varphi_i = 0$, $\int D\varphi_i\varphi_i = (2\pi)^{-1/2}$. The symbols $D\varphi_i$ here are assumed to anticommute with each other and with the generators φ_k, $k \neq i$. We can also have integrals over some rather than all of the generators; the φ_k over which there is no integration can be taken outside the integral, remembering that they anticommute with the symbols $D\varphi_i$.

It follows from the definition (1.149) that for any $f(\varphi)$

$$\int D\varphi\,(\vec{\partial} f(\varphi)/\partial\varphi_i) = \int D\varphi\,(\overleftarrow{\partial} f(\varphi)/\partial\varphi_i) = 0, \tag{1.150}$$

since the integral receives a contribution from only the highest monomial $\varphi_1 ... \varphi_n$, which certainly does not occur in the derivative $\partial f/\partial\varphi_i$.

Let us derive the expression for linear substitution of variables in the integral $I(f)$. Let $\varphi = L\psi$, where L is an arbitrary nondegenerate (i.e., $\det L \neq 0$) matrix. If φ and ψ were ordinary variables rather than generators of a Grassmann algebra, we would be able to write

$$\int D\varphi f(\varphi) = \int J D\psi f(L\psi), \tag{1.151}$$

where $J = D\varphi/D\psi = \det L$ is the Jacobian of the transformation. Equation (1.151) holds for a Grassmann algebra, but now J is an unknown quantity which must be found using (1.151) and the definition (1.149).

The rule (1.151) can be explained as follows. The integral is a linear functional on some space of functions; this space can be understood geometrically, assuming that the original function $f(\varphi)$ and the composite function $f(\varphi(\psi))$ obtained by the variable substitution $\varphi = \varphi(\psi)$ are different representatives of the same point in the space. Then (1.151) states that the integral is also a geometrical object, i.e., the value of $I(f)$ depends only on the choice of the point f, and not its representative. This very requirement determines the transformation law $D\varphi = J D\psi$ of the volume element.

To find J in the special case of a linear substitution we take f to be the highest monomial $\varphi_1\varphi_2...\varphi_n$. Then

$$(2\pi)^{-n/2} = \int D\varphi\varphi_1 ... \varphi_n = \int J D\psi \sum_{i_1...i_n} L_{1i_1} ... L_{ni_n} \psi_{i_1} ... \psi_{i_n}.$$

Assuming that J= const and taking it outside the integral, we obtain

$$1 = J \sum_{i_1 \ldots i_n} \varepsilon_P L_{1i_1} \ldots L_{ni_n}, \tag{1.152}$$

where $\varepsilon_P = \pm 1$ depending on the parity of the permutation $i_1 \ldots i_n \to 1 \ldots n$. The sum entering into (1.152) is $\det L$, from which $J = \det L^{-1}$. Clearly, this form of J guarantees the validity of (1.151) for any function $f(\varphi)$, since only the highest monomial contributes to the integral.

Therefore, for an integral on a Grassmann algebra the role of the Jacobian in (1.151) is played not by the determinant of L, but by the inverse determinant.

Now let $\varphi \equiv \varphi_1 \ldots \varphi_n$ and $a \equiv a_1 \ldots a_n$ be a complete set of generators of a Grassmann algebra, i.e., all the φ_i and the a_k anticommute in pairs. Using the definition of the integral, it is easy to show that

$$\int D\varphi\, f(\varphi) = \int D\varphi\, f(\varphi + a), \tag{1.153}$$

i.e., $D\varphi = D(\varphi + a)$. Equation (1.153) shows that an integral on a Grassmann algebra, like an ordinary integral, is invariant under translations of the integration variable.

Let us conclude by writing out the expression for the direct and inverse Fourier transforms (where n is the number of generators φ):

$$F(a) = \int D\varphi\, f(\varphi) \exp i\varphi a,$$

$$f(\varphi) = (2\pi)^n \int Da\; F(a) \exp(-i\varphi a). \tag{1.154}$$

The first of these equations is the definition of the function $F(a)$, and the second can be proved with the condition that the number of generators n is even and the function $f(\varphi)$ consists only of even monomials.

1.6.3 Gaussian integrals on a Grassmann algebra

Let us consider the Gaussian integral

$$I(K, a) \equiv \int D\varphi \exp\left[\frac{1}{2}\varphi K \varphi + \varphi a\right] \tag{1.155}$$

on a finite-dimensional Grassmann algebra with generators $\varphi \equiv \varphi_1 \ldots \varphi_n$ and $a \equiv a_1 \ldots a_n$. The number n of generators φ is assumed to be even, and the matrix K is assumed to be antisymmetric and nondegenerate.

To isolate the a dependence it is necessary to make a translation $\varphi = \varphi' - K^{-1}a$ of the integration variable. Using the rule (1.153), the antisymmetry of the matrix K, and the equation $\varphi a = -a\varphi$, we obtain

$$I(K,a) = I(K,0)\exp\left(\frac{1}{2}aK^{-1}a\right). \tag{1.156}$$

To calculate the integral $I(K,0)$ we make the replacement $\varphi = L\psi$, for which $K \to K' = L^{\mathrm{T}}KL$. It is well known that for any antisymmetric matrix K of even dimension $n = 2m$ it is possible to find a unimodular (i.e., having unit determinant) matrix L such that $K' = L^{\mathrm{T}}KL$ is a quasidiagonal matrix composed of m two-by-two blocks of the form $\begin{pmatrix} 0 & \lambda_i \\ -\lambda_i & 0 \end{pmatrix}$; the product $\lambda_1\lambda_2\ldots\lambda_m$ is called the *Pfaffian* of the matrix K and denoted as Pf K. The Pfaffian of K coincides up to a sign with the square root of $\det K = \lambda_1^2\lambda_2^2\ldots\lambda_m^2$. In particular, it is easily verified that

$$\mathrm{Pf}\begin{pmatrix} 0 & -B^{\mathrm{T}} \\ B & 0 \end{pmatrix} = (-1)^{m(m+1)/2}\det B, \tag{1.157}$$

where m is the dimension of the matrix B.

Returning to the calculation of the integral $I(K,0)$, we make the substitution $\varphi = L\psi$, choosing the matrix L as described above. Owing to the fact that L is unimodular, we have $I(K,0) = I(K',0)$. We write

$$\exp\left(\frac{1}{2}\varphi K'\varphi\right) = \prod_{k=1}^{m}\exp\left(\lambda_k\varphi_{2k-1}\varphi_{2k}\right) = \prod_{k=1}^{m}\left(1+\lambda_k\varphi_{2k-1}\varphi_{2k}\right).$$

Only the product of the second terms in each set of parentheses contributes to the integral over φ, and we find

$$I(K,0) = (2\pi)^{-m}\prod_{k=1}^{m}\lambda_k = \mathrm{Pf}\left(\frac{K}{2\pi}\right) = \varepsilon\det\left(\frac{K}{2\pi}\right)^{1/2}, \tag{1.158}$$

where $\varepsilon = \pm 1$ is a sign factor distinguishing Pf K from $\det K^{1/2}$.*

---

* We note that in contrast to the bosonic case, the convergence of the integral is not related to the sign of K.

In what follows we assume that Eqs. (1.156)–(1.158) are valid also in the infinite-dimensional case, i.e., for an anticommuting field φ (all the fermionic fields are complex, so that their dimensionality can always be assumed to be even):

$$\int D\varphi \exp\left[\frac{1}{2}\varphi K\varphi + \varphi A\right] = \varepsilon \det\left(\frac{K}{2\pi}\right)^{1/2} \exp\left(\frac{1}{2}AK^{-1}A\right). \qquad (1.159)$$

The integral $\int D\psi D\psi^{+} \exp[\psi^{+}K\psi + \psi^{+}A + A^{+}\psi]$ reduces to the one given above when we transform to universal notation:

$$\psi^{+}K\psi + \psi^{+}A + A^{+}\psi = \frac{1}{2}\begin{pmatrix}\psi \\ \psi^{+}\end{pmatrix}\begin{pmatrix}0 & -K^{\mathrm{T}} \\ K & 0\end{pmatrix}\begin{pmatrix}\psi \\ \psi^{+}\end{pmatrix} + \begin{pmatrix}\psi \\ \psi^{+}\end{pmatrix}\begin{pmatrix}0 & -1 \\ 1 & 0\end{pmatrix}\begin{pmatrix}A \\ A^{+}\end{pmatrix}.$$

Generalizing (1.157), we can express the Pfaffian of a two-by-two matrix in terms of the determinant of K up to a sign and obtain

$$\int D\psi D\psi^{+} \exp\left[\psi^{+}K\psi + \psi^{+}A + A^{+}\psi\right] =$$

$$= \varepsilon \det\left(\frac{K}{2\pi}\right) \exp\left(-A^{+}K^{-1}A\right). \qquad (1.160)$$

Equations (1.145) and (1.159) can be combined to form the single equation

$$\int D\varphi \exp\left[-\frac{1}{2}\varphi K\varphi + \varphi A\right] = \varepsilon \det\left(\frac{K}{2\pi}\right)^{-\varkappa/2} \exp\left[\frac{\varkappa}{2}AK^{-1}A\right]. \qquad (1.161)$$

1.6.4 Gaussian integrals in field theory

If the rule (1.161) is used to formally calculate the integral of $\exp i[\varphi K\varphi/2 + \varphi A]$, where $\varphi K\varphi/2$ is the quadratic form of the free action (1.4), then up to a coefficient independent of A we obtain $\exp[-\varkappa A\Delta A/2]$, where $\Delta = iK^{-1}$. Comparing this expression with (1.89) and taking into account the last equation in (1.31), we find that we have arrived at the generating functional for Green functions of the free theory under consideration.

This is in fact the desired result, but the derivation given above needs to be improved considerably. The point is that K in (1.161) is assumed to be a nondegenerate linear operator on the space of fields φ over which the integration runs; otherwise, the symbol K^{-1} becomes meaningless. In our case K, viewed as an operator on the set of "all fields," is degenerate, since the equation $K\varphi = 0$ has nontrivial solutions: free fields. If the integration space is taken to be, for example, a set of functions which fall off suitably, the operator iK^{-1} will be defined uniquely, but we do not know if its kernel coincides with the time-ordered contraction Δ entering into the functional (1.89) (we shall verify below that this doubt is well founded).

The problem must therefore be posed as follows. Let $\Delta(x, x')$ be some *a priori* selected solution of the equation $K\Delta = i$, i.e., one of the Green functions (the statement "up to a factor of i" is implicit) of the linear operator K. It is necessary to select an integration space $E(\Delta)$ such that on this space K is a nondegenerate operator, and the kernel of the uniquely defined operator iK^{-1} is the function Δ that we have selected.

In applications the role of Δ will be played by the time-ordered contraction which is defined independently of (1.29) and which, according to (1.31), is one of the Green functions of the operator K.

Let us show how to construct the desired space $E(\Delta)$. In calculating the integral of $\exp i[\varphi K\varphi/2+ \varphi A]$ we shift the integration variable $\varphi \to \varphi - \varphi_0$; this, first of all, must not change the integration range, which is equivalent to the requirement $\varphi_0 \in E(\Delta)$, and, second, it must remove the terms linear in φ in the argument of the integrated exponential, which, owing to the condition that K be symmetric on $E(\Delta)$, is equivalent to the requirement $K\varphi_0 = A$. The formal solution of this equation has the form $\varphi_0 = K^{-1}A \equiv -i\Delta A$, and we want the additional condition $\varphi_0 \in E(\Delta)$ to lead uniquely to the choice of the *a priori* specified function Δ.

It is clear from this discussion that the space $E(\Delta)$ must be invariant under shifts by functions of the form $i\Delta A$, where Δ is the selected Green function. The functional argument A plays the role of a parameter in the Gaussian integral, and we shall assume that A belongs to the space E of fields which fall off sufficiently rapidly and which have the required (depending on the theory) reality properties. A natural candidate for $E(\Delta)$ is the space $i\Delta E$, i.e., the set of all functions of the form $\varphi = i\Delta\varphi'$, where φ' is an arbitrary function of E. Let us check that the space $E(\Delta)$ defined in this manner satisfies all the necessary requirements. First, the form $\varphi_1 K\varphi_2$ is defined for any $\varphi_{1,2} \in E(\Delta)$. In fact, representing each of the functions φ_α in the form $i\Delta\varphi'_\alpha$ and using the equation $K\Delta = i$, we arrive at the expression $\Delta\varphi'_1 \cdot \varphi'_2$, the finiteness of which is ensured by the conditions $\varphi'_{1,2} \in E$. Second, if the selected function Δ is symmetric, then K is a symmetric operator on $E(\Delta)$, i.e., $\varphi_1 \cdot K\varphi_2 = K^{\mathrm{T}}\varphi_1 \cdot \varphi_2 = \varkappa K\varphi_1 \cdot \varphi_2$, as is easily verified using the substitution $\varphi_\alpha = i\Delta\varphi'_\alpha$. We note that the symmetry of K is a nontrivial property, since functions from $E(\Delta)$ are not, as we shall see below, well behaved for $t \to \pm\infty$. Integration by parts moves the differential operator K onto a different argument, and the symmetry implies that this does not lead to surface terms. We should emphasize the fact that the above proof of the symmetry applies only to the operator K itself, and not to every differential operator; in general, integration by parts for functions from $E(\Delta)$ leads to the appearance of surface terms.

Third, the operator K on $E(\Delta)$ is nondegenerate, i.e., the homogeneous equation $K\varphi = 0$ has no nontrivial solutions in $E(\Delta)$. In fact, taking $\varphi = i\Delta\varphi'$ and requiring $K\varphi = 0$, we obtain $\varphi' = 0$, from which $\varphi = 0$. Therefore, the inhomogeneous equation $K\varphi = A$ with the condition $\varphi \in E(\Delta)$ has the unique solution $\varphi = -i\Delta A$, which is what we have obtained. We now verify that the requirement $\varphi \in E(\Delta)$ is equivalent to the imposition of certain asymptotic conditions on the equation $K\varphi = A$.

The space $E(\Delta)$ contains the space of functions E which fall off sufficiently rapidly at infinity, as can be verified by taking φ' in the representation $\varphi = i\Delta\varphi'$ to be a function of the form $K\varphi''$, $\varphi'' \in E$, for which $\varphi = -\varphi'' \in E$. However, in the general case functions from $E(\Delta)$ do not fall off well for $t \to \pm\infty$. In fact, from (1.29) and the condition $\varphi' \in E$ it follows that the function $\varphi = i\Delta\varphi'$ has asymptote $\varphi^{(+)} = in\varphi'$ for $t \to \infty$ and asymptote $\varphi^{(-)} = i\varkappa n^{\mathrm{T}}\varphi'$ for $t \to -\infty$. We see from (1.31) that each of these asymptotes is a solution of the free equation: $K\varphi^{(+)} = K\varphi^{(-)} = 0$. The solutions $\varphi^{(\pm)}$ are in general different but not completely independent, since the two asymptotes are determed by the same function $\varphi' \in E$. From the physical point of view the solutions of the free equation do not belong to the class of fields which fall off well at infinity, since such solutions correspond to nonzero flux (of particle number, energy, and so on) for $t = \pm\infty$. We therefore do not have the right to define the Gaussian integral of interest as an integral over a space of fields which fall off rapidly at infinity; in general, shifts by $i\Delta A$, $A \in E$ take us out of this space and it becomes impossible to use (1.161).

The asymptotes of the functions from $E(\Delta)$ for various specific theories are discussed in Sec. 2.5.

In conclusion, we note that in bosonic theories of the oscillator type (see the next chapter) an improper Gaussian integral can be defined by the regularization $K \to K_\varepsilon = K + i\varepsilon$, which makes the operator $\Delta_\varepsilon = iK_\varepsilon^{-1}$ unique. In such theories the contraction Δ defined by (1.29) coincides with the $\varepsilon \to 0$ limit of the operator Δ_ε for the appropriate choice of sign of ε, where this sign always turns out to be such that the addition $i\varepsilon$ introduces a cutoff, rather than growing, factor into the integrand, making the Gaussian integral formally "convergent." The regularized contraction Δ_ε turns out to fall off sufficiently rapidly for $|t - t'| \to \infty$, so that all functions of the form $i\Delta_\varepsilon\varphi'$ forming the space $E(\Delta_\varepsilon)$ will also fall off sufficiently rapidly. However, this procedure is not universal: there are cases where $\Delta \neq \lim \Delta_\varepsilon$ for $\varepsilon \to 0$ (Sec. 2.1.2) or where it is necessary to use different Δ for the same operator K (Sec. 2.2).

1.6.5 *Representations of generating functionals for the* S*-matrix and Green functions by functional integrals*

The arguments of the preceding section make it possible to represent the generating functional of the Green functions of the free theory (1.89) as a Gaussian integral over the space of functions $E(\Delta)$:

$$G^{(0)}(A) = \exp(-\frac{\varkappa}{2}A\Delta A) = c\int_{\Delta} D\varphi \exp i[S_0'(\varphi) + \varphi A]. \quad (1.162)$$

Here $S_0' \equiv \varphi K\varphi/2$ is the quadratic form of the free action, which, as explained in Sec. 1.1.2, differs from the free-action functional S_0 by surface terms. The symbol $\int_{\Delta} D\varphi\ldots$ in (1.162) denotes integration over $E(\Delta)$, and the constant c is defined as

$$c^{-1} = \int_{\Delta} D\varphi \exp iS_0'(\varphi) \ . \quad (1.163)$$

For a theory with an interaction, formally using the relations (1.91), (1.162), and (1.12) we obtain the representation

$$G(A) = \exp iS_v(\varkappa\frac{\vec{\delta}}{\delta iA})G^{(0)}(A) = c\int_{\Delta} D\varphi \exp i[S(\varphi) + \varphi A], \quad (1.164)$$

in which $S(\varphi) = S'_0(\varphi) + S_v(\varphi)$. We shall call this functional the *action*, but it should be remembered that, first, the functional $S'_0(\varphi)$, as mentioned above, differs from the free action and, second, the functional $S_v(\varphi)$ for the case of an interaction containing time derivatives of the fields is an effective interaction, as discussed in Sec. 1.3.4.

Equation (1.164) is equivalent to the following representation of the full Green functions (1.75):

$$G_n(x_1 \ldots x_n) = c\int_{\Delta} D\varphi\ \varphi(x_1)\ldots\varphi(x_n) \exp iS(\varphi). \quad (1.165)$$

Let us now turn to the S-matrix generating functional (1.84). The integral representation for this functional can be obtained by substituting the representation (1.164) into (1.94). It is more instructive to follow another derivation, which is based on the following representation of the differential reduction operator entering into the definition (1.84):

$$\exp\left(\frac{\varkappa}{2}\cdot\frac{\vec{\delta}}{\delta\varphi}\Delta\frac{\vec{\delta}}{\delta\varphi}\right) = c\int_{\Delta} D\psi \exp\left[iS_0'(\psi) + \psi\frac{\vec{\delta}}{\delta\varphi}\right]. \tag{1.166}$$

The integration space and the constant c are the same as in (1.162). For fermions we have used (1.9) to convert to left derivatives so that we can use (1.11), showing that the operator $\exp[\psi\vec{\delta}/\delta\varphi]$ is a shift by ψ, as in the bosonic case. Therefore,

$$\exp\left(\frac{1}{2}\cdot\frac{\delta}{\delta\varphi}\Delta\frac{\delta}{\delta\varphi}\right)F(\varphi) = c\int_{\Delta} D\psi F(\varphi+\psi)\exp iS_0'(\psi) \tag{1.167}$$

for any functional F. In the special case of the S-matrix generating functional (1.84) we obtain

$$R(A) = c\int_{\Delta} D\varphi \exp i[S_0'(\varphi) + S_v(\varphi + A)]. \tag{1.168}$$

If the representations (1.164) and (1.168) are assumed to be known, then (1.93) relating the generating functionals of the S matrix and the Green functions is easily obtained by a shift $\varphi \to \varphi - A$ of the integration variable in (1.168), remaining within $E(\Delta)$ for $A \in E \subset E(\Delta)$.

It would be incorrect to take expressions of the type (1.164) and (1.168) as exact representations, completely free from any connection with perturbation theory. This is not true at all: specification of the integration region $E(\Delta)$, which is determined by only the free part of the action, is equivalent to the implicit assumption that the corresponding perturbation theory is used. One should therefore not be surprised at the appearance of internal inconsistencies in such expressions in cases where perturbation theory is known to be incorrect. In classical language this implies a qualitative difference between the asymptotic behavior of the solutions of the exact and free equations of motion, and in quantum language it implies a qualitative difference between the spectra of the corrresponding Hamiltonians. For example, we can attempt to calculate the functional (1.162) using perturbation theory, by artificially splitting the form $\varphi K\varphi$ into a "free" part $\varphi K'\varphi$ and an "interaction" part $\varphi(K - K')\varphi$. As a result, instead of (1.162) we obtain a Gaussian integral of the same expression, but over a different region $E(\Delta')$, which itself is inconsistent.

It might be thought that in correct expressions like (1.164) and (1.168) the integration space must be determined by the asymptotes of the solutions of the classical equation of motion for the full action, rather than its free part, and that perturbation theory is correct only when the inclusion of the interaction does not change the integration space. We note that this is the case in renormalized perturbation theory [1].

In practice, people do not pay attention to the definition of the integration region in (1.164) and (1.168), because representations of this type are used mainly to simplify the derivation of various closed expressions which in ordinary language are obtained by summing the diagrammatic series of perturbation theory. Experience shows that to reproduce such results it is sufficient to formally manipulate functional integrals without paying any attention to the integration region. As an example, let us consider the calculation of the functional involved in (1.112):

$$I(\Delta, M, \varphi) \equiv \exp\left(\frac{1}{2} \cdot \frac{\delta}{\delta\varphi}\Delta\frac{\delta}{\delta\varphi}\right)\exp\left(\frac{1}{2}\varphi M\varphi\right).$$

Using (1.167), we can write this as a Gaussian integral ($K = i\Delta^{-1}$):

$$c\int D\psi \exp\left[-\frac{1}{2}\psi\Delta^{-1}\psi + \frac{1}{2}(\varphi+\psi)M(\varphi+\psi)\right]. \qquad (1.169)$$

Formally applying the rule (1.161) for calculating the Gaussian integrals (1.163) and (1.169), we obtain

$$I(\Delta, M, \varphi) = \det(1 - M\Delta)^{-\varkappa/2}\exp\left[\frac{1}{2}\varphi(M^{-1} - \Delta)^{-1}\varphi\right],$$

which is the same as the result of direct summation of the graphs of perturbation theory for the functional $I(\Delta, M, \varphi)$.

1.6.6 The stationary-phase method

Let us assume that instead of (1.164) we have the representation

$$G_\lambda(A) = c_\lambda \int_\Delta D\varphi \exp i\lambda^{-1}[S(\varphi) + \varphi A], \qquad (1.170)$$

in which λ is a numerical parameter and the normalization constant c_λ differs from c by the replacement $S'_0 \to \lambda^{-1} S'_0$ in (1.163). In a real situation λ would be Planck's constant $\hbar$, which we have taken to be unity.

In the graphs for G the introduction of λ reduces to multiplication of each line by λ and division of each vertex in (1.123) by λ, so that a graph obtains a factor of λ to the power $N - V$, where N is the number of lines and V is the number of vertices. For a connected graph the quantity $p \equiv N - V + 1$ is the number of loops, i.e., the number of independent closed contours in the graph. In a translationally invariant theory p is the number of independent integration momenta: a momentum is assigned to each line, and the δ function in momentum at each vertex, expressing momentum conservation in the interaction, decreases the number of independent momenta by one. However, the total number of independent momenta is $N - V + 1$ rather than $N - V$, because one (in a connected graph) δ function appears as an overall factor expressing the conservation of the external momenta.

It is thus clear that the expansion of $W_\lambda = \ln G_\lambda$ in powers of λ

$$W_\lambda (A) = \lambda^{-1} \sum_{p=0}^{\infty} \lambda^p W^{(p)} (A) \tag{1.171}$$

is simultaneously an expansion in the number of loops: the general term of the series is the sum of all connected graphs with p loops.

Let us now consider the procedure for calculating the integral (1.170) using the stationary-phase method. We assume that for each fixed $A \in E$ the functional $S(\varphi) + \varphi A$ has a single stationarity point $\psi = \psi(A)$ in the space $E(\Delta)$ over which the integration in (1.170) runs. Then, after expanding the argument of the exponential in (1.170) in a Taylor series about the stationarity point ψ and making the shift $\varphi \to \varphi + \psi$, which owing to the linearity of $E(\Delta)$ does not change the integration region, we obtain

$$G_\lambda (A) = \exp i\lambda^{-1} [S(\psi) + \psi A]\, c_\lambda \int_\Delta D\varphi \times$$

$$\times \exp \{ i\lambda^{-1} \sum_{n=2}^{\infty} \frac{1}{n!} \cdot \frac{\delta^n S(\psi)}{\delta \psi^n} \varphi^n \} . \tag{1.172}$$

The integral involved here, multiplied by the normalization constant d_λ

$$d_\lambda^{-1} = \int_\Delta D\varphi \exp \left[\frac{i}{2} \lambda^{-1} \frac{\delta^2 S(\psi)}{\delta \psi^2} \varphi^2 \right], \tag{1.173}$$

can be represented in the usual manner as the sum of unity and all graphs with lines $\lambda\Delta$ and vertices $\lambda^{-1}\mathcal{M}_n$, $n \geq 3$, where

$$\Delta = i\left(\frac{\delta^2 S(\psi)}{\delta\psi^2}\right)^{-1} = \left[\Delta^{-1} - i\frac{\delta^2 S_v(\psi)}{\delta\psi^2}\right]^{-1}$$

and $\mathcal{M}_n \equiv i\delta^n S_v(\psi)/\delta\psi^n$, $n \geq 3$.

Taking the logarithm of (1.172), we obtain

$$W_\lambda(A) = i\lambda^{-1}[S(\psi) + \psi A] + \ln(c_\lambda/d_\lambda) + \widetilde{W}_\lambda(A), \qquad (1.174)$$

where $\widetilde{W}_\lambda(A)$ denotes the sum of all connected graphs whose lines and vertices are defined above. Classifying these graphs according to the number of loops, we write

$$\widetilde{W}_\lambda(A) = \lambda^{-1}\sum_{p=2}^{\infty}\lambda^p \widetilde{W}^{(p)}(A) . \qquad (1.175)$$

The expansion begins with $p = 2$, since any graph of $\widetilde{W}$ has at least two loops owing to the absence of the vertex $\mathcal{M}_1$; the number of graphs of $\widetilde{W}$ with given p is finite for any p for the same reason.

Calculating the Gaussian integrals (1.163), (1.173) using the rule (1.161), we obtain

$$\ln(c_\lambda/d_\lambda) = -(\varkappa/2)\,\mathrm{tr}\ln(\Delta^{-1}\Delta). \qquad (1.176)$$

Then, comparing representations (1.171) and (1.174)–(1.176), we conclude that

$$W^{(0)}(A) = i[S(\psi) + \psi A], \quad W^{(1)}(A) = -(\varkappa/2)\,\mathrm{tr}\ln(\Delta^{-1}\Delta) \qquad (1.177)$$

and $W^{(p)}(A) = \widetilde{W}^{(p)}(A)$ for $p \geq 2$. Therefore, calculation of the functional integral in (1.170) by the stationary-phase method automatically gives the expansion of the functional $W_\lambda = \ln G_\lambda$ in a series in the number of loops.

The lines and vertices of the graphs of $\widetilde{W}$ contain as a parameter the stationarity point ψ, which is found from the equation

$$\delta S(\psi)/\delta\psi + A = K\psi + \delta S_v(\psi)/\delta\psi + A = 0 \qquad (1.178)$$

with the auxiliary condition $\psi \in E(\Delta)$. We note that the symmetry of the operator K on the space $E(\Delta)$ established in Sec. 1.6.4 ensures the absence of surface terms in the variation of the form $\varphi K\varphi$ for variations $\delta\varphi$ from $E(\Delta)$: $\delta(\varphi K\varphi) = \delta\varphi\cdot K\varphi + \varphi\cdot K\delta\varphi = 2\delta\varphi\cdot K\varphi$.

Equation (1.178) can be solved by iterations in powers of S_v, which leads to representation of ψ as an infinite sum of graphs without loops, the substitution of which into the right-hand side of (1.174) returns us, of course, to the original diagrammatic representation of $W_\lambda(A)$. In field theory graphs without loops are called *tree diagrams*, and in graph theory they are called *Cayley trees*.

The use of the stationary-phase method to calculate the integral (1.168) representing the S-matrix functional leads to the equation

$$K\varphi + \delta S_v(\varphi + A)/\delta\varphi = 0 \qquad (1.179)$$

with the condition $\varphi \in E(\Delta)$. If the argument A is on-shell, i.e., if it is a solution of the free equation $KA = 0$, then in (1.179) it is possible to change to the variable $\psi = \varphi + A$, for which $\delta S(\psi)/\delta\psi = 0$ with the condition $\psi \in E(\Delta) + A$. The substitution $S'_0(\varphi) \to S'_0(\varphi + A)$ directly into (1.168) is not allowed, because $\varphi KA = 0$, but $AK\varphi \neq K^T A\cdot\varphi = 0$ for $\varphi \in E(\Delta)$.

1.6.7 The Dominicis–Englert theorem

The integral (1.164) formally solves the problem of determining the Green functions from a given action functional $S(\varphi)$. The formal solution of the inverse problem can be obtained very easily by noting [16,17] that the integral (1.164) is a functional Fourier transform, and writing the inverse transform as (where ν is the dimensionality of the φ space)

$$(2\pi)^{\nu\varkappa} c \exp iS(\varphi) = \int DA\, G(A) \exp(-i\varphi A)\,. \qquad (1.180)$$

For fermions we have used the generalization of (1.154). This is applicable because the fermionic field is actually always complex and its dimensionality can be taken to be even, and the action functional is always assumed to be bosonic.

Making the substitution $G = \exp W$ in (1.180) and redefining the variables, we obtain

$$(2\pi)^{\nu\varkappa} c \exp\ iS(-A) = \int D\varphi \exp\left[W(\varphi) + iA\varphi\right]. \qquad (1.181)$$

Comparing this with (1.164), we note that up to a normalization factor the right-hand side of (1.181) has the form of the Green-function generating functional of a theory with action $S'(\varphi) = -iW(\varphi)$. This shows that there is a sort of duality between the functionals S and W: for a theory with action S the functional W represents the connected Green functions, and for a theory with action $S' = -iW$ the connected Green functions are represented by the functional iS (up to normalizations).

Perturbation theory for the integral (1.181) gives the diagrammatic technique for calculating the action functional S in terms of the known functional W, i.e., the Green functions. If the functional S is assumed to be known, the equations obtained in this way can be viewed as a system of equations for finding the unknown Green functions.

We multiply both sides of (1.181) by the constant d defined by $d^{-1} = \int D\varphi \exp(-\varkappa\varphi D\varphi/2)$, in which D is the full propagator (see the definitions in Sec. 1.4.9). Then using (1.167), we obtain

$$(2\pi)^{\nu\varkappa} cd \exp\ iS(-A) =$$

$$= \exp\left(\frac{1}{2}\frac{\delta}{\delta\varphi}D^{-1}\frac{\delta}{\delta\varphi}\right)\exp\left\{iA\varphi + \sum_{n\neq 2}\frac{1}{n!}W_n(i\varphi)^n\right\}\Bigg|_{\varphi=0}. \qquad (1.182)$$

According to the general rules of Sec. 1.4.1, the right-hand side of this equation is represented as the sum of unity and all graphs with lines D^{-1} and vertex factors $i^n W_n + iA\delta_{n1}$, where W_n are the coefficients in the expansion (1.127) of the functional $W(A)$, i.e., the connected Green functions of the theory with action S. By taking the logarithm of (1.182) and equating the coefficients of each power of A on both sides of the resulting equation, we can represent bare quantities, i.e., coefficients in the expansion of the functional S, as sums of skeleton graphs with "dressed" lines and vertices as discussed above. We shall not dwell on this, since it is much easier to obtain such relations using the technique of functional Legendre transformations (see Chap. 6).

We note that the logarithm of the preexponential factor in (1.182) is easily found by calculating the Gaussian integrals determining c and d using the rule (1.161): $\ln[cd(2\pi)^{\nu\varkappa}] = -(\varkappa/2)\,\mathrm{tr}\ln(D^{-1}\Delta)$ (we recall that Δ and D are respectively the bare and the full propagators). The factor $(2\pi)^{\nu\varkappa}$ from the

Fourier transform has been cancelled by the analogous factors from the Gaussian integrals.

1.7 Equations in Variational Derivatives

1.7.1 The Schwinger equations

The well known Schwinger equations [18] are exact relations between various Green functions. All these equations can be formulated as a single equation in variational derivatives, which is most simply derived [13] using the equation

$$0 = \int D\varphi\ \vec{\delta}\exp i\,[S(\varphi) + \varphi A]/\delta\varphi(x)\,, \tag{1.183}$$

expressing the invariance of the measure $D\varphi$ under translations $\varphi \to \varphi + \varepsilon$ by a suitably decreasing function $\varepsilon(x)$ (see Sec. 1.7.3 for more details). The justification of (1.183) for fermions can be found in (1.150).

Differentiating the exponential in (1.183), we obtain

$$0 = \int D\varphi\left[\frac{\vec{\delta}S(\varphi)}{\delta\varphi(x)} + A(x)\right]\exp i\,[S(\varphi) + \varphi A]\,.$$

In the bosonic case multiplication of the integrand by φ is equivalent to differentiation of the integral with respect to iA, and in the fermionic case it is equivalent to differentiation with respect to $-iA$, as seen from (1.12). This makes it possible to take the functional $\vec{\delta}S/\delta\varphi$ outside the integral in the form of a differential operator, which gives the desired Schwinger equation:

$$\left\{\left.\frac{\vec{\delta}S(\varphi)}{\delta\varphi(x)}\right|_{\varphi = \varkappa\vec{\delta}/\delta iA} + A(x)\right\}G(A) = 0. \tag{1.184}$$

For the rest of this chapter all equations will be given for the case of a bosonic field, in order to simplify the notation. The generalization to fermions requires only keeping track of the type (left or right) and ordering of the derivatives, and this is easy to do in any particular case.

For a theory with action which is a polynomial in the field, (1.184) is a finite-order equation in variational derivatives. The equation determines a solution only up to a numerical factor; the functional $H(A) = G_0^{-1}G(A)$ is uniquely determined. Making the substitution $G(A) = \exp W(A)$ in (1.184) and, after the differentiation, canceling the factor $\exp W$, we obtain the Schwinger equation for the generating functional of connected Green functions (see the definitions in Sec. 1.4.9). Equating the coefficients of all powers of iA in the resulting equations to zero term by term, we arrive at an infinite set of coupled equations for the Green functions.

As an example, let us consider the equations for the connected Green functions in the theory with action $S(\varphi) = \varphi K\varphi/2 + \lambda\int dx\varphi^3(x)/6$. The Schwinger equation (1.184) for this theory takes the form

$$\{\int dx' K(x,x')\,\delta/\delta iA(x') + \frac{\lambda}{2}\delta^2/\delta(iA(x))^2 + A(x)\}\,G(A) = 0. \tag{1.185}$$

Substituting $G = \exp W$ into this, we obtain the equation for W, which we write in abbreviated form as [the notation can be understood by comparing with (1.185)]:

$$K\frac{\delta W}{\delta iA} + \frac{\lambda}{2}\left[\frac{\delta^2 W}{\delta(iA)^2} + \left(\frac{\delta W}{\delta iA}\right)^2\right] + A = 0.$$

It is convenient to contract this equation with the function $[A\Delta](x) \equiv \int dx' A(x')\Delta(x', x)$:

$$A\frac{\delta W}{\delta A} = \frac{\lambda}{2}\left[A\Delta\frac{\delta^2 W}{\delta A^2} + A\Delta\left(\frac{\delta W}{\delta A}\right)^2\right] - A\Delta A. \tag{1.186}$$

We used the fact that $\Delta = iK^{-1}$. Substituting $W(A)$ in the form of the series (1.127) and equating the coefficients of all powers of iA, we obtain a system of coupled equations for the connected Green functions W_n. These equations are conveniently written out graphically using the notation $W_n \equiv$ , $i\lambda \equiv$ , and $\Delta \equiv$ . In this notation the system of coupled equations obtained from (1.186) becomes ($n = 0, 1, 2,\ldots$):

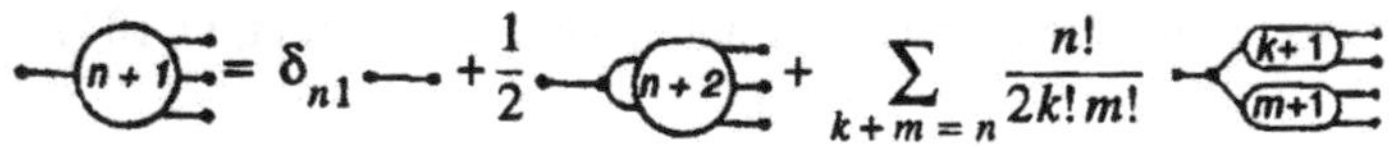

(1.187)

Here δ_{n1} is the Kronecker delta. The first two equations in (1.187) are

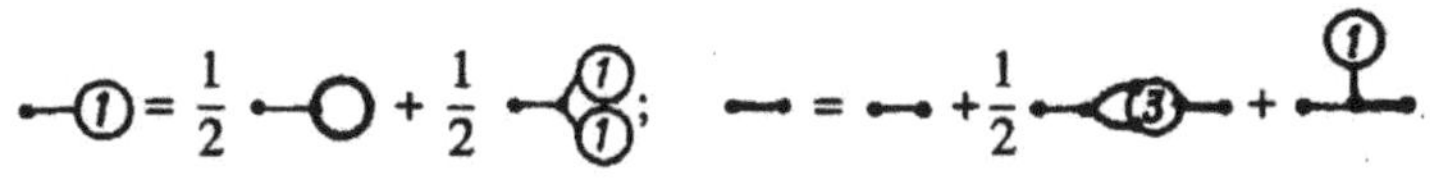
(1.188)

The heavy line denotes the full propagator $D \equiv W_2 \equiv$ 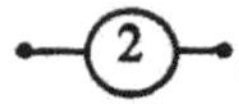, and the circle 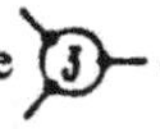 denotes the amputated function (see the definition in Sec. 1.4.9). Multiplying the second of Eqs. (1.188) on the right by D^{-1} and on the left by Δ^{-1}, we obtain

$$\Delta^{-1} - D^{-1} = \frac{1}{2}$$ 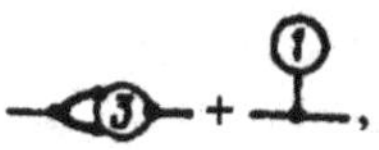 (1.189)

which determines the quantity $\Sigma \equiv \Delta^{-1} - D^{-1}$, called the *self-energy* or the *mass operator* [1].

The Schwinger equations for any other specific (polynomial) theory are just as easy to obtain. The starting point can always be the general expression (1.183), which, if universal notation is not used, becomes a system of equations, one for each independent field φ.

1.7.2 *Linear equations for connected Green functions*

The Schwinger equation (1.184) is linear in the full Green functions G_n, but nonlinear in their connected parts W_n. Linear equations for the connected functions can be obtained from the Dominicis–Englert formula (1.181) in exactly the same way as the Schwinger equations are obtained from the integral (1.164):

$$0 = \int D\varphi \frac{\delta}{\delta\varphi(x)} \exp\left[W(\varphi) + iA\varphi\right].$$

From this taking into account (1.181) we obtain the desired equation:

$$\left\{ \frac{\delta W(\varphi)}{\delta\varphi(x)}\bigg|_{\varphi = \delta/\delta iA} + iA(x) \right\} \exp iS(-A) = 0. \tag{1.190}$$

Multiplying it by $\exp(-iS(-A))$ and writing

$$Q_n(x_1...x_n;A) \equiv \exp[-iS(-A)]\frac{\delta^n}{\delta A(x_1)...\delta A(x_n)}\exp[iS(-A)],$$

we find

$$A(x_1) + \sum_{n=1}^{\infty}\frac{1}{(n-1)!}\times$$

$$\times\int...\int dx_2...dx_n W_n(x_1x_2...x_n)Q_{n-1}(x_2...x_n;A) = 0. \qquad (1.191)$$

Equating the coefficients of all powers of A to zero, we obtain an infinite set of equations for the connected Green functions W_n. These equations, in contrast to the Schwinger equation (1.184), are linear in each of the functions W_n, but they couple together all the functions W_n, even for a polynomial action S. The Schwinger equations in this case couple together only a finite number of connected functions. This is the price paid for the gain in linearity.

1.7.3 General method of deriving the equations

Let us consider the most general form of the action functional:

$$iS(\varphi) \equiv A(\varphi) = \sum_{n=0}^{\infty}\frac{1}{n!}A_n\varphi^n \equiv$$

$$\equiv \sum_{n=0}^{\infty}\frac{1}{n!}\int...\int dx_1...dx_n A_n(x_1...x_n)\varphi(x_n)...\varphi(x_1). \qquad (1.192)$$

The symmetric functions $A_n(x_1...x_n)$, which will be referred to as *potentials*, can be either ordinary functions or generalized functions. A particular theory corresponds to a choice of certain values of the potentials $A \equiv \{A_0A_1...\}$. In the fermionic case odd potentials A_{2k+1} must be taken to be fermionic and even ones bosonic, so that the functional (1.192) is always bosonic. The integral

$$G(A) = \int D\varphi \exp iS(\varphi) = \int D\varphi \exp A(\varphi) \qquad (1.193)$$

is a functional of all the potentials A. This functional obviously satisfies the following system of equations:

$$\frac{\delta G}{\delta A_n(x_1 \ldots x_n)} = \frac{1}{n!}\frac{\delta^n G}{\delta A_1(x_1)\ldots\delta A_1(x_n)}, \quad n \neq 1 \tag{1.194}$$

(for $n = 1$ we obtain an identity), which we shall refer to as the *connection equations* and write in abbreviated notation as

$$\frac{\delta G}{\delta A_n} = \frac{1}{n!}\left(\frac{\delta}{\delta A_1}\right)^n G, \quad n \neq 1.$$

In addition, the functional (1.193) satisfies the Schwinger equation (1.184):

$$0 = \int D\varphi \frac{\delta}{\delta\varphi} \exp A(\varphi) = \left\{ \sum_{n=1}^{\infty} \frac{1}{(n-1)!} A_n \left(\frac{\delta}{\delta A_1}\right)^{n-1} \right\} G(A). \tag{1.195}$$

Here and below we do not write the argument x explicitly, understanding that

$$[A_n \varphi^{n-1}](x_1) \equiv \int \ldots \int dx_2 \ldots dx_n A_n(x_1 \ldots x_n)\, \varphi(x_n) \ldots \varphi(x_2).$$

The Schwinger equation determines $G(A)$ up to a factor independent of the potential A_1, while the entire system (1.194), (1.195) determines $G(A)$ up to a numerical factor.

Let us assume that the functional $A(\varphi)$ contains only even potentials A_{2n}, and ask the question: how can the complete system of equations for $G(A)$ be written in terms of only even potentials? This formulation of the problem is meaningful for a fermionic theory if we want to completely avoid the use of anticommuting variables.

The connection equations are obvious in this case:

$$\frac{\delta G}{\delta A_{2n}} = \frac{2^n}{(2n)!}\left(\frac{\delta}{\delta A_2}\right)^n G, \quad n \neq 1,$$

and the analog of the Schwinger equation can be derived from the expression

$$0 = \int D\varphi \frac{\delta}{\delta\varphi(x)} [\varphi(x') \exp A(\varphi)] =$$

$$= \int D\varphi \left[\delta(x - x') + \varphi(x') \frac{\delta A(\varphi)}{\delta\varphi(x)} \right] \exp A(\varphi).$$

The expression inside the square brackets in the last integral contains only even powers of the field, so it can be taken outside the integral in the form of a functional of the derivative $\delta/\delta A_2$.

Let us now discuss equations of the type (1.184), (1.195) from a more general point of view. We assume that the substitution $\varphi(x) = F(x; \varphi')$ has been made in the integral (1.193), where F is some functional of φ' depending parametrically on x. We assume that in the substitution $\varphi \to \varphi'$ the space of fields over which the integration runs in (1.193) is transformed into itself. In the fermionic case we would restrict ourselves to linear inhomogeneous transformations of the field, which were studied in Sec. 1.6.2.

The Jacobian of the substitution $J = D\varphi/D\varphi'$ is equal to $\det M^{\times}$, where M is a linear operator with kernel $M(x, x') = \delta\varphi(x)/\delta\varphi'(x')$. This kernel, and also J, has a functional dependence on φ' if the substitution is nonlinear. In the rest of our discussion an important role will be played by transformations $\varphi \to \varphi'$ for which the Jacobian $D\varphi/D\varphi'$ is equal to unity. Such transformations obviously form a group, which we shall call the *group of motions of the* φ *space.*

Each regular substitution $\varphi = \varphi(\varphi')$ induces a potential transformation $A \to A'$:

$$A(\varphi) = A(\varphi(\varphi')) \equiv A'(\varphi'). \tag{1.196}$$

The integral (1.193) is clearly invariant under all transformations $A \to A'$ induced by the group of motions of the φ space:

$$G(A) = \int D\varphi \exp A(\varphi) = \int D\varphi' \exp A'(\varphi') = G(A'). \tag{1.197}$$

The simplest elements of the group of motions are translations by some fixed function ε which falls off suitably*: $\varphi(x) = \varphi'(x) + \varepsilon(x)$. These transformations form a group, and the Schwinger equation (1.195), as we now verify, expresses the invariance of the functional $G(A)$ under the induced group of potential transformations. Actually, this invariance is expressed by the equation

$$G(A) = \int D\varphi \exp A(\varphi) = \int D\varphi \exp A(\varphi + \varepsilon) \equiv G(A^{\varepsilon}),$$

* We recall that the spaces $E(\Delta)$ defined in Sec. 1.6.4 are invariant under such translations.

which shows that the functional $G(A^{\varepsilon})$ is actually independent of ε. Its first variation with respect to ε in the neighborhood of $\varepsilon = 0$ is

$$\delta_{\varepsilon} G = \int dx\ \varepsilon(x) \int D\varphi \delta / \delta\varphi(x) \cdot \exp A(\varphi),$$

from which we see that (1.195) is equivalent to the requirement that $G(A)$ be invariant under infinitesimal transformations of the induced group $A \to A^{\varepsilon}$.

Now let $\varphi(x) = F(x, \varepsilon; \varphi')$ be some Lie group with parameters (numerical or functional) ε which is a subgroup of the group of motions. We shall assume that $\varphi = \varphi'$ for $\varepsilon = 0$. The invariance of the functional (1.193) under infinitesimal transformations of the induced group $A \to A^{\varepsilon}$ is expressed by the equation

$$0 = \int D\varphi \delta_{\varepsilon} \exp A(\varphi) = \int D\varphi \delta_{\varepsilon}\varphi \frac{\delta A(\varphi)}{\delta\varphi} \exp A(\varphi), \tag{1.198}$$

where $\delta_{\varepsilon}\ldots$ denotes the first variation with respect to ε at the point $\varepsilon = 0$. It is easily seen that under normal conditions (the meaning of this will become clear below) the resulting equation for $G(A)$ is a consequence of the Schwinger equation. Actually, in obtaining the equation from (1.198) the factor $\delta_{\varepsilon}\varphi[\delta A(\varphi)/\delta\varphi]$ is taken outside the integral in the form of a differential operator with the substitution $\varphi \to \delta/\delta A_1$, and we obtain the equation $LL'G = 0$, where L and L' are operators generated respectively by the factors $\delta_{\varepsilon}\varphi$ and $\delta A(\varphi)/\delta\varphi$. The Schwinger equation in this notation has the form $L'G = 0$, so that the equation $LL'G=0$ is an automatic consequence of it.

If L and L' are differential operators of finite order, it can happen that the order of the operator LL' is lower than that of L', because the symbol $\delta_{\varepsilon}\varphi \cdot [\delta A(\varphi)/\delta\varphi]$ stands not for a simple product, but a convolution in the arguments of $\delta_{\varepsilon}\varphi(x)$ and $\delta/\delta\varphi(x)$. In particular, this will be the case when the highest power of the polynomial action $A(\varphi)$ is invariant under this transformation and therefore does not contribute to the variation $\delta_{\varepsilon}A(\varphi)$. In such cases an equation of order lower than that of the Schwinger equation is obtained from (1.198); equations of this type are called *Ward identities* (see the examples in Chap. 3).

The above discussion does not pertain to the so-called anomalous situation in which spontaneous symmetry breaking occurs. In our language this phenomenon arises because the equation $LL'G = 0$ is not true in spite of the fact that the Schwinger equation $L'G = 0$ is satisfied. This is possible, roughly speaking, when the contraction of the operator L with $L'G$ gives rise to divergences, and this "extra infinity" cancels the "zero" of $L'G$. The usual mechanism for the appearance of such an infinity is Goldstone particles (see Chap. 6).

1.7.4 Iteration solution of the equations

In this section we analyze the iteration solution of the equations of motion (1.194), (1.195), and on the basis of this analysis we give another version of the proof of the theorem about the connectedness of the logarithm (the first version is given in Sec. 1.4.8).

Multiplying (1.194) and (1.195) by $G^{-1} = \exp(-W)$, we obtain

$$\delta W/\delta A_n = H_n/n!, \quad n \neq 1; \tag{1.199}$$

$$A_1 + \sum_{n=2}^{\infty} A_n H_{n-1}/(n-1)! = 0. \tag{1.200}$$

We have introduced the functions H_n (the arguments $x_1 \ldots x_n$ are understood):

$$H_n = \exp(-W)\left(\frac{\delta}{\delta A_1}\right)^n \exp W = \left(\frac{\delta W}{\delta A_1} + \frac{\delta}{\delta A_1}\right)^n \cdot 1, \tag{1.201}$$

which are polynomial forms of the derivatives $W_n \equiv \delta^n W/\delta A_1^n$, $n \geq 1$. The derivatives W_n have the meaning of connected Green functions, and the H_n are the full Green functions without vacuum loops for the theory with the action (1.192).

Equations (1.199) and (1.200) can be iterated, constructing the solution in the form of a series in multiple powers of the vertices A_n, $n \geq 3$. For finding the zeroth-order approximation $W^{(0)}$ we obtain from (1.199) and (1.200) the system of equations

$$A_1 + A_2 W_1 = 0, \; \delta W/\delta A_0 = 1, \; \delta W/\delta A_2 = (W_2 + W_1^2)/2 \tag{1.202}$$

(we have substituted $H_0 = 1$, $H_1 = W_1$, and $H_2 = W_2 + W_1^2$), the solution of which has the form

$$W^{(0)} = \frac{1}{2}\,\mathrm{tr}\ln\Delta + \frac{1}{2}A_1\Delta A_1 + A_0 \tag{1.203}$$

up to an unimportant additive constant. The quantity $\Delta = -A_2^{-1}$ in (1.203) is the bare propagator. Using the equation $A_2^{-1} = \delta\,\mathrm{tr}\ln A_2/\delta A_2$, it is easy to verify that (1.203) does satisfy (1.202). We note that in the fermionic case $\mathrm{tr}\ln\Delta$ would enter into (1.203) with the other sign.

For determining the difference $W - W^{(0)}$ it is sufficient to use Eqs. (1.199) for $n \geq 3$, which are conveniently contracted with the corresponding potentials A_n:

$$A_n \frac{\delta W}{\delta A_n} = \frac{1}{n!} A_n H_n = \sum_{n_1 + \ldots + n_k = n} \ldots \tag{1.204}$$

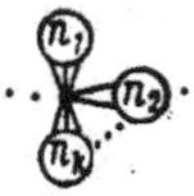

We have given the graphical expression for the general structure of the right-hand side, denoting the derivative W_k by a circle with k external lines. The vertex A_n is associated with the point of intersection of all n lines in (1.204), and numerical factors unimportant for the rest of the discussion have been omitted.

In the construction of the iteration solution the blocks W_k on the right-hand side of (1.204) are calculated using the approximation of W found earlier. The starting point is the zeroth-order approximation (1.203), in which

$$W_1 \equiv \delta W/\delta A_1 = \Delta A_1 \equiv \quad ; \quad W_2 \equiv \delta^2 W/\delta A_1^2 = \Delta \equiv$$

and $W_k = 0$ for $k \geq 3$. Discarding the contributions of W_k with $k \geq 3$ in (1.201), we obtain $H_3 = 3W_2W_1 + W_1^3$, $H_4 = 3W_2^2 + 6W_2W_1^2 + W_1^4$, and so on. Substituting these quantities into the right-hand side of (1.204), we find

$$A_3 \frac{\delta W}{\delta A_3} = \frac{1}{6} \quad + \frac{1}{2} \quad ; \quad A_4 \frac{\delta W}{\delta A_4} = \frac{1}{24} \quad + \frac{1}{4} \quad + \frac{1}{8} \tag{1.205}$$

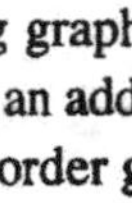
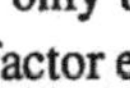
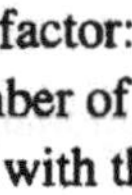
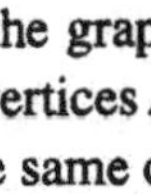
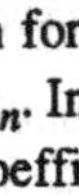

and so on [theories with $A_n = 0$ for $n > 4$ are usually considered, because then only two equations of the system (1.204) survive]. Determination of the functionals $A_n \delta W/\delta A_n$ is equivalent to determination of W, since the corresponding graphs differ only by a numerical factor: the graph for $A_n \delta W/\delta A_n$ contains an additional factor equal to the number of vertices A_n. In particular, the first-order graphs of (1.205) enter into W with the same coefficients as in (1.205).

In the next step of the iteration procedure we determine the second-order graphs of W, and so on. Knowing W, we can then find the functional $G = \exp W = \det \Delta^{1/2} \exp \tilde{W}$ (we have isolated from W the term involving $\mathrm{tr}\ln\Delta$, which is not represented by a graph). Clearly, in the end we arrive at the same diagrammatic representation which can be obtained directly from

(1.193) using (1.167) and the general technique of Sec. 1.4. We shall therefore not dwell on the analysis of the graphs and only discuss one problem: the proof of the theorem about the connectedness of the logarithm, in this case the functional $\widetilde{W} = \ln \widetilde{G}$. We proved this theorem in Sec. 1.4.8 using knowledge of the symmetry coefficients of the graphs. Now we verify that it is much simpler to do this using the equations of motion (1.204). The proof reduces to the observation that the equations possess the *property of conservation of connectedness*, namely, if all the graphs of $\widetilde{W}$ up to some order are connected, the graphs of the following order obtained in iterations of Eqs. (1.204) will also be connected. This is obvious from the general structure of the right-hand side of (1.204): the blocks $\widetilde{W}_k$ involved here are multiple derivatives with respect to A_1 of graphs of $\widetilde{W}$ which are connected by assumption, and the differentiation with respect to A_1 does not spoil the connectedness. Joining the connected blocks W_k to the vertex A_n, as shown in (1.204), we of course obtain a connected graph. The connectedness of all the graphs of $\widetilde{W}$ is thereby proved by induction.

We still need to check that $\widetilde{W}$ contains all the connected graphs which are contained in $\widetilde{G}$, and with the same coefficients. For this we write

$$\widetilde{G} = \exp \widetilde{W} = 1 + \widetilde{W} + \widetilde{W}^2/2 + \ldots$$

and then separate the connected part of this equation. We have shown that the connected part of $\widetilde{W}$ coincides with $\widetilde{W}$ itself, and all graphs of any power $\widetilde{W}^n$, $n > 1$, are unconnected (they are products of graphs). From this we immediately find the desired statement: $\widetilde{W}$ is the connected part of $\widetilde{G}$.

This proof has the advantage that it does not require knowledge of the symmetry coefficients of the graphs and has an obvious generalization to any theory with fermions.

1.8 One-Irreducible Green Functions

1.8.1 Definitions

Let us consider the functional $W(A) = \ln G(A)$, where $G(A)$ is the integral (1.193). The derivative

$$\delta W(A) / \delta A_1(x) \equiv \alpha_1(x) \tag{1.206}$$

has the meaning of the first connected Green function of the theory with action (1.192). The equation (1.206) expresses α_1 as a functional of all the potentials A.

The Legendre transform of W with respect to A_1 is the functional

$$\Gamma(\alpha_1, A'') = W(A) - \alpha_1 A_1, \tag{1.207}$$

where we have used the notation $\alpha_1 A_1 = \int dx\, \alpha_1(x) A_1(x)$, and we assume that the potential A_1 on the right-hand side is expressed as a functional of α_1 and $A'' \equiv \{A_n, n \neq 1\}$ by means of (1.206).

The functional (1.207) is sometimes referred to as the effective action, but we prefer the term *first Legendre transform*, meaning *transform with respect to the first potential*. More complicated transforms will be studied in Chap. 6.

Differentiating (1.207) with respect to α_1 and taking into account (1.206), we obtain the equation

$$\frac{\delta\Gamma}{\delta\alpha_1(x)} = \int dx' \frac{\delta W}{\delta A_1(x')} \frac{\delta A_1(x')}{\delta\alpha_1(x)} - \frac{\delta}{\delta\alpha_1(x)}(\alpha_1 A_1) = -A_1(x), \tag{1.208}$$

explicitly determining (for known Γ) the potential A_1 as a functional of α_1 and A'' and implicitly determining α_1 as a functional of the potentials A. We see from (1.208) that the functional

$$\Phi(\alpha_1, A) = \Gamma(\alpha_1, A'') + \alpha_1 A_1 \tag{1.209}$$

at the point $\alpha_1(A) = \delta W(A)/\delta A_1$ is stationary with respect to variations of α_1 for fixed potentials A:

$$\delta\Phi(\alpha_1, A)/\delta\alpha_1\big|_{\alpha_1(A)} = 0. \tag{1.210}$$

It is also clear that the value of Φ at the stationarity point coincides with $W(A)$.

Using (1.206) and (1.208), we obtain the equation

$$\delta(x-y) = \frac{\delta A_1(x)}{\delta A_1(y)} = \int dz \frac{\delta A_1(x)}{\delta\alpha_1(z)} \cdot \frac{\delta\alpha_1(z)}{\delta A_1(y)} =$$

$$= -\int dz \frac{\delta^2\Gamma}{\delta\alpha_1(x)\,\delta\alpha_1(z)} \cdot \frac{\delta^2 W}{\delta A_1(z)\,\delta A_1(y)}, \tag{1.211}$$

showing that the second derivatives of Γ and W are the kernels of operators which are reciprocals of each other up to a sign. Henceforth, expressions like (1.206), (1.208), and (1.211) will be written in abbreviated form:

$$\frac{\delta W}{\delta A_1} = \alpha_1, \quad \frac{\delta \Gamma}{\delta \alpha_1} = -A_1, \quad \frac{\delta^2 \Gamma}{\delta \alpha_1^2} \cdot \frac{\delta^2 W}{\delta A_1^2} = -1. \tag{1.212}$$

Multiple derivatives $W_n \equiv \delta^n W/\delta A_1{}^n$ (the arguments $x_1 \ldots x_n$ are understood) are the connected Green functions of the theory with action (1.192). In terms of Γ, the first connected function $W_1 = \alpha_1$ becomes an independent variable, and higher connected functions W_n, $n > 1$, can be represented as functionals of α_1 and A'' by using the recursion relation $\delta W_n/\delta A_1 = W_{n+1}$ following from the definition of W_n and expressing the *raising operator* $\mathcal{D} \equiv \delta/\delta A_1$ involved in it in the variables α_1 and A'':

$$\mathcal{D} = \frac{\delta \alpha_1}{\delta A_1} \cdot \frac{\delta}{\delta \alpha_1} = \frac{\delta^2 W}{\delta A_1^2} \cdot \frac{\delta}{\delta \alpha_1} = -\left(\frac{\delta^2 \Gamma}{\delta \alpha_1^2}\right)^{-1} \frac{\delta}{\delta \alpha_1} \tag{1.213}$$

[we have taken into account (1.212)]. For $n \geq 1$ we have

$$W_n = \mathcal{D}^{n-1} W_1 = \left[-\left(\frac{\delta^2 \Gamma}{\delta \alpha_1^2}\right)^{-1} \frac{\delta}{\delta \alpha_1}\right]^{n-1} \cdot \alpha_1. \tag{1.214}$$

The functions W_n are thereby expressed in the variables α_1 and A''.

Multiple derivatives $\delta^n \Gamma/\delta \alpha_1{}^n \equiv \Gamma_n$ (the arguments $x_1 \ldots x_n$ are understood) are referred to as the *1-irreducible Green functions* (in the variables α_1 and A'') of the theory with the action (1.192). In order to express these functions in the variables A, it is necessary to make the substitution $\alpha_1 \to \alpha_1(A)$.

Equations (1.214) allow us to explicitly express the connected functions W_n in terms of the 1-irreducible functions Γ_n. For $n = 2$ Eq. (1.214) coincides with (1.211) and serves to define the full propagator $D \equiv W_2 = -\Gamma_2^{-1}$. For calculating higher-order functions W_n, $n > 2$, it is necessary to use the following rule for differenting an inverse operator: $(L^{-1})' = -L^{-1}L'L^{-1}$. With this we obtain

$$\mathcal{D}[-\Gamma_2^{-1}] = \Gamma_3[-\Gamma_2^{-1}]^3 \equiv$$  $. \qquad (1.215)$

Here and below the heavy line denotes the full propagator $D = -\Gamma_2^{-1}$, the shaded circle with $n \geq 3$ lines leaving it denotes the 1-irreducible function Γ_n, and the unshaded circle will denote the amputated function $W_n^{\text{am}} = D^{-n}W_n$ (see the definition in Sec. 1.4.9).

From (1.214) and (1.215) we obtain

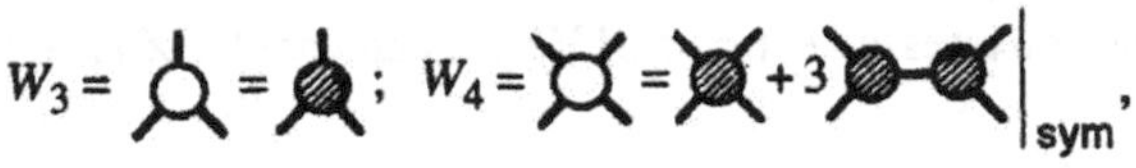

(1.216)

where sym stands for complete symmetrization in the arguments of the function W_4. The raising operator $\mathcal{D}$ acting on an equation of the type (1.216) differentiates in succession each line according to the rule (1.215) and each vertex Γ_n, connecting to it an additional line D. Differentiating (1.216) for W_4 in this manner, we obtain

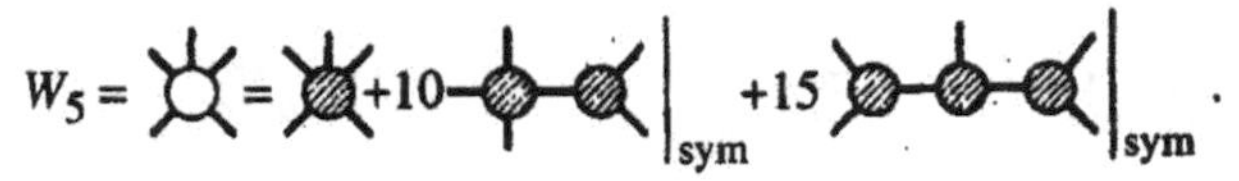

The equations given above can be solved for the 1-irreducible vertices Γ_n, expressing them in terms of amputated vertices:

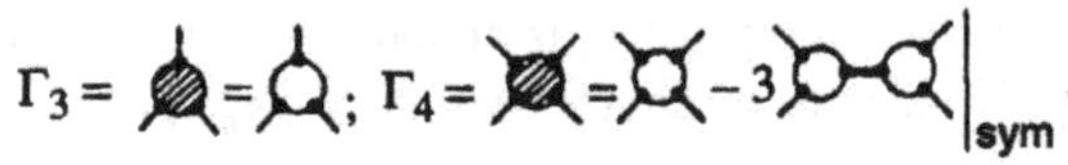

(1.217)

and so on. The general expression is easily obtained using (1.212):

$$\Gamma_n = \left(\frac{\delta A_1}{\delta\alpha_1}\frac{\delta}{\delta A_1}\right)^{n-1}\frac{\delta\Gamma}{\delta\alpha_1} = -\left[\left(\frac{\delta^2 W}{\delta A_1^2}\right)^{-1}\frac{\delta}{\delta A_1}\right]^{n-1} A_1. \tag{1.218}$$

The *raising operator for the 1-irreducible functions* Γ_n is

$$\mathcal{D}_\gamma \equiv \frac{\delta}{\delta\alpha_1} = \left(\frac{\delta^2 W}{\delta A_1^2}\right)^{-1}\frac{\delta}{\delta A_1} = D^{-1}\mathcal{D}. \tag{1.219}$$

1.8.2 The equations of motion for Γ

The problem of constructing the connected functions W_n from a known functional Γ becomes important when we can determine Γ independently of

W. For this it is necessary to write down the full set of equations of motion for Γ, which would determine this functional exactly as the system (1.199), (1.200) determines the functional W. We shall now do this.

The needed equations are obtained from (1.199) and (1.200) by the replacements $A_1 \to \alpha_1$ and $W \to \Gamma$ of the independent variables and the desired functional. All the equations involve the functions H_n defined by (1.201), which now must be expressed in terms of Γ by means of (1.206) and (1.213):

$$H_n = [\alpha_1 + \mathcal{D}]^n \cdot 1 = \left[\alpha_1 - \Gamma_2^{-1} \frac{\delta}{\delta\alpha_1}\right]^{n-1} \cdot \alpha_1. \tag{1.220}$$

We note that H_n contains the term $\alpha_1{}^n$.

Expressing the potential A_1 in the Schwinger equation (1.200) in terms of Γ using (1.208), we obtain (we recall that $\Delta = -A_2{}^{-1}$)

$$\delta\Gamma/\delta\alpha_1 = -\Delta^{-1}\alpha_1 + \sum_{n=3}^{\infty} [1/(n-1)!] A_n H_{n-1}. \tag{1.221}$$

The connection equations (1.199) involve partial derivatives of W with respect to the potentials A_n, $n \neq 1$, in addition to H_n. To express these derivatives in terms of Γ we differentiate (1.207) with respect to some potential A_n with $n \neq 1$, assuming that α_1 is an independent variable and that A_1 is a functional of α_1 and the other potentials $A'' = \{A_n, n \neq 1\}$:

$$\frac{\delta\Gamma}{\delta A_n} = \frac{\delta W}{\delta A_n} + \frac{\delta W}{\delta A_1} \cdot \frac{\delta A_1}{\delta A_n} - \alpha_1 \frac{\delta A_1}{\delta A_n} = \frac{\delta W}{\delta A_n}. \tag{1.222}$$

On the left-hand side $\delta/\delta A_n$ is the partial derivative in the variables α_1, A'' and on the right-hand side it is the partial derivative in the variables $A = \{A_1, A''\}$. The equation (1.222) allows the connection equations (1.199) to be rewritten as

$$\delta\Gamma/\delta A_n = H_n/n!, \quad n \neq 1. \tag{1.223}$$

The system (1.221), (1.223) together with (1.220) determines the functional Γ up to an additive constant.

1.8.3 *Iteration solution of the equations and proof of 1-irreducibility*

The system (1.221), (1.223) can be iterated and the solution can be constructed in the form of a power series in the vertices A_n, $n \geq 3$. For the zeroth-order approximation from (1.220), (1.221), and (1.223) we obtain $\delta\Gamma/\delta\alpha_1 = -\Delta^{-1}\alpha_1$; $\delta\Gamma/\delta A_0 = 1$; $\delta\Gamma/\delta A_2 = (\alpha_1{}^2 - \Gamma_2{}^{-1})/2$. This system is easily solved [cf. (1.203)]:

$$\Gamma^{(0)} = \frac{1}{2}\,\mathrm{tr}\ln\Delta + A_0 - \frac{1}{2}\alpha_1\Delta^{-1}\alpha_1, \tag{1.224}$$

where $\Delta = -A_2{}^{-1}$. The equations (1.223) for $n \geq 3$ are sufficient for determining the difference $\Gamma' \equiv \Gamma - \Gamma^{(0)}$. Let us study the first steps of the iteration procedure for the simple example of a φ^3 theory with a single nonzero vertex A_3. For convenience we contract (1.223) for $n = 3$ with A_3, which gives

$$A_3\frac{\delta\Gamma'}{\delta A_3} = \frac{1}{6}A_3H_3 = \frac{1}{6}\ \ \ +\frac{1}{2}\ \ \ +\frac{1}{6}\ \ \ \ldots \tag{1.225}$$

Here we have used the diagrammatic notation of the preceding section: the heavy line is the full propagator $D = -\Gamma_2^{-1}$ and the shaded circle is Γ_3. In addition, we have introduced the leg  to denote the variable α_1, and the point of intersection of three lines or legs denotes A_3.

In the zeroth-order approximation from (1.224) we find $D = \Delta$ and $\Gamma_3 = 0$. Substituting these into the right-hand side of (1.225), we find the first-order graphs of Γ: $\Gamma \ni \frac{1}{6}$  $+ \frac{1}{2}$ . The line in these graphs denotes the bare propagator Δ. In calculating higher orders D must be represented as a series $D = -\Gamma_2{}^{-1} = (\Delta^{-1} - \Gamma'_2)^{-1} = \Delta + \Delta\Gamma'_2\Delta + \ldots$, where $\Gamma'_2 \equiv \delta^2\Gamma'/\delta\alpha_1{}^2$. Comparing this equation to the definition of the self-energy $\Sigma = \Delta^{-1} - D^{-1}$, we see that $\Gamma'_2 = \Sigma$ in the variables α_1, A″.

Calculating D and Γ_3 taking into account the first-order graphs found above, we easily find the second-order graphs for Γ to be $\frac{1}{12}$  $+ \frac{1}{4}$  and so on.

All these graphs of Γ are *1-irreducible*, i.e., they remain connected when any one line is cut. It is easy to understand that in the process of iterating (1.225) 1-reducible graphs cannot appear at all, because this equation possesses the property of *conservation of 1-irreducibility*, similar to the way (1.204) possesses the property of conservation of connectedness: if all the graphs of Γ up to some order are 1-irreducible, then the graphs of next highest order arising in iterations of (1.225) also will be 1-irreducible. The reason is simple: in the analogous equations (1.204) for W, 1-reducible graphs were generated in the iterations by the terms of the polynomials H_n containing factors of W_1, shown in (1.204) by the block connected to the vertex A_n by a single line. In transforming to Γ the entire block W_1 becomes an independent variable α_1 and ceases to be involved in the iterations. This argument is valid for an interaction involving any potentials A.

Having thus proved the 1-irreducibility of all the graphs of Γ, let us now discuss the question of the coefficients of these graphs. For this we rewrite the definition (1.207), isolating from the functionals W and Γ the corresponding zeroth-order approximations (1.203), (1.224) [$W= W^{(0)} + W'$, $\Gamma = \Gamma^{(0)} + \Gamma'$]:

$$-\frac{1}{2}\alpha_1\Delta^{-1}\alpha_1 + \Gamma' = \frac{1}{2}A_1\Delta A_1 + W' - \alpha_1 A_1. \tag{1.226}$$

We shall take α_1 to be an independent variable. Then A_1 is a functional of α_1 defined by (1.208):

$$A_1 = -\frac{\delta\Gamma}{\delta\alpha_1} = \Delta^{-1}\alpha_1 - \frac{\delta\Gamma'}{\delta\alpha_1} \equiv \Delta^{-1}\alpha_1 - \Gamma'_1. \tag{1.227}$$

Substituting A_1 in this form into (1.226), we obtain

$$\Gamma' = \frac{1}{2}\Gamma'_1\Delta\Gamma'_1 + W'. \tag{1.228}$$

Now let us take the 1-irreducible part of this equation. The first term on the right-hand side consists of two blocks Γ'_1 joined by a single line Δ, so that all these graphs are 1-reducible. Using the 1-irreducibility of Γ' proved earlier, we conclude that Γ' coincides with the 1-irreducible part of the functional W', in which the variable A_1 is expressed in terms of α_1 using (1.227). The potential A_1 enters into the graphs of W' in the form of a vertex to which a single line Δ is connected. Clearly, if in substituting (1.227) we take into account the addition Γ'_1 even one time, the resulting graph is 1-reducible, because it will contain the block Γ'_1 connected by a single line to the rest of the graph. From this it follows that

$$\Gamma'(\alpha_1, A'') = \text{1–irreducible part of} \quad W'(A_1 = \Delta^{-1}\alpha_1,\ A''). \tag{1.229}$$

The statement we have proved makes the iteration construction of the functional Γ' redundant: it is sufficient to take all the graphs of W', make the replacement $A_1 = \Delta^{-1}\alpha_1$ in them, and then discard the resulting graphs which turn out to be 1-reducible. It is useful to recall that W' contains all possible connected graphs with the standard symmetry coefficients.*

The graphs of Γ can be classified not according to the number of vertices, but according to some other feature, for example, the number of loops. In the tree approximation only the contributions of graphs of first order in the vertices A_n, $n \geq 3$, are kept on the right-hand side of (1.229). Together with the zeroth-order approximation (1.224) this gives

$$\Gamma_{\text{tree}} = \frac{1}{2} \operatorname{tr} \ln \Delta + \sum_{n \neq 1} \frac{1}{n!} A_n \alpha_1^n . \tag{1.230}$$

In transforming to the variational functional (1.209) the linear term $A_1\alpha_1$ is added and the sum over n is grouped together to form the full action (1.192):

$$\Phi(\alpha_1, A)_{\text{tree}} = \frac{1}{2} \operatorname{tr} \ln \Delta + iS(\alpha_1) . \tag{1.231}$$

This approximation for Γ corresponds to the leading term of the stationary-phase approximation (see Sec. 1.6.6) in the functional integral.

1.9 Renormalization Transformations

We shall term *any nondegenerate linear inhomogeneous transformation of the Heisenberg field operator* $\hat{\varphi}_{\text{H}}(x)$ *a renormalization transformation*:

$$\hat{\varphi}_{\text{H}} = Z\hat{\varphi}'_{\text{H}} + a, \quad \hat{\varphi}'_{\text{H}} = Z^{-1}(\hat{\varphi}_{\text{H}} - a), \tag{1.232}$$

where Z is a linear operator with $\det Z \neq 0$ (Z is usually written as $Z^{1/2}$) and $a(x)$ is a translation parameter similar to the classical field $\varphi(x)$, i.e., an ordinary function for bosons and an anticommuting function for fermions.

The goal of this section is to derive expressions for the transformations induced by (1.232) for various functionals. The starting point is Eq. (1.58) determining the Green functions without vacuum loops of the field $\hat{\varphi}_{\text{H}}$. Replacing $\hat{\varphi}_{\text{H}}$ in (1.58) by $\hat{\varphi}'_{\text{H}}$, we obtain the Green functions of the field $\hat{\varphi}'_{\text{H}}$. Introducing the corresponding generating functional

---

* The graphs for several orders of the functionals Γ and W are given in Appendix 2.

$$H'(A) = \sum_{n=0}^{\infty} \frac{1}{n!} H_n' (iA)^n = (0| T_D \exp i(\hat{\varphi}_H' A) |0)$$

(we have returned to the notation of Sec. 1.4.9) and expressing $\hat{\varphi}_H'$ in it in terms of $\hat{\varphi}_H$ using (1.232), taking into account the equation $Z^{-1}\hat{\varphi}_H = \hat{\varphi}_H Z^{-1T}$, we obtain

$$H'(A) = H(Z^{-1T}A) \exp[-iaZ^{-1T}A]. \tag{1.233}$$

According to (1.86), the generating functional of the full Green functions $G(A)$ differs from $H(A)$ by the factor G_0 representing vacuum loops. Let us define $G'_0 = G_0$; then the functional G will transform exactly like H. Taking the logarithm, we find the transformation rule for the generating functional of connected Green functions:

$$W'(A) = W(Z^{-1T}A) - iaZ^{-1T}A. \tag{1.234}$$

From this we see that the shift $\hat{\varphi}_H' = \hat{\varphi}_H - a$ leads to a change of only the first connected function ($W_1' = W_1 - a$), while all the other W_n, $n \neq 1$, remain unchanged. In the dilatation $\hat{\varphi}_H' = Z^{-1}\hat{\varphi}_H$ any function W_n gets a factor of Z^{-1} for each argument x: $W_n' = Z^{-n}W_n$ [see the comment after (1.131)]. In particular, the full propagator $D = W_2$ transforms according to the rule $D' = Z^{-2}D$. If D is understood as the kernel of a linear operator, the transformation formula can be written in matrix form: $D' = Z^{-1}DZ^{-1T}$.

Using the definition (1.132) of the generating functional of amputated Green functions and the known transformation rules for connected functions, it is easy to obtain

$$W^{\mathrm{am}\prime}(A) = W^{\mathrm{am}}(ZA) - iaD^{-1T}ZA. \tag{1.235}$$

Finally, the transformation rule for the generating functional of 1-irreducible functions must be found. In the notation of this section the definitions (1.206), (1.207) take the form $\Gamma(\alpha) = W(A) - i\alpha A$, $i\alpha = \delta W/\delta A$. The transformation rule for $\Gamma(\alpha)$ is easily found from the known transformation rule for $W(A)$. We give only the result, leaving its proof to the reader:

$$\Gamma'(\alpha) = \Gamma(Z\alpha + a). \tag{1.236}$$

These expressions can be used to rewrite any equation in variational derivatives for any of the functionals studied above in terms of the transformed functional.

In various physical problems the renormalization transformation (1.232) is used to go from the original Heisenberg field $\hat{\varphi}_H$ to the *renormalized field* $\hat{\varphi}'_H \equiv \hat{\varphi}_{\text{ren}}$ satisfying certain additional requirements which fix the form of the transformation (1.232). If the theory possesses a symmetry (translational, relativistic, or some other invariance), it is usually required that the renormalization transformation not violate this symmetry. The translation parameter a is usually fixed by the requirement $W'_1 = (0|\hat{\varphi}_{\text{ren}}|0) = 0$, which taking into account (1.234) gives $a = (0|\hat{\varphi}_H|0) = W_1$; the dilatation operator Z is fixed by some condition imposed on the renormalized propagator $D_{\text{ren}} = Z^{-1}DZ^{-1T}$. For relativistic interactions the renormalization is accompanied by rearrangement of the perturbation series and elimination of ultraviolet divergences [1], but these subjects lie outside the scope of this book.

Let us conclude by obtaining the expression for the renormalization transformation for the action functional. The starting point will be the representation (1.164). We write $G'(A) = c'\int D\varphi \exp i[S'(\varphi) + \varphi A]$ and seek the renormalized action functional $S'(\varphi)$ and normalization constant c' in terms of the known functional $G'(A)$. From (1.164) and the transformation formula for G analogous to (1.233) we obtain

$$G'(A) = c\int D\varphi \exp i[S(\varphi) + (\varphi - a)Z^{-1}A], \tag{1.237}$$

where c is a constant defined by (1.163). Making the change of integration variable $\varphi = Z\varphi' + a$ in (1.237), we find

$$G'(A) = c \det Z\int D\varphi \exp i[S(Z\varphi + a) + \varphi A].$$

The constant $\det Z$ is the Jacobian of the substitution. Therefore,

$$c' = c \det Z, \quad S'(\varphi) = S(Z\varphi + a). \tag{1.238}$$

The factor $\det Z$ can, of course, be included as an addition to the transformed action S', but we prefer the form in (1.238).

Therefore, the renormalization transformation (1.232) of the Heisenberg field is equivalent to the transformation (1.238) of the action functional, so that the transformed (renormalized) Green functions can be calculated according to the usual rules starting from the appropriately transformed action functional. The transformation (1.238) in the language of [1] corresponds to the introduction of counterterms into the Lagrangian.

1.10 Anomalous Green Functions and Spontaneous Symmetry Breaking

Up to now it has always been assumed that the Green functions are uniquely determined by the action functional, and if the perturbation series converges, this of course must be true. However, there is a large class of problems of great practical importance (for example, the entire theory of phase transitions in statistical physics) for which the perturbation series certainly diverges. Here the objection might be made that the original expression (1.58) determines the Green functions uniquely and independently of the perturbation theory. However, this is true only when all the quantities involved in (1.58) are mathematically well defined, and this is far from always so. In the case of immediate interest one is usually speaking of translationally invariant systems studied in infinite volume, and for such systems the usual operator formalism of quantum theory cannot be completely correct, at least because the translational invariance makes the ground-state energy proportional to the system volume and therefore infinite.

It thus becomes necessary to have a new definition of the Green functions. This can be obtained using the set of equations of motion (1.194), (1.195), stipulating that these equations describe the dynamics of a system with a given action, and that the desired Green functions, or, more precisely, their generating functional, is by definition a solution of the equations of motion.* Such a definition no longer assumes the uniqueness of the Green functions—there may be many solutions.

Perturbation theory corresponds to the iteration solution of the equations; this is called the *normal solution*. Every other solution that exists is called an *anomalous solution*. An anomalous solution cannot be represented by a power series in the vertices A_n (we shall use the notation of Sec. 1.7.3), since the equations of motion determine the power series uniquely: it is the series of ordinary perturbation theory and none other. Therefore, each anomalous solution must contain a "nonanalyticity" in the vertices A_n, $n \geq 3$.

The normal solution is represented as a functional integral (1.193) and is invariant under all potential transformations induced by the group of motions

---

* This approach provides the basis for determining the Green functions in, for example, [19].

of the field. The anomalous solution may have no such property owing to spontaneous symmetry breaking [see (1.197), (1.198) and the discussion following them]. Spontaneous symmetry breaking is always accompanied by degeneracy of the solution (the term "degeneracy" is understood as "nonuniqueness"): if $G(A)$ is some solution of the equations of motion and $A \to A^{\varepsilon}$ is some potential transformation induced by the element ε of the group of motions of the field, then the "shifted functional" $G^{\varepsilon}(A) \equiv G(A^{\varepsilon})$ is also a solution of the equations of motion. This is equivalent to the statement that the equations of motion are invariant under all such shifts: "a solution is transformed into a solution." We shall not dwell on the proof here, because all these questions are discussed in detail in Chap. 6.

The practical construction of anomalous solutions is a difficult task, because the only effective calculational technique in field theory is perturbation theory, whereas here one is interested in a solution which cannot be expanded in a perturbation series. Usually some approximations are made in the equations of motion in order to simplify these equations enough that they can be solved exactly. When proceeding in this manner it is necessary to be sure that the approximations made only simplify the problem and do not distort it to the point where it becomes unrecognizable.

A more regular method of constructing anomalous solutions is the variational method, which is essentially what Chap. 6 is devoted to. The idea of this method is the following: the original problem of solving the equations of motion is reduced to the variational problem by constructing the functional for which the equations of motion play the role of the Euler equations determining the stationarity point. In this language, degeneracy of the solution implies nonuniqueness of the stationarity point, and spontaneous symmetry breaking corresponds to the case where the varied functional is invariant under some symmetry group, but has a noninvariant (and therefore necessarily degenerate) stationarity point.

From the computational point of view the advantage of this method is that the variational functional can be constructed using perturbation theory without losing the possibility of finding anomalous solutions. This is easy to understand: the coordinates of a stationarity point analytic in some parameters of a function, for example, a polynomial, can depend on these parameters nonanalytically.

This general idea can be illustrated by the method discussed in Sec. 1.8.1 for calculating connected Green functions W_n using the known functional $\Gamma(\alpha_1, A'')$ (all the notation is the same as in Sec. 1.8.1). In this method the first connected function W_1 is found as the stationarity point of the functional (1.209), the 1-irreducible Green functions are determined by the values of the derivatives of the functional (1.209) at the stationarity point, and the connected functions W_n with $n \geq 2$ are constructed from the 1-irreducible functions using the simple rule (1.214). This method makes it possible to find an anomalous solution for $W_1 = \alpha_1$ even when the variational functional (1.209) is used in some finite order of perturbation theory. The presence or absence of such solutions depends, of course, on the explicit form of the variational functional, and it in turn is determined by the action functional. In specific calculations, for example, in the Goldstone model [20], the variational functional is constructed in the form of a loop expansion, and the anomalous solution, if it exists, appears already in the tree approximation (1.231).

We end this discussion for now, because Chap. 6 is devoted to spontaneous symmetry breaking and variational methods, and here we want only to define the relevant concepts and show that the Legendre transform introduced in the preceding sections is directly related to the problem of spontaneous symmetry breaking.

We conclude with a remark which should be born in mind when reading Chaps. 4 and 5. In the present chapter all the expressions have been written down for ordinary (pseudo-Euclidean) quantum field theory, but the technical methods themselves—diagrammatic representations, functional integrals, variational equations—are completely universal. When studying Euclidean field theory and statistical physics in the following chapters, we shall not consider it necessary to repeat the standard constructions of this chapter each time. We shall discuss only the particular features of the relevent representations. The reader must therefore clearly understand and remember the following: *if a representation of the type (1.84) is obtained for some object, no matter what theory it pertains to, it automatically follows that the standard diagrammatic representations of Sec. 1.4 and the representation as a functional integral found using (1.166) and (1.167) are valid for it. Finally, it is always possible to write down equations for it in terms of variational derivatives (Schwinger equations) using the standard recipe (1.183).* All these constructions are elementary and, as a rule, we shall not dwell on them further.

Chapter 2

SPECIFIC SYSTEMS

The first three sections of this chapter contain a brief compilation of information about various fields and contractions, and the last three sections are auxiliary to Chap. 1. The material is arranged such that the later sections make extensive use of the material in the first three.

2.1 Quantum Mechanics

The field formalism described in the preceding chapter is fully applicable also to the ordinary quantum mechanics of systems with a finite number of degrees of freedom. The simplest example is that of a particle in a given external field. The corresponding classical Lagrangian has the form $\mathscr{L} = m\dot{q}^2/2 - \mathscr{V}(q)$, where q is the particle coordinate, $\dot{q} \equiv \partial_t q$ is the velocity, and $\mathscr{V}$ is a given potential.

Depending on the form of the potential, in quantum mechanics one has either the problem of finding the energy levels and the corresponding wave functions, or the scattering problem. They can both be solved using field objects: the generating functionals of the S matrix and Green functions.

From the technical point of view, the manner in which the full Lagrangian $\mathscr{L}$ is split into the free part $\mathscr{L}_0$ and the interaction is important. The situation is most similar to that of the traditional field theoretic techniques when $\mathscr{L}_0$ is taken to be the oscillator Lagrangian, and the rest of the potential is taken to be the interaction Lagrangian (anharmonic oscillator). This splitting is justified physically for problems with a purely discrete spectrum (it can be hoped that the perturbation series converge only when the interaction does not lead to a qualitative change of the spectrum). The second type of splitting, when $\mathscr{L}_0$ is taken to be only the kinetic term $m\dot{q}^2/2$, is convenient for the scattering problem.

We shall study these two cases separately.

2.1.1 *The oscillator*

The classical free action for the oscillator with frequency ω has the form

$$S_0(q) = (m/2)\int dt\,(\dot{q}^2 - \omega^2 q^2)\,, \tag{2.1}$$

and the kernel of the corresponding quadratic form (1.4) is the operator $K = -m(\partial_t^2 + \omega^2)$. The action (2.1) is associated with the quantum Hamiltonian $\mathbf{H}_0 = \hat{p}^2/2m + m\omega^2\hat{q}^2/2 = \omega(a^+a + 1/2)$, where $\hat{q}$ and $\hat{p}$ are the coordinate and momentum operators in the Schrödinger picture and a^+ and $a \equiv a^-$ are the creation and annihilation operators in the same picture

$$a^{\pm} = (\hat{q}\sqrt{\omega m} \mp i\hat{p}/\sqrt{\omega m})/\sqrt{2}, \tag{2.2}$$

satisfying the canonical commutation relation $[a, a^+] = 1$.

The ground state of $\mathbf{H}_0$ is the "vacuum" $|0\rangle$, defined by $a|0\rangle = 0$; the excited ("n-particle") states $|n\rangle = (n!)^{-1/2}(a^+)^n|0\rangle$ in the one-dimensional theory are nondegenerate and have energies $\omega(n + 1/2)$. The coordinate operator $\hat{\varphi}(t) \equiv \hat{q}(t)$ playing the role of the free field has the following form in the interaction picture:

$$\hat{\varphi}(t) = e^{i\mathbf{H}_0 t}\hat{q}e^{-i\mathbf{H}_0 t} = (ae^{-i\omega t} + a^+e^{i\omega t})/\sqrt{2\omega m}. \tag{2.3}$$

The normal-ordered product of the fields is defined by the usual rules in Sec. 1.1.4, and the roles of the operators $\hat{a}(t)$ and $\hat{b}(t)$ in the decomposition of the field are respectively played by the first and second terms on the right-hand side of (2.3) (i.e., a^+ on the left and a on the right). The simple and time-ordered contractions defined by (1.13) and (1.29) have the form

$$n(t, t') = (2\omega m)^{-1}\exp i\omega(t' - t), \tag{2.4}$$

$$\Delta(t, t') = (2\omega m)^{-1}[\theta_{tt'}\exp i\omega(t' - t) + \theta_{t't}\exp i\omega(t - t')]. \tag{2.5}$$

The contraction Δ is continuous at $t = t'$, so that (1.30) is satisfied automatically. In the energy representation

$$\Delta(t, t') = \frac{i}{2\pi m}\int dE\frac{\exp iE(t' - t)}{E^2 - \omega^2 + i0}, \tag{2.6}$$

from which we see that the contraction Δ is actually the Green function for the operator K given above, in accordance with the general rule (1.31). We also note that the normal-ordered product possesses the properties (1.87) and (1.134).

The generating functionals of the S matrix and the Green functions for the anharmonic oscillator are determined by the usual expressions in Sec. 1.3.5.

Obviously, for a problem with a purely discrete spectrum which is shifted when the interaction is switched on, the S-matrix operator does not exist in the strict sense of the word. We are actually interested in the shifts of the level energies and the exact wave functions. They can be found from the asymptotic formulas of nonstationary perturbation theory for a discrete level (see Appendix 1) using the representation (1.64) for the evolution operator. In particular, the shift of the ground-state energy is, according to (1.74), expressed in terms of the logarithm of the vacuum expectation value of the S matrix, which in diagrammatic perturbation theory is represented as the sum of all connected vacuum loops.

We note that the Green functions of the field, in contrast to the S matrix, are well defined and can also be used to find the exact energies and exact wave functions (the former can be found from the locations of the poles in spectral representations and the latter from the residues at these poles).

2.1.2 *The free particle*

In this case $\mathbf{H}_0 = \hat{p}^2/2m$, and the free field – the coordinate operator in the interaction picture – has the form

$$\hat{\varphi}(t) = e^{i\mathbf{H}_0 t}\hat{q}e^{-i\mathbf{H}_0 t} = \hat{q}+\hat{p}t/m. \tag{2.7}$$

The usual field-theoretic technique of creation and annihilation operators is inapplicable here and we must give a different definition of the normal-ordered product. This can be done using the general rules of Sec. 1.1.4, taking $\hat{a}$ and $\hat{b}$ in the decomposition $\hat{\varphi}(t) = \hat{a}(t) + \hat{b}(t)$ to be the first and second terms on the right-hand side of (2.7), respectively (i.e., momenta on the left, coordinates on the right). This decomposition obviously satisfies all the necessary requirements of Sec. 1.1.4, but, of course, it is not the only possible one.

Using (1.13) and (1.29) we find the contractions for the N-product:

$$n(t,t') = [\hat{a}(t),\hat{b}(t')] = im^{-1}t', \tag{2.8}$$

$$\Delta(t,t') = im^{-1}[t'\theta_{tt'}+t\theta_{t't}]. \tag{2.9}$$

In contrast to (2.4) and (2.5), these contractions are not functions of the difference $t-t'$, because the operators $\hat{a}$ and $\hat{b}$ that we have selected are mixed in the time evolution.

The contraction Δ is continuous at $t = t'$ and therefore automatically satisfies the condition (1.30). The operator K in the free equation of motion (1.2) has the form $-m\partial_t^2$ in our case, and the general equation (1.31) is satisfied: $K\Delta = i$, as is easily verified.

We use $|\mathbf{x}\rangle$ and $|\mathbf{k}\rangle$ to denote the eigenstates of the coordinate and momentum operators, normalized to a δ function, with eigenvalues $\mathbf{x}$ and $\mathbf{k}$, respectively. The ground state of $\mathbf{H}_0$ is the plane wave $|\mathbf{k}\rangle$ with $\mathbf{k} = 0$; this state belongs to the continuum rather than to the discrete spectrum, so that the usual definitions of the Green functions lose their meaning and these objects need not be considered. However, this does not prevent us from using the technique of the Wick theorem for reducing various operators to normal-ordered form and calculating the matrix elements; in particular, the S-matrix operator in normal-ordered form is defined according to the general equation (1.64).

In the calculation of the matrix elements it is more convenient to deal not with the field operator itself $\hat{\varphi}(t) \equiv \hat{q}(t)$, but with the exponential $\exp i\hat{\varphi}A \equiv \exp i\int dt\,\hat{\varphi}(t)A(t)$. From the definition of the N-product we obtain

$$N\exp i\hat{\varphi}A = \exp(i\hat{p}\overline{tA}/m)\exp(i\hat{q}\bar{A}), \qquad (2.10)$$

where $\bar{A} \equiv \int dtA(t)$ and $\overline{tA} \equiv \int dt\,tA(t)$. Using the fact that in coordinate space $\exp i\hat{q}c$ is a multiplication operator and $\exp i\hat{p}c$ is a translation by c, we obtain

$$\langle \mathbf{k}_1 | N\exp i\hat{\varphi}A | \mathbf{k}_2 \rangle = \delta(\mathbf{k}_1 - \mathbf{k}_2 - \bar{A})\exp(i\mathbf{k}_1\overline{tA}/m), \qquad (2.11)$$

$$\langle \mathbf{x}_1 | N\exp i\hat{\varphi}A | \mathbf{x}_2 \rangle = \delta(\mathbf{x}_1 - \mathbf{x}_2 + \overline{tA}/m)\exp(i\mathbf{x}_2\bar{A}). \qquad (2.12)$$

These expressions replace (1.87), which is not satisfied for the N-product we are considering [like (1.134)].

In calculating the elements of the S matrix for scattering in a potential V, it is convenient to write the latter as a Fourier integral: $V(q) = \int d\lambda\,\tilde{V}(\lambda)\exp i\lambda q$. Substituting the interaction functional $S_v(\varphi) = -\iint dt d\lambda\,\tilde{V}(\lambda)\exp i\lambda\varphi(t)$ into (1.64), we arrive at the following representation for the normal-ordered S-matrix functional:

$$R(\varphi) = \exp\left(\frac{1}{2}\cdot\frac{\delta}{\delta\varphi}\Delta\frac{\delta}{\delta\varphi}\right)\exp iS_v(\varphi) =$$

$$= \sum_{n=0}^{\infty} \frac{(-i)^n}{n!} \int \ldots \int \prod_i (\tilde{\mathcal{V}}(\lambda_i)\, d\lambda_i dt_i) \exp\Big[-\frac{1}{2}\sum_{ks} \lambda_k \lambda_s \Delta(t_k, t_s) + $$
$$+ i \sum_k \lambda_k \varphi(t_k)\Big]. \tag{2.13}$$

Equation (2.11) must be used to calculate the matrix element of the S-matrix operator $\mathbf{U} \equiv NR(\hat{\varphi})$ in momentum space:

$$\langle \mathbf{k}_1 | \mathbf{U} | \mathbf{k}_2 \rangle = \sum_{n=0}^{\infty} \frac{(-i)^n}{n!} \int \ldots \int \prod_i (\tilde{\mathcal{V}}(\lambda_i)\, d\lambda_i dt_i) \times$$
$$\times \delta(\mathbf{k}_1 - \mathbf{k}_2 - \sum_s \lambda_s) \times$$
$$\times \exp\Big[-(1/2) \sum_{ks} \lambda_k \lambda_s \Delta(t_k, t_s) + i(\mathbf{k}_1/m) \sum_k \lambda_k t_k\Big]. \tag{2.14}$$

Performing the integrations over $t_1 \ldots t_n$, we arrive at the usual resolvent formulas of perturbation theory.

Although, as noted above, the Green functions for this theory are meaningless, their generating functional (1.86) can be calculated formally. Using (2.11), for the free theory we obtain

$$G^{(0)}(A) \equiv \langle 0 | T \exp i\hat{\varphi} A | 0 \rangle = \delta(\bar{A}) \exp\Big[-\frac{1}{2} A \Delta A\Big], \tag{2.15}$$

where $|0\rangle$ is the state $|\mathbf{k}\rangle$ with $\mathbf{k} = 0$. For a theory with an interaction it is necessary to use (1.90). The explicit nonanalyticity of the functional (2.15) in the argument A indicates the impossibility of correctly defining the Green functions.

Let us conclude by deriving a simple relation between the on-shell S-matrix functional and the S-matrix elements in momentum space. First we use (2.11) to calculate the matrix element of the operator in (2.15) between states of arbitrary momentum:

$$\langle \mathbf{k}_1 | T \exp i\hat{\varphi} A | \mathbf{k}_2 \rangle = \delta(\mathbf{k}_1 - \mathbf{k}_2 - \bar{A}) \exp[-A\Delta A/2 + i\mathbf{v}_1 \overline{tA}] \tag{2.16}$$

(here and below $\mathbf{v} \equiv \mathbf{k}/m$). Representing the δ function in (2.16) by a Fourier integral and the factor $\exp[-A\Delta A/2]$ by a Gaussian functional integral (1.162), we reduce the right-hand side of (2.16) to the form

$$c\,(2\pi)^{-d}\int d\mathbf{a}\int_{\Delta} D\varphi \exp i\,[S_0'(\varphi) + \mathbf{a}\,(\mathbf{k}_1 - \mathbf{k}_2) + A\,(\varphi + f)]\,, \qquad (2.17)$$

where d is the dimensionality of $\mathbf{k}$, $S'_0(\varphi) = \varphi K\varphi/2$ is the quadratic form of the free action, and $f(t) = \mathbf{v}_1 t - \mathbf{a}$ is a function on the mass shell (for us $\varphi K\varphi = -m\int dt\,\varphi\ddot{\varphi}$, and the mass shell is the set of all linear functions). According to (1.90), to transform to an element of the S-matrix $\mathbf{U} = T\exp iS_v(\hat{\varphi})$ it is necessary to act on (2.17) with the operator $\exp iS_v(\delta/\delta iA)$ and then set $A = 0$. This trick is apparently valid only for potentials for which the classical trajectories go into free asymptotes sufficiently rapidly, which is what will be assumed here (the Coulomb potential is thereby excluded). After transforming to the S matrix, the form $A(\varphi + f)$ in (2.17) is replaced by $S_v(\varphi + f)$, the integral over φ becomes the S-matrix generating functional (1.168), and we finally obtain $\langle \mathbf{k}_1 | \mathbf{U} | \mathbf{k}_2 \rangle = (2\pi)^{-d}\int d\mathbf{a} R(f) \exp i\mathbf{a}(\mathbf{k}_1 - \mathbf{k}_2)$.

If the interaction $S_v(\varphi)$ is not explicitly time-dependent, the energy δ function must be separated from the matrix element: $\langle \mathbf{k}_1 | \mathbf{U} | \mathbf{k}_2 \rangle = \delta(\mathbf{k}_1 - \mathbf{k}_2) - i2\pi\delta(E_1 - E_2)T(\mathbf{k}_1, \mathbf{k}_2)$, where $E \equiv \mathbf{k}^2/2m$ and $T(\mathbf{k}_1, \mathbf{k}_2)$ is the scattering amplitude. The representation obtained above for the matrix element is not convenient for isolating the δ function, since the latter is associated with time translations, and the functional space over which the integration in (2.17) runs is not invariant under such translations (see Sec. 2.5).

The representation found by A. V. Kuz'menko and the author is more convenient (see also [87]):

$$\langle \mathbf{k}_1 | \mathbf{U} | \mathbf{k}_2 \rangle = c \int_{(\mathbf{k}_1 \mathbf{k}_2)} D\varphi \exp iF(\varphi)\,. \qquad (2.18)$$

The integration space $(\mathbf{k}_1\, \mathbf{k}_2)$ consists of functions with asymptotes $\varphi_+(t) = \mathbf{v}_1 t + \mathbf{a}_1$, $\varphi_-(t) = \mathbf{v}_2 t + \mathbf{a}_2$ (the velocities $\mathbf{v}_i = \mathbf{k}_i/m$ are fixed, and the parameters $\mathbf{a}_i$ are arbitrary), and $F(\varphi) = S'_0(\varphi) + S_v(\varphi) + (\mathbf{k}_2\mathbf{a}_2 - \mathbf{k}_1\mathbf{a}_1)/2$. For potentials which fall off sufficiently rapidly, the functional $F(\varphi)$ is finite on functions from the space $(\mathbf{k}_1\mathbf{k}_2)$ and stationary on the classical trajectory with given $\mathbf{k}_{1,2}$ to all variations with $\delta\mathbf{k}_1 = \delta\mathbf{k}_2 = 0$ (the surface terms in the variation of the action cancel with the variation of the piece $\mathbf{k}_2\mathbf{a}_2 - \mathbf{k}_1\mathbf{a}_1$). The normalization constant c in (2.18) will be determined below.

For deriving the representation (2.18) it is sufficient to show that its right-hand side for the functional $F_A(\varphi) = S'_0(\varphi) + \varphi A + (\mathbf{k}_2\mathbf{a}_2 - \mathbf{k}_1\mathbf{a}_1)/2$

coincides with the right-hand side of (2.16). First, we separate the δ function contained in (2.16) from the integral (2.18). For this we consider the group of translations $\varphi(t) \to \varphi_{\mathbf{a}}(t) = \varphi(t) + \mathbf{a}$, under which, as is easily checked, $F_A(\varphi_{\mathbf{a}}) = F_A(\varphi) + \mathbf{a}[\mathbf{k}_2 - \mathbf{k}_1 + \bar{A}]$, and the space $(\mathbf{k}_1\mathbf{k}_2)$ is transformed into itself. We choose the constraint $\Phi(\varphi) = \mathbf{a}_1 + \mathbf{a}_2$ and introduce the expression $2\int d\mathbf{a}\delta[\Phi(\varphi_{-\mathbf{a}})] = 2\int d\mathbf{a}\delta\,[\mathbf{a}_1 + \mathbf{a}_2 - 2\mathbf{a}] = 1$ inside the integral in (2.18). Then, shifting the integration variable $\varphi \to \varphi_{\mathbf{a}}$, we can explicitly perform the integration over $\mathbf{a}$, which gives the desired δ function, and we obtain

$$2\,(2\pi)^d\delta\,[\mathbf{k}_2-\mathbf{k}_1+\bar{A}]\,c\int_{(\mathbf{k}_1\mathbf{k}_2)} D\varphi\delta\,[\mathbf{a}_1+\mathbf{a}_2]\exp\,iF_A(\varphi)\,.$$

The remaining integral is calculated using the substitution $\varphi(t) = \varphi'(t) + h(t) + \mathbf{v}_1 t + t\bar{A}/2m$, where $h = i\Delta A$ and Δ is the contraction (2.9). This substitution changes the integration space: knowing the asymptotes $h_+(t) = -t\bar{A}/m$ and $h_-(t) = -t\bar{A}/m$, we find $\varphi'_+(t) = \mathbf{a}_1 + t\bar{A}/2m$ and $\varphi'_-(t) = \mathbf{a}_2 - t\bar{A}/2m$ (for $\mathbf{k}_2 - \mathbf{k}_1 + \bar{A} = 0$). It is easily verified (taking into account all the surface terms and the equations $\mathbf{k}_2 - \mathbf{k}_1 + \bar{A} = 0$, $\mathbf{a}_1 + \mathbf{a}_2 = 0$) that after the substitution the functional $iF_A(\varphi)$ takes the form $iS'_0(\varphi') + i\mathbf{v}_1 t\bar{A} - A\Delta A/2$. The last two terms are taken outside the integral and give an exponential on the right-hand side of (2.16). The remaining integral does not depend on either A or the momenta $\mathbf{k}_{1,2}$ and determines the normalization constant c in (2.18):

$$2c\,(2\pi)^d\int_{(00)} D\varphi\delta\,[\mathbf{a}_1+\mathbf{a}_2]\exp\,iS_0{}'(\varphi) \;=\; 1,$$

where (00) is the space $(\mathbf{k}_1\mathbf{k}_2)$ for $\mathbf{k}_1 = \mathbf{k}_2 = 0$.

We have therefore proved the representation (2.18). It is convenient because the integration space is specified by the parameters numbering the quantum asymptotic states, and it is invariant under time translations $\varphi(t) \to \varphi_\tau(t) = \varphi(t + \tau)$. This makes it possible to separate from (2.18) all the δ functions when necessary. The first term of the perturbation series for $\exp iF(\varphi) = \exp i[F_0(\varphi) + S_v(\varphi)]$ generates the term $\delta(\mathbf{k}_1 - \mathbf{k}_2)$ on the right-hand side of (2.18), and the contributions of the other terms contain only a δ function in the energy. As already mentioned, the latter is related to time translations, for which $\mathbf{a}_i \to \mathbf{a}_i + \mathbf{v}_i\tau$ and $F(\varphi_\tau) = F(\varphi) + i\tau(E_2 - E_1)$ owing to the invariance of the action. Choosing, for example, the constraint $\Phi(\varphi) = \mathbf{v}_1\mathbf{a}_1$ (the scalar product of two vectors), we introduce unity in the form $\mathbf{v}_1{}^2\int d\tau\delta[\Phi(\varphi_{-\tau})] = \mathbf{v}_1{}^2\int d\tau\delta(\mathbf{v}_1\mathbf{a}_1 - \tau\mathbf{v}_1{}^2)$ inside the integral. Making the shift $\varphi \to \varphi_\tau$, we can calculate the integral over τ, which gives the desired

δ function, and obtain an integral representation directly for the scattering amplitude ($\mathbf{v}_1^2 = \mathbf{v}_2^2$):

$$T(\mathbf{k}_1, \mathbf{k}_2) = ic\mathbf{v}_1^2 \int_{(\mathbf{k}_1\mathbf{k}_2)} D\varphi\delta(\mathbf{v}_1\mathbf{a}_1)\,[\exp iF(\varphi) - \exp iF_0(\varphi)]. \quad (2.19)$$

Preliminary separation of the δ function is necessary, for example, when calculating the amplitude by the stationary-phase method (the semiclassical approximation), since the conservation laws expressed by the δ functions automatically make the kernel of the quadratic form of the second variation of the action degenerate in the neighborhood of a classical trajectory. The degeneracy makes the kernel noninvertible and prevents the construction of the standard perturbation theory in powers of $\hbar$ (the loop expansion). Similar difficulties are encountered in gauge field theory (Chap. 3), from which we have borrowed the trick of overcoming them by introducing a constraint [see Eq. (3.6)]. The constraint lifts the degeneracy of the kernel and makes it possible to correctly determine the corresponding propagator (the inverse kernel) and the determinant, i.e., the preexponential in the semiclassical approximation.

2.2 Nonrelativistic Field Theory

Nonrelativistic field theory is the quantum mechanics of systems with an arbitrary number of identical particles in the second-quantized formalism. In the classical theory such systems are described by a complex field ψ, ψ^+ which is an ordinary or anticommuting function, depending on the statistics of the particles obtained in the quantization. The latter is done by means of the usual canonical commutation relations, and in the fermionic case by anticommutation relations.

The action functional of the free theory has the form

$$S(\psi^+, \psi) = \int dx \psi^+(x)\,[i\partial_t - \mathscr{E}]\,\psi(x) \equiv \psi^+ K \psi, \quad (2.20)$$

where $x \equiv (t, \mathbf{x})$, and $\mathscr{E}$ is the given *single-particle Hamiltonian*, a linear operator acting only on the argument $\mathbf{x}$ and interpreted as the ordinary quantum-mechanical Hamiltonian for a single particle of the system: $\mathscr{E} = \hat{p}^2/2m + V(\mathbf{x}) - \mu$. The first term is the kinetic energy of the particle, and the second is the potential of the external field (for example, the lattice potential for electrons in a solid). The constant μ introduced for convenience as an arbitrary parameter is called the chemical potential. For spinless particles the argument **x** is a vector in the space on which the field is studied, and for particles with spin **x** includes a spin index in accordance with the conventions of Sec. 1.1.1.

We see from (2.20) that $i\psi^+$ is the momentum canonically conjugate to the coordinate ψ, i.e., in this case the unified field $\varphi \equiv (\psi, \psi^+)$ represents not the generalized coordinates, but the entire phase space: coordinates and momenta. The canonical commutation relations for operators in the Schrödinger picture take the form $\hat{\psi}(\mathbf{x})\hat{\psi}^+(\mathbf{x}') - \varkappa\hat{\psi}^+(\mathbf{x}')\hat{\psi}(\mathbf{x}) = \delta(\mathbf{x} - \mathbf{x}')$, where, as usual, $\varkappa = \pm 1$ depending on the statistics. If the classical field $\psi(\mathbf{x})$ is expanded in a complete orthonormal set $\Phi_\alpha(\mathbf{x})$ of eigenfunctions of the operator $\mathscr{E}$, for the coefficients of the expansion $\psi(\mathbf{x}) = \Sigma_\alpha a_\alpha \Phi_\alpha(\mathbf{x})$, $\psi^+(\mathbf{x}) = \Sigma_\alpha \overset{+}{a}_\alpha \Phi^*_\alpha(\mathbf{x})$, upon quantization we obtain the standard commutation relations for the creation–annihilation operators: $a_\alpha a_\beta - \varkappa a_\beta a_\alpha = \overset{+}{a}_\alpha \overset{+}{a}_\beta - \varkappa \overset{+}{a}_\beta \overset{+}{a}_\alpha = 0$, $a_\alpha \overset{+}{a}_\beta - \varkappa \overset{+}{a}_\beta a_\alpha = \delta_{\alpha\beta}$. The realization of a_α and $\overset{+}{a}_\alpha$ by operators in the Fock space is well known, and we shall not discuss it here.

The Hamiltonian corresponding to the action (2.20) has the form

$$\mathbf{H}_0 = \int d\mathbf{x}\hat{\psi}^+(\mathbf{x})\mathscr{E}\hat{\psi}(\mathbf{x}) = \sum_\alpha \varepsilon_\alpha \overset{+}{a}_\alpha a_\alpha, \tag{2.21}$$

where ε_α are the eigenvalues of $\mathscr{E}$ on Φ_α: $\mathscr{E}\Phi_\alpha = \varepsilon_\alpha \Phi_\alpha$. For a discrete level α the particle number operator $\overset{+}{a}_\alpha a_\alpha$ in the state α takes only the two eigenvalues 0 and 1 in the case of a fermionic field, and in the case of a bosonic field its eigenvalues can be any nonnegative integer. For the continuum $\overset{+}{a}_\alpha a_\alpha$ is interpreted as the particle number density operator for the state α.

If the field is bosonic, all single-particle energies ε_α are always assumed to be nonnegative. Otherwise, the Hamiltonian (2.21) will be unbounded below, which physically corresponds to instability of the system to condensation of an infinitely large number of particles in levels with negative energies. If there are no discrete levels with $\varepsilon = 0$, the ground state of $\mathbf{H}_0$ for a bosonic field will be the Fock vacuum $|0a\rangle$, for which $a_\alpha|0a\rangle = 0$. If there are discrete levels with $\varepsilon = 0$, the ground state of $\mathbf{H}_0$ is infinitely degenerate.

In the fermionic case the system is also stable when there are levels with negative ε_α. If there are no discrete levels with $\varepsilon = 0$, the ground state of $\mathbf{H}_0$ will be the state in which all levels with negative energy are filled and all levels with positive energy are empty. If there are discrete levels with $\varepsilon = 0$, the ground state of $\mathbf{H}_0$ is 2^n-fold degenerate, where n is the number of levels with zero ε.

In the interaction picture $a^{\pm}_{\alpha}(t) = a^{\pm}_{\alpha}\exp(\pm i\varepsilon_{\alpha}t)$, where $a^{-}_{\alpha} \equiv a_{\alpha}$, and the free-field operators take the form

$$\hat{\psi}(x) = \sum_{\alpha} a_{\alpha}\Phi_{\alpha}(\mathbf{x})\exp(-i\varepsilon_{\alpha}t), \tag{2.22}$$

$$\hat{\psi}^{+}(x) = \sum_{\alpha} a^{+}_{\alpha}\Phi^{*}_{\alpha}(\mathbf{x})\exp(i\varepsilon_{\alpha}t),$$

where $x \equiv t, \mathbf{x}$.

The canonical "particle–hole" transformation is often done for fermionic particles. Here the set of all levels α is split into two groups 1, 2 and the operators b_{α}, b^{+}_{α} are introduced [21]:

$$b_{\alpha} = a_{\alpha},\ b^{+}_{\alpha} = a^{+}_{\alpha} \text{ for } \alpha \in 1; \quad b_{\alpha} = a^{+}_{\alpha},\ b^{+}_{\alpha} = a_{\alpha} \text{ for } \alpha \in 2. \tag{2.23}$$

The split is usually done according to the sign of the energy: $\alpha \in 1$ if $\varepsilon_{\alpha} \geq 0$ and $\alpha \in 2$ if $\varepsilon_{\alpha} < 0$. However, this is not necessary: the transformation (2.23) is canonical (only for fermions) for any splitting into sets 1 and 2.

The expansion (2.22) for $\hat{\psi}$ can be rewritten as

$$\hat{\psi}(x) = \sum_{\alpha}[\chi_1(\alpha)b_{\alpha} + \chi_2(\alpha)b^{+}_{\alpha}]\Phi_{\alpha}(\mathbf{x})\exp(-i\varepsilon_{\alpha}t), \tag{2.24}$$

where $\chi_{1,2}$ are the characteristic functions of sets 1 and 2: $\chi_1(\alpha) = 1$ if $\alpha \in 1$ and $\chi_1(\alpha) = 0$ if $\alpha \in 2$, and similarly for χ_2.

The normal-ordered product of free-field operators is determined according to the general rules of Sec. 1.1.4, with the contributions of all the annihilation and creation operators respectively playing the role of $\hat{a}(t)$ and $\hat{b}(t)$. Different N-products, N_a and N_b, are possible for fermions. We shall give the formulas for the most commonly used N_b-product, assuming arbitrary splitting into sets 1 and 2. Set 2 is empty in the special case of the N_a-product.

Assuming, as usual, that ψ and ψ^{+} are respectively the first and second components of the unified field $\varphi \equiv (\psi, \psi^{+})$, for the matrix of simple contractions (1.16) we obtain $n^{b}_{11} = n^{b}_{22} = 0$ and

$$n^{b}_{12}(x, x') \equiv \hat{\psi}(x)\hat{\psi}^{+}(x') - N_b[\hat{\psi}(x)\hat{\psi}^{+}(x')] =$$

$$= \sum_{\alpha}\chi_1(\alpha)\Phi_{\alpha}(\mathbf{x})\Phi^{*}_{\alpha}(\mathbf{x}')\exp i\varepsilon_{\alpha}(t' - t), \tag{2.25}$$

$$n_{21}^{b}(x,x') \equiv \dot{\psi}^{+}(x)\,\dot{\psi}(x') - N_b[\dot{\psi}^{+}(x)\,\dot{\psi}(x')] =$$

$$= \sum_{\alpha}\chi_2(\alpha)\,\Phi_{\alpha}^{*}(\mathbf{x})\,\Phi_{\alpha}(\mathbf{x}')\exp i\varepsilon_{\alpha}(t-t'),$$

and for the time-ordered contractions from (1.29) we find $\Delta_{11}^{b} = \Delta_{22}^{b} = 0$ and

$$\Delta_{12}^{b}(x,x') = \sum_{\alpha}[\chi_1\theta_{tt'} + \varkappa\chi_2\theta_{t't}]\,\Phi_{\alpha}(\mathbf{x})\,\Phi_{\alpha}^{*}(\mathbf{x}')\exp i\varepsilon_{\alpha}(t'-t),$$

$$\Delta_{21}^{b}(x,x') = \sum_{\alpha}[\chi_2\theta_{tt'} + \varkappa\chi_1\theta_{t't}]\,\Phi_{\alpha}^{*}(\mathbf{x})\,\Phi_{\alpha}(\mathbf{x}')\exp i\varepsilon_{\alpha}(t-t'). \quad (2.26)$$

To make the notation more compact we drop the argument α of $\chi_{1,2}$ and, as always, write $\theta(t-t') \equiv \theta_{tt'}$. The symmetry of the matrix Δ is expressed as $\Delta_{21} = \varkappa\Delta_{12}{}^{\mathrm{T}}$.

In the energy representation

$$\Delta_{12}^{b}(x,x') = \frac{i}{2\pi}\int dE\sum_{\alpha}\left[\frac{\chi_1}{E-\varepsilon_{\alpha}+i0} - \frac{\varkappa\chi_2}{E-\varepsilon_{\alpha}-i0}\right]\times$$

$$\Phi_{\alpha}(\mathbf{x})\,\Phi_{\alpha}^{*}(\mathbf{x}')\exp iE(t'-t),$$

$$\Delta_{21}^{b}(x,x') = \frac{i}{2\pi}\int dE\sum_{\alpha}\left[\frac{\varkappa\chi_2}{E+\varepsilon_{\alpha}+i0} - \frac{\chi_1}{E+\varepsilon_{\alpha}-i0}\right]\times$$

$$\Phi_{\alpha}^{*}(\mathbf{x})\,\Phi_{\alpha}(\mathbf{x}')\exp iE(t'-t). \quad (2.27)$$

We can set $\varkappa = -1$ in (2.27), because $\varkappa$ enters only in the combination $\varkappa\chi_2$, and χ_2 can be nonzero only for a fermionic field. Setting $\varkappa = -1$ in (2.27), we see that the choice of splitting into sets 1 and 2 affects only the signs of the additions $\pm i0$ in the energy denominators.

Equation (1.31) now takes the matrix form (1.109), and the role of L in (1.109) is played in our case by the kernel of the free action (2.20), i.e., $L = i\partial_t - \mathscr{E}$. According to (1.109), $\Delta_{12}^{b} = iL^{-1}$; in the fermionic case different splittings into sets 1 and 2, which lead to different Δ_{12}^{b}, correspond to different choices of the Green function L^{-1}.

The contractions Δ have a discontinuity at $t = t'$ and are defined according to the usual rule (1.30) in matrix form.

The normal-ordered product possesses the property (1.134), and Eq. (1.87) for the N_b-product is valid for the state $|0b\rangle \equiv \Pi_2 a^+_\alpha |0a\rangle$, in which all levels of group 2 are filled and the levels of group 1 are empty: $b_\alpha |0b\rangle = 0$ for any α. For the N_a-product Eq. (1.87) is valid for the Fock vacuum $|0a\rangle$.

Let us discuss this topic in more detail.

2.2.1 *The quantum Bose and Fermi gases*

In this case the single-particle Hamiltonian does not contain a potential term: $\mathscr{E} = \hat{p}^2/2m - \mu$. The eigenfunctions of $\mathscr{E}$, which are plane waves, are labeled by the momentum $\mathbf{p}$ and spin index s if the particles have spin: $\Phi_\alpha(\mathbf{x}) = (2\pi)^{-d/2} e_s \exp i\mathbf{px}$, where d is the dimensionality of the space and e_s is a complete orthonormal set of spin states. For a bosonic field the condition that $\varepsilon(\mathbf{p}) = \mathbf{p}^2/2m - \mu$ be positive requires $\mu \leq 0$, and for a fermionic field any sign of μ is allowed. For $\mu > 0$ the ground state of $\mathbf{H}_0$ is the state in which all levels with $\mathbf{p}^2/2m < \mu$ are filled and levels with $\mathbf{p}^2/2m > \mu$ are empty, i.e., μ is interpreted as the Fermi energy.

If the splitting into sets 1 and 2 is done according to the sign of $\varepsilon(\mathbf{p})$, the contraction Δ^b_{12} in (2.27) takes the form

$$\Delta^b_{12}(x, x') = \frac{i\delta_{ss'}}{(2\pi)^{d+1}} \int dE \int d\mathbf{p} \frac{\exp i[E(t'-t) - \mathbf{p}(\mathbf{x}'-\mathbf{x})]}{E - \varepsilon(\mathbf{p}) + i0 \cdot \mathrm{Sgn}\,\varepsilon(\mathbf{p})}, \tag{2.28}$$

where $\mathrm{Sgn}\,\varepsilon(\mathbf{p}) = \pm 1$ depending on the sign of $\varepsilon(\mathbf{p})$, and $x \equiv t, \mathbf{x}, s$.

For such systems it is usual to study a two-body interaction, the Hamiltonian of which in the Schrödinger picture has the form

$$\mathbf{V} = \frac{1}{2} \iint d\mathbf{x} d\mathbf{x}' \hat{\psi}^+(\mathbf{x}) \hat{\psi}^+(\mathbf{x}') \mathcal{V}(\mathbf{x}, \mathbf{x}') \hat{\psi}(\mathbf{x}') \hat{\psi}(\mathbf{x}), \tag{2.29}$$

where $\mathcal{V}(\mathbf{x}, \mathbf{x}')$ is a symmetric *two-body interaction potential*. The operator (2.29) is written in N_a form, and the corresponding classical interaction functional (1.59) represents, as discussed in Sec. 1.3.2, the Sym-form of the operator $-\int dt\, \mathbf{V}(t)$, which is easily found using (1.45) from the known N_a form of this operator. We note that (1.45) leads to the following rule for going from the N_a form to the N_b form:

$$N_a F(\hat{\varphi}) = N_b \left\{ \exp\left[\frac{1}{2} \frac{\delta}{\delta\varphi} (n^b - n^a) \frac{\delta}{\delta\varphi}\right] F(\varphi) \right\} \Big|_{\varphi = \hat{\varphi}}. \tag{2.30}$$

Here $\varphi \equiv (\psi, \psi^+)$, and n^a and n^b are the corresponding matrices of contractions (2.25). If we work with the N_b-product, the role of the reduced generating vertex (see Sec. 1.4.1) in the graphs of perturbation theory will be played by the N_b form of the operator $-i\int dt\mathbf{V}(t)$. The leading term in the interaction functional has the form (1.116) and can be reduced to a Yukawa-type interaction, as shown in Sec. 1.4.7.

2.2.2 *The atom*

The atom is a system containing any number of electrons in the Coulomb field of the nucleus: $\mathscr{E} = \hat{p}^2/2m + \mathscr{V}(\mathbf{x})$, $\Phi_\alpha(\mathbf{x})$ are the electron wave functions in the Coulomb field, and ε_α are the corresponding energies. The index α runs over both the discrete and the continuous spectrum. The discrete energy levels are degenerate, and the set of all states with the same energy is called a shell.

An important problem is the determination of the energy of the ground and excited states of an atom with a given number of electrons interacting with each other via the two-body Coulomb interaction (2.29).

This problem can be solved by field-theoretic methods using (1.74), which is valid for any nondegenerate eigenstate of the free Hamiltonian. Only states in which some of the shells are completely filled and others are completely empty will be nondegenerate for the operator (2.21). If at least one shell is only partially filled, the corresponding level of $\mathbf{H}_0$ is certainly degenerate, because the energy does not change when the order of putting the electrons in the partially filled shell is changed.

For the state $|0b\rangle$ with completely filled shells it is possible to introduce the corresponding N_b-product and to use the standard representation of the logarithm of the expectation value of the S matrix involved in (1.74) in the form of a sum of connected vacuum loops. Without going into detail, we note that the energies of states differing by one particle or hole from the state $|0b\rangle$ with completely filled shells can be found from the spectral representation of the Green function $(0b|T_D[\hat{\psi}_H(x)\hat{\psi}_H^+(x')]|0b)$, where $|0b)$ is the eigenstate of the full Hamiltonian which becomes the state $|0b\rangle$ when the interaction between the electrons is switched on. This Green function can be calculated by the standard diagrammatic techniques using Eq. (1.73), which is valid for any nondegenerate state.

For a Coulomb field the nondegeneracy condition is an important restriction, because the degree of degeneracy is very high. However, it must be stated that in practice the level energies are usually calculated not using ordinary perturbation theory in the interelectron interaction, but by the

Hartree–Fock self-consistent field method [22,23], about which more will be said in Sec. 6.2.7. The formulas of nonstationary perturbation theory for a degenerate level generalizing (1.74) are given in Appendix 1.

2.2.3 *Electrons in solids and phonons*

In this case $\mathscr{E} = \hat{p}^2/2m + \mathscr{V} - \mu$, where $\mathscr{V}$ is the periodic lattice potential; $\Phi_\alpha(\mathbf{x})$ are the electron wave functions in the given periodic field; E_α are the corresponding energies (which, as is well known, form bands), and $\varepsilon_\alpha = E_\alpha - \mu$. In the ground state of $\mathbf{H}_0$ all levels with $E_\alpha < \mu$ (where μ is the Fermi energy) are filled, and by a suitable choice of μ it is possible to obtain any *a priori* given electron concentration. In simple models the lattice potential is often neglected and the conduction electrons are treated as a free gas.

In addition to the Coulomb two-body interaction (2.29) between electrons, in a solid an important role is played by the interaction of the electrons with lattice vibrations: phonons. Its Hamiltonian in the Schrödinger picture is written as [21] $\mathbf{V} = g\int d\mathbf{x}\Sigma_s \hat{\psi}_s^+(\mathbf{x})\hat{\psi}_s(\mathbf{x})\hat{\varphi}(\mathbf{x})$. In this expression, $\mathbf{x}$ denotes only the space coordinate, s is the spin index, g is the coupling constant, and $\hat{\varphi}(\mathbf{x})$ is the Hermitian phonon field operator. In the interaction picture

$$\hat{\varphi}(x) = (2\pi)^{-d/2}\int d\mathbf{k}\sqrt{\frac{\varepsilon_k}{2}}\,[a^+(\mathbf{k})\exp(ikx) + a(\mathbf{k})\exp(-ikx)], \quad (2.31)$$

where $x \equiv (t, \mathbf{x})$; d is the dimensionality of the $\mathbf{x}$ space; $kx \equiv \varepsilon_k t - \mathbf{kx}$; $\varepsilon_k = \lambda|\mathbf{k}|$; the constant λ is the speed of sound; and $a^+(\mathbf{k})$ and $a(\mathbf{k})$ are the bosonic creation and annihilation operators of a phonon of momentum $\mathbf{k}$. The field $\varphi(x)$ is interpreted as the density of bound charge, which is proportional to the divergence of the vector field of the lattice deformations. Of the three acoustical branches of the deformation field, only the longitudinal one contributes to the divergence. The Hamiltonian $\mathbf{V}$ given above models the strongly screened Coulomb interaction of the electrons with the bound charge arising in lattice deformations.

The normal-ordered product of fields (2.31) is determined by the usual splitting into creation and annihilation operators and possesses the properties (1.87) and (1.134). The corresponding contractions are determined by the general rules (1.13) and (1.29). Here we give only the time-ordered one:

$$\Delta(x, x') = i(2\pi)^{-d-1}\iint dE d\mathbf{k}\frac{\varepsilon_k^2 \exp i[E(t'-t) - \mathbf{k}(\mathbf{x}'-\mathbf{x})]}{E^2 - \varepsilon_k^2 + i0}. \quad (2.32)$$

The electron–phonon interaction is a special case of the Yukawa interaction, the diagrammatic technique for which was discussed in Sec. 1.4.6.

2.3 Relativistic Field Theory

The quantization of relativistically covariant fields is discussed in detail in [1], so here we shall restrict ourselves to only a listing of the corresponding free systems with their action functional and contractions. The normal-ordered product for all relativistic fields is defined in the usual manner in terms of creation–annihilation operators and possesses the properties (1.87) and (1.134), where the state $|0\rangle$ in (1.87) is always the Fock vacuum and simultaneously the ground state of the free Hamiltonian. We shall not need the explicit form of the expansions of fields in creation and annihilation operators, so we shall not give them here.

In this section we use the usual notation of relativistic theories: $x \equiv \{x^\alpha, \alpha = 0, 1, 2, 3\}, x^0 \equiv t, \{x^i, i = 1, 2, 3\} \equiv \mathbf{x}$. Repeated Greek indices are summed from 0 to 3, and repeated Latin indices are summed from 1 to 3. The metric tensor $g_{\alpha\beta} = g^{\alpha\beta}$ is defined as $g_{11} = g_{22} = g_{33} = -g_{00} = -1$, $g_{\alpha\beta} = 0$ for $\alpha \neq \beta$; the tensor $g_{\alpha\lambda}g^{\lambda\beta} = g_\alpha{}^\beta$ is the Kronecker δ. The covariant vector B_α is constructed from the contravariant vector B^α according to the rule $B_\alpha = g_{\alpha\beta}B^\beta$; the invariant scalar product $x_\alpha y^\alpha = g_{\alpha\beta}x^\alpha y^\beta = x^0 y^0 - \mathbf{xy}$ is denoted xy. Finally, $\partial_\alpha \equiv \partial/\partial x^\alpha$ and $\int dx \ldots \equiv \int dx^0 dx^1 dx^2 dx^3 \ldots$, where the integration over each x^α runs from $-\infty$ to ∞.

The simplest relativistically invariant system is a real scalar field $\varphi(x)$, for which

$$S_0(\varphi) = \frac{1}{2}\int dx\, [\partial_\alpha \varphi(x) \cdot \partial^\alpha \varphi(x) - m^2 \varphi^2(x)]\,, \tag{2.33}$$

and the kernel of the corresponding quadratic form (1.4) is the Klein–Gordon operator $K = -\partial_\alpha \partial^\alpha - m^2 \equiv -\Box - m^2$. The field contractions have the form

$$n(x, x') = (2\pi)^{-3} \int dp\, \theta(p_0)\, \delta(p^2 - m^2) \exp ip(x' - x)\,, \tag{2.34}$$

$$\Delta(x, x') = i(2\pi)^{-4} \int dp \frac{1}{p^2 - m^2 + i0} \exp ip(x' - x)\,, \tag{2.35}$$

where $p^2 = p_0^2 - \mathbf{p}^2$ by convention. The contractions are explicitly translationally invariant, i.e., they depend only on the difference $y \equiv x - x'$. In contrast to quantum mechanics and nonrelativistic theory, Eqs. (2.34) and (2.35) are not ordinary but generalized functions having singularities on the surface $y^2 = 0$ (the light cone). The strongest singularity is located at $y = 0$.

The following representations of the contractions are useful:

$$\Delta(x, x') = d(y^2) + if(y^2), n(x, x') = d(y^2) + i\varepsilon(y_0) f(y^2), \quad (2.36)$$

where $y \equiv x - x'$; d and f are real, generalized functions of the variable y^2; $\varepsilon(y_0) = \theta(y_0) - \theta(-y_0)$ is the sign function. The function f is nonzero only inside the light cone, i.e., in the region $y^2 \geq 0$. The representation (2.36) for n is formal, because f is singular on the surface $y^2 = 0$ having the point $y = 0$ in common with the plane $y_0 = 0$, where ε undergoes a jump. The representation (2.36) becomes completely correct only after regularization, which will be discussed below.

For a complex scalar field φ, φ^+ the free action $S_0(\varphi, \varphi^+)$ is written in the form $\int dx[\partial_\alpha\varphi^+(x)\partial^\alpha\varphi(x) - m^2\varphi^+(x)\varphi(x)]$. Contractions of φ with φ^+, like those of φ^+ with φ, coincide with the contractions (2.34) and (2.35) of a real field.

For a real vector field $B^\alpha(x)$

$$S_0(B) = -\frac{1}{4}\int dx\,[F_{\alpha\beta}(x) F^{\alpha\beta}(x) - 2m^2 B_\alpha(x) B^\alpha(x)], \quad (2.37)$$

where $F_{\alpha\beta}(x) \equiv \partial_\alpha B_\beta(x) - \partial_\beta B_\alpha(x)$, and Eq. (1.4) has the form $B^\alpha K_{\alpha\beta} B^\beta/2$ with the kernel

$$K_{\alpha\beta} = (\Box + m^2)\, g_{\alpha\beta} - \partial_\alpha \partial_\beta. \quad (2.38)$$

The corresponding equations of motion are equivalent to the system $(\Box + m^2)B^\alpha = 0$, $m^2\partial_\alpha B^\alpha = 0$. We see from the latter equation that for a massive field any solution is four-dimensionally transverse: $\partial_\alpha B^\alpha = 0$.

The contractions of the field $B^\alpha(x)$ have the form

$$n^{\alpha\beta}(x, x') = \mathcal{D}_x^{\alpha\beta} n(x, x'), \quad \Delta^{\alpha\beta}(x, x') = \mathcal{D}_x^{\alpha\beta}\Delta(x, x'), \quad (2.39)$$

where n and Δ are the contractions (2.34) and (2.35) and $\mathcal{D}$ is a differential operator:

$$\mathcal{D}_x^{\alpha\beta} = -g^{\alpha\beta} - m^{-2}\partial^\alpha\partial^\beta. \quad (2.40)$$

We note that in this case the general relation (1.31) has covariant form: $K_{\alpha\lambda}\Delta^{\lambda\beta} = ig_\alpha{}^\beta$. It does not admit passage to the case of a massless field, since the operator $K_{\alpha\beta}$ then becomes essentially degenerate: for $m = 0$ we have $K_{\alpha\beta}\partial^\beta\varphi = 0$ for any function $\varphi(x)$. A massless vector field is studied in the next chapter.

Scalar and vector fields are bosonic and describe (after quantization) particles with spin 0 and 1, respectively. Particles with spin 1/2 are described by a fermionic Dirac field, i.e., by a four-component complex spinor field $\psi \equiv \psi_1$, $\bar{\psi} = \psi^+\gamma^0 \equiv \psi_2$ (γ^0 is one of the Dirac matrices γ^α acting on the spin indices of the field). The free action has the form

$$S_0(\psi_1, \psi_2) = \int dx\, \psi_2(x)\, [i\gamma^\alpha\partial_\alpha - m]\, \psi_1(x)\,, \tag{2.41}$$

[this is the special case (2.20)] and the nonzero elements of the contraction matrix of the unified field $(\psi_1, \psi_2) \equiv (\psi, \bar{\psi})$ are given by

$$\begin{aligned} n_{12}(x,x') &= \mathcal{D}_x n(x,x'), \quad n_{21}(x,x') = -\mathcal{D}_{x'} n(x,x'), \\ \Delta_{12}(x,x') &= \mathcal{D}_x \Delta(x,x'), \quad \Delta_{21}(x,x') = -\mathcal{D}_{x'} \Delta(x,x'), \end{aligned} \tag{2.42}$$

in which n and Δ are the contractions (2.34) and (2.35) and $\mathcal{D}_x \equiv i\gamma^\alpha\partial_\alpha + m$.

Let us conclude with the following remark on regularization. As already mentioned, the field contractions defined above are not ordinary but generalized functions. The product of generalized functions with superposed singularities is meaningless even as a generalized function, and for this reason the contributions of the graphs of perturbation theory containing multiple products of propagators Δ often turn out to be meaningless in relativistic theories. In momentum space this is manifested as the well known "ultraviolet" divergences of integrals at large momenta.

Any procedure which rigorously defines the contribution of each graph of perturbation theory is referred to as regularization. A typical example is Pauli–Villars regularization [1], in which n and Δ in each expression of this section are replaced by the regularized contractions n_{reg} and Δ_{reg}. The latter are defined as

$$n_{\text{reg}} = n + \sum_{i=1}^{N} c_i n_i\,, \quad \Delta_{\text{reg}} = \Delta + \sum_{i=1}^{N} c_i \Delta_i\,, \tag{2.43}$$

in which n_i and Δ_i are the usual contractions (2.34) and (2.35) with the replacement $m \to M_i$; $M_1 \ldots M_N$ are arbitrary parameters, and the numerical coefficients c_i in (2.43) are determined by the following system of equations:

$$m^{2k} + \sum_{i=1}^{N} c_i M_i^{2k} = 0, \quad k = 0, 1, \ldots, N-1. \tag{2.44}$$

In momentum space

$$\tilde{\Delta}_{\text{reg}} = i\left[\frac{1}{p^2 - m^2 + i0} + \sum_{i=1}^{N} \frac{c_i}{p^2 - M_i^2 + i0}\right], \tag{2.45}$$

and the conditions (2.44) imply that after the sum (2.45) is reduced to a common denominator, the coefficients of all powers of p^2 except the zeroth vanish in the numerator. Therefore, $\tilde{\Delta}_{\text{reg}}(p^2)$ falls off sufficiently rapidly at infinity and its Fourier transform $\Delta_{\text{reg}} = d_{\text{reg}} + if_{\text{reg}}$ [see (2.36)] will now be an ordinary, rather than generalized, function having a certain number of smooth derivatives which depends on N. We see from (2.36) and (2.43) that the contraction n_{reg} is related to Δ_{reg} by the same expression (2.36), which is now completely correct, since the function f_{reg}, which vanishes outside the light cone, must have a zero of finite order owing to its smoothness on the light cone itself. From this, in particular, it follows that Eq. (1.30) is satisfied automatically for regularized contractions. We also note that the proof of unitarity of the S matrix given in Sec.1.5.2 will be valid in each order of perturbation theory for the regularized theory independently of the presence of time derivatives of the field in the Hermitian interaction functional.

The transformation to the regularized theory in expressions of the type (1.164) is done using the substitution $K \to K_{\text{reg}}$, where $K_{\text{reg}} = i\Delta_{\text{reg}}^{-1}$ in accordance with (1.31).

2.4 Integral Representations of the Transition Amplitude

For ordinary quantum mechanics with Hamiltonian $\mathbf{H}$ the quantity

$$\mathcal{A}(\mathbf{x}_1, \tau_1; \mathbf{x}_2, \tau_2) \equiv \langle \mathbf{x}_1 | \exp i\mathbf{H}(\tau_2 - \tau_1) | \mathbf{x}_2 \rangle \tag{2.46}$$

has the meaning of the transition amplitude in coordinate space for a finite time interval $\tau_2 \leq t \leq \tau_1$, and $\theta(\tau_1 - \tau_2)\mathcal{A}(\mathbf{x}_1, \tau_1; \mathbf{x}_2, \tau_2)$ is the retarded Green function of the nonstationary Schrödinger equation (the kernel of the operator $i[i\partial_t - \mathbf{H}]^{-1}$, with the retardation condition) in the same space.

Historically, such amplitudes (2.46) were the first objects for which Feynman obtained the functional integral representation [24]. The original derivation of Feynman is based on the definition of the functional integral in terms of the well known interpolation procedure (see [14,15]), which we shall not use owing to its ambiguity. We shall derive the Feynman formula using only the rules for calculating Gaussian integrals formulated in Sec. 1.6.1.

As in Sec. 2.1.2, as $\mathbf{H}_0$ we take only the kinetic term $\hat{p}^2/2m$. Representing $\exp i\mathbf{H}_0\tau$ by a Gaussian integral

$$\exp\, i\mathbf{H}_0\tau = \exp\left[i\frac{\hat{p}^2\tau}{2m}\right] = \left(\frac{im}{2\pi\tau}\right)^{d/2}\int d\lambda\, \exp\left[-\frac{im\lambda^2}{2\tau}+i\hat{p}\lambda\right], \quad (2.47)$$

in which d is the dimensionality of p, and using the fact that the shift operator $\exp i\,\hat{p}\,\lambda$ takes $|\mathbf{x}\rangle$ into $|\mathbf{x}-\lambda\rangle$, for the free theory we obtain

$$\mathcal{A}_0(\mathbf{x}_1,\tau_1;\mathbf{x}_2,\tau_2) = \left[\frac{m}{2\pi i(\tau_1-\tau_2)}\right]^{d/2}\exp\left[\frac{im(\mathbf{x}_1-\mathbf{x}_2)^2}{2(\tau_1-\tau_2)}\right]. \quad (2.48)$$

For a theory with an interaction the evolution operator in the Schrödinger picture entering into (2.46) can be expressed using (1.55) in terms of the evolution operator in the interaction picture, which is represented by a T-exponential (1.64). With the intention of using Eq. (1.90) later, we calculate the functional

$$\mathcal{A}(\mathbf{x}_1,\tau_1;\mathbf{x}_2,\tau_2;A) \equiv \langle\mathbf{x}_1|\exp(-i\mathbf{H}_0\tau_1)\,T\exp(i\hat{\varphi}A)\exp(i\mathbf{H}_0\tau_2)\,|\mathbf{x}_2\rangle, \quad (2.49)$$

interpreted as the transition amplitude of a free particle in an arbitrary external force field $A(t)$. The free field $\hat{\varphi}$ in (2.49) denotes the coordinate operator in the interaction picture (2.7), and the integration over time in the linear form $\hat{\varphi}A$ runs, as everywhere below, only over a finite interval from τ_2 to τ_1.

To calculate the amplitude (2.49) it is necessary to write the T-exponential in N-ordered form using the rule (1.46), and the result of acting with the operator $\exp(i\mathbf{H}_0\tau)$ on $|\mathbf{x}\rangle$ must be represented as a Gaussian integral using (2.47). The matrix element of the N-ordered exponential can then be found from (2.12). Omitting the elementary calculations,* we present only the result. The amplitude (2.49) differs from (2.48) by the replacement of the argument of the exponential in (2.48) by the expression

---

* The calculations simplify considerably if we set $\tau_1=\tau$ and $\tau_2=0$, which does not lead to loss of generality.

$$\frac{im\tau_2}{2\tau_1(\tau_1-\tau_2)}\left(\mathbf{x}_1-\mathbf{x}_2+\frac{\overline{tA}}{m}-\frac{\tau_1\bar{A}}{m}\right)^2+\frac{im}{2\tau_1}\left(\mathbf{x}_1-\mathbf{x}_2+\frac{\overline{tA}}{m}\right)^2+$$

$$+i\mathbf{x}_2\bar{A}-\frac{1}{2}A\Delta A, \tag{2.50}$$

in which Δ denotes the contraction (2.9) and $\bar{A}\equiv\int dtA(t)$ and $\overline{tA}\equiv\int dt\, tA(t)$; both here and in the form $A\Delta A$, the time integration runs only over a given finite interval.

Collecting the terms quadratic and linear in A in (2.50), the result can be written as

$$\mathcal{A}(\mathbf{x}_1,\tau_1;\mathbf{x}_2,\tau_2;A) = \mathcal{A}_0(\mathbf{x}_1,\tau_1;\mathbf{x}_2,\tau_2)\exp\left[-\frac{1}{2}A\tilde{\Delta}A+iAf\right]. \tag{2.51}$$

Here f is a linear function: $f(t)=\mathbf{x}_2+\mathbf{v}(t-\tau_2)$, $\mathbf{v}\equiv(\mathbf{x}_1-\mathbf{x}_2)/(\tau_1-\tau_2)$, and

$$\tilde{\Delta}(t,t') = \Delta(t,t')+i[\tau_2(t+t')-tt'-\tau_1\tau_2]/m(\tau_1-\tau_2). \tag{2.52}$$

We note that it is necessary to symmetrize $t\rightleftarrows t'$ in finding the kernel of the quadratic form $\tilde{\Delta}(t,t')$ from (2.50). The function (2.52) differs from $\Delta(t,t')$ by the solution of the classical free equation $K\varphi=0$ in the variable t (for us, $K=-m\partial_t^2$). From this it follows that $\tilde{\Delta}$, like Δ, is the Green function of the free equation: $K\tilde{\Delta}=i$. This Green function is special in that it satisfies, as is easily seen from (2.9) and (2.52), the conditions $\tilde{\Delta}(\tau_1,t')=\tilde{\Delta}(\tau_2,t')=0$, from which it is clear that $\tilde{\Delta}$ is the Green function for the equation $K\varphi=\psi$ with the auxiliary conditions $\varphi(\tau_1)=\varphi(\tau_2)=0$. This property uniquely distinguishes $\tilde{\Delta}$ from the set of all Green functions of the free equation.

In what follows it is also important to note that the function f entering into (2.51) is a solution of the homogeneous equation $Kf=0$ satisfying the conditions $f(\tau_i)=\mathbf{x}_i$, $i=1,2$, which determine this solution uniquely.

Let us now digress from the amplitude (2.49) and consider the integral

$$\int\limits_{(\mathbf{x}_1\mathbf{x}_2)} D\varphi\exp\, i\int_{\tau_2}^{\tau_1}dt\left[\frac{1}{2}m\dot{\varphi}^2+\varphi A\right] = \int\limits_{(\mathbf{x}_1\mathbf{x}_2)} D\varphi\exp\, i[S_0(\varphi)+\varphi A]. \tag{2.53}$$

The symbol $(\mathbf{x}_1\mathbf{x}_2)$ indicates that the integration runs over the set (which is not a linear space) of real functions $\varphi(t)$ on the interval $\tau_2\le t\le\tau_1$ satisfying the boundary conditions $\varphi(\tau_i) = \mathbf{x}_i$. We note that the free action itself $S_0(\varphi)$ enters into (2.53), rather than its quadratic form $S_0' = \varphi K \varphi/2$.

In (2.53) we make the change of variable $\varphi \to \varphi + f$, where f is the same linear function as in (2.51). The new variable φ will then satisfy zero boundary conditions, from which, in particular, it follows that the cross terms of the form $(\dot{\varphi} + \dot{f})^2$ do not contribute to the action. Therefore, the integral (2.53) can be rewritten as

$$\exp i\,[S_0(f) + Af] \int_{(00)} D\varphi \exp i\,[S_0(\varphi) + \varphi A]\,. \tag{2.54}$$

The zero boundary conditions on φ also cause the surface terms to vanish in the integration by parts in the form $S_0(\varphi)$, so that it can be written as $\varphi K \varphi/2$. Therefore, the Gaussian integral in (2.54) can be performed using the shift $\varphi \to \varphi + i\tilde{\Delta}A$, which does not lead outside the class of real functions satisfying zero boundary conditions. After the shift the factor $\exp[-A\tilde{\Delta}A/2]$ is isolated from the integral (2.54), and the remaining integral $J \equiv \int_{(00)} D\varphi \exp iS_0(\varphi)$ no longer depends on A. Comparing the expression obtained from (2.54) with (2.51) and using the fact that $S_0(f)$ coincides with the argument of the exponential in (2.48), as is easily checked, we conclude that the quantities (2.54) and (2.51) are proportional:

$$\mathcal{A}(\mathbf{x}_1, \tau_1; \mathbf{x}_2, \tau_2; A) = \text{const} \int_{(\mathbf{x}_1 \mathbf{x}_2)} D\varphi \exp i\,[S_0(\varphi) + \varphi A]\,. \tag{2.55}$$

The normalization constant is numerically equal to the ratio of the pre-exponential factor in (2.48) and the Gaussian integral J defined above and depends only on m and $\tau_1 - \tau_2$.

In order to find the transition amplitude for a theory with arbitrary interaction $S_v(\varphi)$, according to (1.90) we must operate on the right-hand side of (2.55) with $\exp iS_v(\delta/\delta iA)$ and then set $A = 0$. We find

$$\mathcal{A}(\mathbf{x}_1, \tau_1; \mathbf{x}_2, \tau_2) = \text{const} \int_{(\mathbf{x}_1 \mathbf{x}_2)} D\varphi \exp iS(\varphi)\,, \tag{2.56}$$

where $S(\varphi) = S_0(\varphi) + S_v(\varphi)$ is the complete action functional for the interval $[\tau_1, \tau_2]$. It is worth noting that the range of integration over t in the functional $S_v(\varphi)$ is finite because it is finite in the linear form φA on which the operator $S_v(\delta/\delta iA)$ acts.

As already noted repeatedly, for a theory with time derivatives in the interaction the role of $S_v(\varphi)$ is played by the effective interaction defined in Sec. 1.3.4 and explicitly depending on the choice of method of quantizing the classical theory. We note that when the functional integral is determined using the interpolation procedure, the functional S in (2.56) is always assumed to be classical, and the quantum amplitudes (2.56) for different methods of quantization differ owing to the use of different interpolation variants. This has been checked [14,15] for several simple quantization recipes, but there is no general rule for associating a certain interpolation method with any particular choice of quantum interaction operator $\mathbf{V}$ in the way we associate a functional $S_v^{\text{eff}}(\varphi)$ with it.

The generalization of the quantum-mechanical expression (2.56) to bosonic field theories will be straightforward in "coordinate space," where the field operator $\hat{\varphi}(\mathbf{x})$ is diagonal in the Schrödinger picture: the arguments $\mathbf{x}$ in (2.56) are replaced by "configurations" $\alpha(\mathbf{x})$, which number the eigenstates of the operator $\hat{\varphi}(\mathbf{x})$ like $\mathbf{x}$ numbers the eigenstates of the coordinate operator in quantum mechanics. In this representation the role of the wave functions $\psi(\mathbf{x})$ is played by the functionals $F(\alpha)$, the scalar product of which is determined by the integral $\int D\alpha F_1{}^*(\alpha)F_2(\alpha)$. The coordinate and momentum operators $\hat{\varphi}(\mathbf{x})$ and $\hat{p}(\mathbf{x})$ in the Schrödinger picture are respectively realized as the operations of multiplication by $\alpha(\mathbf{x})$ and differentiation with respect to $i\alpha(\mathbf{x})$.

We conclude by noting two important differences between Eqs. (2.56) and (1.164). First, (2.56) involves the actual free action $S_0(\varphi)$, and not the quadratic form corresponding to it, as in (1.164). Second and more important, the space of functions over which the integration runs in (2.56) is, in contrast to (1.164), independent of the method of splitting the full action into the free part and the interaction. This advantageous feature of the representation (2.56) allows it to be viewed as a truly exact relation which is not even implicitly based on perturbation theory. This cannot be said of (1.164).

Calculating the functional integral (2.56) by the stationary-phase method (see Sec. 1.6.6) and restricting ourselves to the first two terms in the expansion analogous to (1.171), we obtain the well known semiclassical approximation for the transition amplitude. The condition that the action $S(\varphi) = \int dt\,[m\dot{\varphi}^2/2 - \mathcal{V}(\varphi)]$ be stationary is the equation of the classical trajectory $m\ddot{\varphi} + \mathcal{V}_\varphi = 0$ (here and below $f_z \equiv \partial_z f \equiv \partial f/\partial z$), supplemented by the boundary conditions $\varphi(\tau_1) = \mathbf{x}_1$, $\varphi(\tau_2) = \mathbf{x}_2$. We assume that the desired trajectory is unique (otherwise the result is the sum over all trajectories) and denote it as q. The first term of the expansion analogous to (1.171) gives the factor $\exp iS(q)$ in the amplitude (2.56), and the second determines the preexponential factor, which will be proportional to $\det K^{-1/2}$, where $K = -m\partial_t^2 - \mathcal{V}_{qq} \equiv K_0 + u$ is the kernel of the second variation of the action on the trajectory. Here K_0 and K are treated as linear operators in the space of square-integrable functions on the interval $\tau_2 \le t \le \tau_1$, defined on the dense set of doubly differential functions with zero boundary values and square-integrable second derivative. It is well known that the symmetric operator $K_0 = -m\partial_t^2$ is self-adjoint in this region. The operation of multiplication by $u = -\mathcal{V}_{qq}$ is restricted (it is assumed that the trajectory does not pass through the singular points of the potential), so $K = K_0 + u$ is also self-adjoint.

Strictly speaking, the determinant exists only for operators of the form $1 + B$, where B is a kernel operator (the sum of the moduli of the eigenvalues is finite). The operator K_0^{-1} [integral operator with kernel $-i\,\widetilde{\Delta}(t, t')$] will be a kernel operator, because its eigenvalues fall off as $1/n^2$. The product of a kernel operator by a continuous operator is a kernel operator, so $\det K_0^{-1}K = \det(1 + K_0^{-1}u)$ is well defined. It can be calculated explicitly, with the result expressed in terms of the classical trajectory.

For simplicity, let us consider the one-dimensional case and use Eq. (1.147) to calculate the determinant, considering the variation $\delta K(t) = -\delta\mathcal{V}_{qq}(q(t))$ for an arbitrary variation of the parameters $x_{1,2}$ specifying the trajectory $q(t;\, x_1,\, x_2)$: $\delta \ln\det K = \mathrm{tr}\,(G\delta K) = \int dt G(t,\, t)\delta K(t)$, where $G(t,\, t')$ is the kernel of the integral operator $G = K^{-1}$ (the Green function) and the integration runs over the interval $\tau_2 \le t \le \tau_1$. The Green function of our boundary-value problem is expressed in terms of suitable solutions $y_{1,2}$ of the homogeneous equation: $G(t,\, t') = [\theta_{tt'}y_1(t)y_2(t') + \theta_{t't}y_2(t)y_1(t')]/mW$, where $y_{1,2}$ is any pair of independent solutions of the equation $Ky = 0$ satisfying the conditions $y_1(\tau_1) = y_2(\tau_2) = 0$, and the t-independent constant $W = y_1\dot{y}_2 - \dot{y}_1y_2$ is their Wronskian. Taking into account the explicit form of G, we obtain $\delta \ln\det K = (mW)^{-1}\int dt y_1(t)y_2(t)\delta K(t)$. This integral can be done as follows. Varying the equation $Ky_1 = 0$, we obtain $K\delta y_1 + \delta K\cdot y_1 = 0$, from which we find $y_1y_2\delta K = -y_2K\delta y_1$, and the integral of the last expression completely reduces to surface terms owing to the equation $Ky_2 = 0$ (integration by parts). In this manner we obtain $\delta \ln\det K = (W)^{-1}[y_2\delta\dot{y}_1 - \dot{y}_2\delta y_1]|_{\ldots}$, where the symbol $|_{\ldots}$ denotes the difference of the values at the ends of the interval $\tau_{1,2}$.

The functions $y_{1,2}$ can be related to the classical trajectory q. Differentiating the equation for q, $m\ddot{q} + \mathcal{V}_q(q) = 0$, we easily verify that the derivative $\partial q/\partial\alpha$ with respect to any parameter α is a solution of the homogeneous equation $K[\partial q/\partial\alpha] = 0$. For a trajectory $q(t;\, x_1,\, x_2)$ with independent parameters $x_{1,2}$ we can take $y_1 = \partial q/\partial x_2$ and $y_2 = \partial q/\partial x_1$. Then for $t = \tau_1$ we shall have $y_1 = \partial x_1/\partial x_2 = 0$, $\dot{y}_1 = \partial v_1/\partial x_2$, $y_2 = \partial x_1/\partial x_1 = 1$, $\dot{y}_2 = \partial v_1/\partial x_1$, and for $t = \tau_2$ we obtain $y_1 = \partial x_2/\partial x_2 = 1$, $\dot{y}_1 = \partial v_2/\partial x_2$, $y_2 = \partial x_2/\partial x_1 = 0$, $\dot{y}_2 = \partial v_2/\partial x_1$, where $v_i \equiv \dot{q}(\tau_i)$ are the velocities at the ends of the interval. For this choice the required boundary conditions $y_i(\tau_i) = 0$ are satisfied, $W = y_1\dot{y}_2 - \dot{y}_1y_2 = -\partial v_1/\partial x_2 = \partial v_2/\partial x_1$ (the values at the two ends of the interval) and $[y_2\delta\dot{y}_1 - \dot{y}_2\delta y_1]|\ldots = \delta[\partial v_1/\partial x_2]$. Finally, we obtain $\delta \ln\det K = -\delta \ln[\partial v_1/\partial x_2]$, from which up to a normalization factor we have $\det K \sim (\partial v_1/\partial x_2)^{-1} \sim (\partial^2 S(q)/\partial x_1\partial x_2)^{-1}$. The last representation was obtained taking into account the equations $\partial S(q)/\partial x_1 = mv_1$ and $\partial S(q)/\partial x_2 = -mv_2$ for the action on the classical trajectory $q(t;\, x_1,\, x_2)$.

Taking into account the normalization of the amplitude (2.48) for the free theory, the final result can be written as

$$\mathscr{A}(x_1, \tau_1; x_2, \tau_2)\,|_{sc} = \left[\frac{i}{2\pi}\cdot\frac{\partial^2 S(q)}{\partial x_1 \partial x_2}\right]^{1/2} \exp iS(q)\,.$$

In the generalization to the multidimensional case the expression in the square brackets becomes a matrix in the indices labeling x_1 and x_2, and [...] is understood as the determinant of this matrix. It can also be expressed in terms of the corresponding trajectory $q(t; x_2, v_2)$ with given initial values x_2, v_2 as independent variables: $\partial^2 S(q)/\partial x_1 \partial x_2 = m\partial v_1(x_1, x_2)/\partial x_2 = -m\partial v_2(x_1, x_2)/\partial x_1 = -m[\partial x_1(x_2, v_2)/\partial v_2]^{-1}$.

The semiclassical approximation is also used when $K_0^{-1}K$ has negative eigenvalues. Each of them gives a contribution $-\pi/2$ to the phase of det $(K_0^{-1}K)^{-1/2}$. Therefore, the preexponential factor of the amplitude $\mathscr{A}$ then acquires the phase factor $\exp(-i\pi\nu/2)$, where ν is the number of negative eigenvalues (the Morse index). According to the Morse theorem, ν is the number of focal [with respect to the family $q(t; x_2, v_2)$] points of the trajectory, i.e., the number of values of t in the interval $\tau_2 \leq t \leq \tau_1$ for which $\partial q(t; x_2, v_2)/\partial v_2 = 0$ (in the multidimensional case, the zero of the determinant taking into account its multiplicity). Actually, if we study the problem on the interval $\tau_2 \leq t \leq t'$ with variable upper limit $t' \leq \tau_1$, the eigenvalues of $K_0^{-1}K$ become functions of t', and the focal points by definition are points at which one of the eigenvalues vanishes (in the multidimensional case the zero can be a multiple one). It is known from classical mechanics that for sufficiently small $t' - \tau_2$ all eigenvalues are positive, and as t' increases they vary strictly monotonically. Therefore, the number of negative eigenvalues for $t' = \tau_1$ is equal to the number of "passages through zero" for the problem with variable upper limit $t' \leq \tau_1$.

2.5 The Space $E(\Delta)$ for Various Systems

In the calculation of the Gaussian integral in (2.54), the Green function iK^{-1} was determined uniquely by imposing boundary conditions. In the case of an infinite time interval the analogous role is played by the asymptotic conditions at $t \to \pm\infty$ on the functions which are integrated. As explained in Sec. 1.6.4, these conditions are contained in the requirement $\varphi \in E(\Delta)$, where $E(\Delta)$ is a linear space consisting of functions of the form $i\Delta A$, $A \in E$, where E is a space of suitably decreasing functions with the needed reality properties, and Δ is the *a priori* chosen Green function, which we wish to obtain by computing the Gaussian integral. As Δ we shall always take the time-ordered contraction defined by (1.28) and in this case the asymptotes of the function $\varphi = i\Delta A$ for $t \to \pm\infty$ are respectively $\varphi_{(+)} = inA$ and $\varphi_{(-)} = i\varkappa n^{\mathrm{T}}A$, where n is a simple contraction and $\varkappa = \pm 1$, depending on the statistics. Both asymptotes $\varphi_{(\pm)}$ are solutions of the equation $K\varphi = 0$.

We now wish to describe these asymptotes explicitly for various specific theories. We begin with the quantum-mechanical oscillator studied in

Sec. 2.1.1. In this case the two basis solutions of the free equation $K\varphi = 0$ are the harmonics $\exp(\pm i\omega t)$, and the asymptotes $\varphi_{(\pm)}$ have the form $ic_{(\pm)}\exp(\mp i\omega t)$, as seen from (2.4), where the constants $c_{(+)}$ and $c_{(-)}$ are the conjugates of each other owing to the reality of the function $A \in E$ entering into the definition of the two asymptotes [but here the functions $\varphi = i\Delta A \in E(\Delta)$ are complex; see the note regarding this in Sec. 4.2]. The situation with other bosonic theories of the oscillator type, for example, the theory of a real scalar field (Sec. 2.3) is similar: the asymptote $\varphi_{(+)}$ is an arbitrary negative-frequency solution of the Klein–Gordon equation and $\varphi_{(-)}$ is the conjugate positive-frequency solution.

Let us now turn to the quantum mechanics of a free particle, studied in Sec. 2.1.2, which is not a theory of the oscillator type. In this case the two basis solutions of the free equation are 1 and t, and for the asymptotes $\varphi_{(\pm)}$ from (2.8) we obtain $\varphi_{(+)} = c_{(+)}$, $\varphi_{(-)} = c_{(-)}t$, where $c_{(\pm)}$ are arbitrary real constants. This theory is distinguished from the oscillator theory by the fact that all the functions of the space $E(\Delta)$ are real, and the asymptotes for $t \to \pm\infty$ are completely independent.

In the case of the nonrelativistic theory considered in Sec. 2.2, the free equation of motion is an equation of first, rather than second, order in time derivatives. Therefore, for each of the functions ψ, ψ^+ there is only one solution per degree of freedom, namely, $\exp(-i\varepsilon t)$ for ψ and $\exp(i\varepsilon t)$ for ψ^+. In such theories one of the asymptotes of each of the functions ψ, ψ^+ is zero, and the other is a solution of the free equation; the zero asymptote will occur either for $t \to \infty$ or for $t \to -\infty$, depending on whether or not the level in question is filled, and the asymptotes $\psi_{(+)}$ and $\psi_{(-)}{}^+$ are conjugates of each other. In fermionic theories different spaces $E(\Delta)$ correspond to different variants of the level filling, and the choice of $E(\Delta)$ uniquely determines the Green function $\Delta = iK^{-1}$ in the integral (1.162).

It is useful to note that the asymptotic behavior of the functions from the space $E(\Delta)$ is very important from the viewpoint of the stationary-phase method discussed in Sec. 1.6.6. As an example, let us consider the use of this method to calculate the on-shell S-matrix generating functional. As explained at the end of Sec. 1.6.6, to determine the stationarity point it is necessary to find the solution of the classical equation of motion belonging to the manifold $E(\Delta) + f$, where f is some fixed solution of the free equation of motion. To be specific, we assume that we are dealing with the quantum mechanics of Sec. 2.1.2, where the solutions of the free equation are linear functions $a + bt$, and the space $E(\Delta)$ consists of functions with asymptote c for $t \to \infty$ and $c't$ for $t \to -\infty$, where c and c' are arbitrary. For a well behaved potential the solution of the exact classical equation has free asymptotes $a_{(\pm)} + b_{(\pm)}t$ and the condition $\varphi \in E(\Delta) + f$ fixes only two of the four parameters of the asymptotes, namely, $b_{(+)}$ and $a_{(-)}$. If we had replaced $E(\Delta)$ by a space E of suitably decreasing functions, we would thereby have attempted to fix all four parameters of the asymptotes, and the desired trajectory would not exist, except for rare exceptions (absence of scattering).

We also note that this example clearly indicates the unsatisfactoriness of the situation where the exact and free classical equations have asymptotes of different types,* as discussed in Sec. 1.6.5.

2.6 Functional Integrals Over Phase Space

For systems with second-order (in time derivatives) equations of motion, the unified field φ which is integrated over in equations like (2.56) and (1.164) represents only the generalized coordinates of the system. If desired, the number of variables can always be doubled by changing to integrals over trajectories in the phase space of coordinates and momenta. We shall show how this is done for the example of quantum mechanics; the corresponding constructions will simultaneously serve as a good illustration and supplement of the material of Sec. 1.3.4, which is devoted to theories with an interaction containing time derivatives of the field.

Let us consider a classical system with Hamiltonian of the general form $\mathcal{H}(q,p) = p^2/2m + \mathcal{V}(q,p)$. Upon quantization, q and p become operators, and the quantum interaction Hamiltonian $\mathbf{V}$ is not determined uniquely owing to the arbitrariness in the arrangement of the noncommuting factors $\hat{q}$ and $\hat{p}$. One of the possible quantization variants is the choice $\mathbf{V} = \text{Sym}\,\mathcal{V}(\hat{q}, \hat{p})$, which ensures that $\mathbf{V}$ is Hermitian.

In going to the interaction picture, $\hat{q}$ and $\hat{p}$ in the Hamiltonian $\mathbf{V}$ are replaced by the corresponding operators $\hat{q}(t)$ and $\hat{p}(t)$. The operator $\hat{q}(t)$ has the form (2.7), and $\hat{p}(t)$ does not actually depend on time in the interaction picture: $\hat{p}(t) = \hat{p} = m\partial_t\hat{q}(t)$.

In the language of a single field $\varphi \equiv q$, the operator $\text{Sym}\,\mathcal{V}(\hat{\varphi}, m\partial_t\hat{\varphi})$ contains time derivatives, and, as explained in Sec. 1.3.4, it is necessary to construct the corresponding effective interaction, which plays the role of $S_v(\varphi)$ in Eqs. (2.56) and (1.164). However, it is possible to proceed differently, assuming that $\hat{q}(t)$ and $\hat{p}(t)$ are two independent fields. In the language of the two-component field $\varphi = (\varphi_0 \equiv q, \varphi_1 \equiv p)$, the interaction Hamiltonian $\mathbf{V}$ no longer contains time derivatives and so the interaction functional (1.59) corresponding to it coincides with the classical expression: $S_v(\varphi_0, \varphi_1) =$

---

* More precisely, this is an indication of the inapplicability of perturbation theory under such conditions.

$-\int dt\, \mathcal{V}(\varphi_0, \varphi_1)$. The second route naturally leads us to integrals over the entire phase space.

Let us first use the standard rules (the N-product is defined in Sec. 2.1.2) to calculate the matrix of simple (n) and time-ordered (Δ) contractions for the two-component field $\varphi = (\varphi_0, \varphi_1) \equiv (q, p)$:

$$n(t, t') = \frac{i}{m}\begin{pmatrix} t', m \\ 0, 0 \end{pmatrix}, \quad \Delta(t, t') = \frac{i}{m}\begin{pmatrix} t'\theta_{tt'} + t\theta_{t't}, & m\theta_{tt'} \\ m\theta_{t't}, & 0 \end{pmatrix}. \tag{2.57}$$

As always, $\theta_{tt'} \equiv \theta(t - t')$.

For the classical free action for φ we shall use the expression

$$S_0(\varphi) = \int dt\,[p\dot{q} - p^2/2m] \equiv \int dt\,[\varphi_1\dot{\varphi}_0 - \varphi_1^2/2m]. \tag{2.58}$$

It corresponds to the free equation $K\varphi = 0$ with matrix operator K: $K_{00} = 0$, $K_{10} = -K_{01} = \partial_t$, $K_{11} = -1/m$. It is easily checked that the matrix of time-ordered contractions (2.57) is, as usual, the Green function of the free equation: $K\Delta = i$. Let us also give the quadratic form (1.4) corresponding to the action (2.58):

$$S_0'(\varphi) = \frac{1}{2}\varphi K\varphi = \frac{1}{2}\int dt\,[p\dot{q} - q\dot{p} - p^2/m]. \tag{2.59}$$

All the expressions of Secs. 1.6.4–1.6.6 are valid, of course, also for the two-component field φ. The integration space $E(\Delta)$ is defined according to the general rules of Sec. 1.6.4. Using the known matrix of contractions (2.57) it is not difficult to find the asymptotes of two-component functions $\varphi \in E(\Delta)$: $\varphi_0 = c$, $\varphi_1 = 0$ for $t \to \infty$ and $\varphi_0 = c't$, $\varphi_1 = mc'$ for $t \to -\infty$; c and c' are arbitrary real constants.

If we are only interested in the Green function of the component φ_0 (i.e., the coordinate operator), in (1.164) we can take the linear form to have the particular form $\varphi_0 A_0$ and perform the functional integration over φ_1, i.e., over the momentum. We then arrive at the ordinary path integral in coordinate space, but instead of the simple classical interaction functional $S_v(\varphi_0, \varphi_1)$ we obtain the effective interaction discussed in Sec. 1.3.4 (if S_v actually depends on φ_1).

In finding the analog of Eq. (2.56) it is necessary to first calculate the amplitude (2.49) for the two-component field φ, setting $\hat{\varphi}A = \hat{\varphi}_0 A_0 + \hat{\varphi}_1 A_1$ in (2.49). The result, which we present without derivation, is the direct generalization of (2.51) to the case of a two-component field: the linear form

$Af \equiv A_0f_0 + A_1f_1$ involves the solution of the free equation $Kf = 0$ with $f_0 = \mathbf{x}_2 + \mathbf{v}(t - \tau_2), f_1 = m\mathbf{v}$, where $\mathbf{v} \equiv (\mathbf{x}_1 - \mathbf{x}_2)/(\tau_1 - \tau_2)$ is the velocity on a straight-line trajectory. The matrix $\tilde{\Delta}$ is the Green function of the problem with zero boundary conditions for the component φ_0 and no restrictions on the boundary values of the component φ_1.

We use E_0 to denote the space of two-component functions φ satisfying these conditions, and consider the functional integral of $\exp i[S_0(\varphi) + A\varphi]$ over the manifold $E_0 + f$. It is easily verified that for any $\varphi \in E_0$ the values of $S_0(\varphi)$ and $S'_0(\varphi)$ coincide and $S_0(\varphi + f) = S_0(\varphi) + S_0(f)$, where $S_0(f)$ is the argument of the exponential in (2.48). Therefore, the proof of the proportionality of Eqs. (2.51) and (2.53) applies also to the present case, and we obtain

$$\mathcal{A}(\mathbf{x}_1, \tau_1; \mathbf{x}_2, \tau_2) = \text{const} \int_{\text{All}} Dp \int_{(\mathbf{x}_1\mathbf{x}_2)} Dq \exp iS_{\text{cl}}(q, p). \tag{2.60}$$

This integral is the transition amplitude for the quantum theory with Hamiltonian $\mathbf{H} = \hat{p}^2/2m + \text{Sym}\, V(\hat{q}, \hat{p})$; the functional $S_{\text{cl}} = S_0 + S_v = \int dt[p\dot{q} - \mathcal{H}(q, p)]$ entering into (2.60) is the classical action on the phase space of this theory.

Equation (2.60) is sometimes viewed as a universal recipe for "correct quantization" because it automatically ensures the Hermiticity of the quantum interaction Hamiltonian and, consequently, the unitarity of the evolution operator. However, this is, of course, only the one recipe in which $\text{Sym}\, V(\hat{q}, \hat{p})$ is taken as $\mathbf{V}$. If the quantization is done differently, for example, by setting $\mathbf{V} = \text{Re}NV(\hat{q}, \hat{p})$, where Re denotes the Hermitian part of the operator, then in (2.60) instead of S_{cl} we have a different functional, which can be found explicitly from the Sym-form of $\mathbf{V}$.

In conclusion, let us discuss the question of the canonical invariance of representations of the type (2.60). Canonical invariance implies that (2.60) remains true if the pair q, p is replaced by any other pair of canonical variables q', p', taking $S'_{\text{cl}}(q', p') = \int dt[p'\dot{q}' - \mathcal{H}'(q', p')]$, where $\mathcal{H}'(q', p') = \mathcal{H}(q, p)$. The special case of the transformation $q' = p$, $p' = -q$ corresponds to going from coordinate to momentum space in (2.60). Here $S'_{\text{cl}}(q', p') = \int dt[-q\dot{p} - \mathcal{H}(q, p)] = S_{\text{cl}}(q, p) - p_1q_1 + p_2q_2$, where q_i and p_i are the boundary values of q and p.

Without assuming that the canonical invariance is self-evident, we note that the momentum-space analog of the representation (2.60) [and (2.56)] can actually be obtained by the same method as the representation (2.18): the problem with a linear source is studied, the δ function is isolated by means of a constraint, and the remaining Gaussian integral with the constraint is calculated using a suitable shift. From this it obviously follows that the representation (2.60) is actually invariant under linear canonical transformations of the Cartesian variables p and q, but in other cases (for example, in transforming from p and q to action–angle variables for the oscillator) equations of the type (2.60) must be proved anew each time (if this is possible at all).

Chapter 3

THE MASSLESS YANG–MILLS FIELD

In this chapter we give a brief discussion of the specific features of the theory of massless vector (gauge) fields.

3.1 Quantization of the Yang–Mills Field

3.1.1 *The classical theory*

Let Γ be a group of unitary matrices Ω acting in some finite-dimensional space, and $T_1 \ldots T_N$ be the set of corresponding Hermitian generators normalized by the condition $\mathrm{tr}[T_a T_b] = \delta_{ab}$. The Yang–Mills field [25] associated with this group is a set of N real vector fields $B_a^\alpha(x)$ (everywhere in this chapter we use the standard relativistic notation of Sec. 2.3), which are conveniently described by a single matrix function $B^\alpha(x) = \Sigma_a T_a B_a^\alpha(x)$. A group of *gauge transformations* Γ_x is specified on the set of matrices $B^\alpha(x)$, and the elements of this group are parametrized by the function $\Omega(x)$ with values in Γ:

$$B^\alpha(x) \to B^\alpha_\Omega(x) = \Omega(x) B^\alpha(x) \Omega^{-1}(x) + i\varepsilon^{-1}\partial^\alpha \Omega(x) \cdot \Omega^{-1}(x) . \quad (3.1)$$

The real parameter ε plays the role of the coupling constant in the theory.

For an infinitesimal transformation of the form $\Omega(x) = 1 - i\varepsilon u(x)$ from (3.1) we obtain $B^\alpha_u(x) = B^\alpha(x) + \mathcal{D}^\alpha(x) u(x)$, where $\mathcal{D}^\alpha$ is a linear operator on the space of matrix functions (the covariant derivative) :

$$\mathcal{D}^\alpha \equiv \partial^\alpha + i\varepsilon [B^\alpha, \ldots] . \quad (3.2)$$

The classical action for the Yang–Mills field is taken to be the functional

$$S(B) = -\frac{1}{4}\int dx \, \mathrm{tr}\, [F_{\alpha\beta}(x) F^{\alpha\beta}(x)] , \quad (3.3)$$

in which $F^{\alpha\beta} = \partial^\alpha B^\beta - \partial^\beta B^\alpha + i\varepsilon[B^\alpha, B^\beta]$ is the field tensor. From (3.1) we easily obtain the transformation law for F: $F_\Omega{}^{\alpha\beta} = \Omega F^{\alpha\beta} \Omega^{-1}$. From this we see that the action (3.3) is gauge-invariant. For reference, we note that the gauge-invariant interaction of the Yang–Mills field with a multiplet of other fields φ is introduced via the replacement of the derivatives $\partial^\alpha \to \nabla^\alpha \equiv \partial^\alpha + i\varepsilon B^\alpha$ in the Lagrangian of the field φ (then $\nabla\varphi \to \Omega\nabla\varphi$ upon the simultaneous transformation $\varphi \to \Omega\varphi$, $B \to B_\Omega$).

The classical equation of motion for the action (3.3) has the form $\mathcal{D}_\alpha F^{\alpha\beta} = 0$. It is also easily verified that the gauge invariance of the action leads to the equation $\mathcal{D}_\alpha \mathcal{D}_\beta F^{\alpha\beta} = 0$ for any B.

If the original group Γ is Abelian, all the commutators vanish, the gauge transformation (3.1) becomes a ***gradient transformation*** $B^\alpha \to B^\alpha + \partial^\alpha u$, and the action (3.3) becomes the usual quadratic form (2.37) for a system of massless vector fields.

3.1.2 A general recipe for quantization

The canonical quantization of the theory (3.3) is nontrivial, because the corresponding Lagrangian is degenerate (or singular), i.e., for it there is no one-to-one relation between the velocities and momenta. Actually, from the definition of $S(B)$ it is clear that this quantity does not contain the velocity $\dot{B}^0 = \partial_0 B^0$ at all, and the corresponding momentum is identically equal to zero.

The quantization of systems with degenerate Lagrangians is a special problem which has been solved (see, for example, [26]), and the corresponding constructions for the case of the Yang–Mills field are described in detail in [4]. We shall therefore not discuss this topic here, and limit ourselves, following [27,28] to simple heuristic arguments which lead to the correct expressions.

In an ordinary theory with nondegenerate Lagrangian the Green functions of a field are represented by functional integrals of a product of fields with weight $\exp iS(\varphi)$ (see Sec. 1.6.5). At the level of rigor of our present discussion, it makes no sense to accurately specify the integration region for the functional integrals or to distinguish the free action from its quadratic form (1.4). The free action for (3.3) is the term quadratic in the field B, which has the form $S_0(B) = BKB/2$, where K is a matrix with indices a, α (a is the "isotopic" label and α is the vector label). This matrix has the form $\delta_{ab}K_{\alpha\beta}$, where $K_{\alpha\beta}$ is the operator (2.38) for $m = 0$. As already noted, this operator is essentially degenerate: in momentum space the form $K_{\alpha\beta} = -p^2 g_{\alpha\beta} + p_\alpha p_\beta$ is transverse, i.e., four-dimensionally orthogonal to the vector p: $K_{\alpha\beta}p^\beta = p^\alpha K_{\alpha\beta} = 0$. This feature presents an immediate obstacle to using the standard technique of perturbation theory, based on the expression

$$\int DB\ F(B) \exp iS(B) =$$

$$= \{ \int DB \exp \left[iS_0(B) + B\frac{\delta}{\delta A} \right] \} F(A) \exp iS_v(A) \Big|_{A=0}, \qquad (3.4)$$

in which $S(B) = S_0(B) + S_v(B)$ is the full action (3.3) and F is an arbitrary functional. Under ordinary conditions the Gaussian integral on the right-hand side of (3.4) can be done and gives the reduction operator $\exp[1/2\cdot(\delta/\delta A)\Delta(\delta/\delta A)]$ with $\Delta = iK^{-1}$, which makes it possible to construct the standard diagrammatic representations of Sec 1.4. In our case, owing to the degeneracy of the kernel K the operator $\Delta = iK^{-1}$ does not exist and the Gaussian integral in (3.4) loses meaning.

A technical trick was suggested in [27,28] for transforming the integral on the left-hand side of (3.4) for a gauge-invariant functional F (assuming that only gauge-invariant quantities can have direct physical meaning) such that there are no difficulties later on with the construction of perturbation theory. This trick is based on the observation that the integral of an invariant quantity is proportional to the volume of the invariance group, in this case the gauge group. The infinite factor $\int D\Omega(x)$ must first be isolated explicitly and only then is the perturbation theory constructed.

The trick consists of the following. The surface $f(x; B) = 0$ in the space of all fields $B(x)$ is studied. The surface is specified by the matrix function $f(x; B) = \Sigma_a T_a f_a(x; B)$, the coefficients of which are certain functionals of the field B (henceforth we omit the argument x of f). Of course, equality of the matrix function to zero is understood as equality of the coefficients of all the generators T_a to zero. The surface f must be chosen such that for any $B(x)$ an element $\Omega(x) \in \Gamma_x$ is found, and found uniquely, such that $f(B_\Omega) = 0$. Geometrically, this implies that any group orbit (i.e., set of points B_Ω, where Ω runs through Γ_x) has one and only one point in common with the surface f. For what follows it is also important that this common point will always be a point of intersection and not a point of tangency. These conditions are always satisfied in perturbation theory for the relatively simple surfaces which are dealt with in practice.

The following equation determines the functional $\overline{\omega}_f(B)$ up to a coefficient:

$$\overline{\omega}_f(B) \int D\Omega\delta\left[f(B_\Omega)\right] = \text{const}\,. \qquad (3.5)$$

The functional integral $\int D\Omega\ldots$ should be understood as an integral over the invariant measure on the group Γ_x, and the symbol δ denotes the functional δ

function, i.e., $\prod_{a,x}\delta\,[f_a(x;B_\Omega)]$. Owing to the invariance of the measure $D\Omega$, the functional $\overline{\omega}_f$ defined by (3.5) is gauge-invariant; the constant on the right-hand side of (3.5) determines the normalization of $\overline{\omega}_f$, which will be fixed later.

To isolate the group volume from the integral $\int DBF(B) \equiv J$ of some gauge-invariant functional F, it is necessary to introduce into the integral the left-hand side of (3.5), which according to (3.5) is equivalent to the introduction of a constant. As a result we find that $J = \text{const}\int DB\int D\Omega F(B)\,\overline{\omega}_f(B)\,\delta\,[f(B_\Omega)]$. We change the order of integration in this integral and then make the variable substitution $B_\Omega \to B$. The measure DB is invariant under this substitution, because the gauge transformation (3.1) is a combination of a translation and a unitary rotation. The functionals $F(B)$ and $\overline{\omega}_f(B)$ are also invariant, and we obtain the desired representation with the group volume isolated:

$$\int DB\ F(B) \;=\; \text{const}\ [\int D\Omega]\int DB\ F(B)\,\overline{\omega}_f(B)\,\delta\,[f(B)]\,, \tag{3.6}$$

which is valid for any gauge-invariant functional $F(B)$ for any choice of surface f (one says "for any choice of gauge").

The introduction of the δ function inside the functional integral in (3.4) makes it possible to use perturbation theory, since a Gaussian integral containing a δ function

$$G_f^{(0)}(A) \equiv \text{const}\int DB\ \delta\,[f(B)]\exp i\left[\frac{1}{2}BKB + AB\right] \tag{3.7}$$

is well defined (see the next section). We shall fix the normalization constant in (3.7) by the condition $G_f^{(0)}(0) = 1$.

These arguments suggest the possibility of defining the Green-function generating functional of the field B as

$$G_f(A) \;=\; \text{const}\int DB\ \delta\,[f(B)]\,\overline{\omega}_f(B)\exp i\,[S(B) + AB] \tag{3.8}$$

with normalization $G_f(0) = 1$ for the free theory. The functional (3.8) depends on the choice of surface f, since the addition AB to the exponent is not gauge-invariant. We must reconcile ourselves to this, because the important point is not the definition, but the nature of the object in question: the Green functions of a field are not gauge-invariant quantities. Only the on-shell S matrix (more precisely, renormalized S matrix) has any direct physical meaning. It describes real scattering processes, and it is the only object that we can expect to be gauge-invariant, i.e., independent of gauge. This does actually occur.

Equation (3.8) differs from the ordinary representation (1.164) by, first, the δ function, which reduces the integral over all fields B to an integral over the surface f, and, second, by the appearance of the additional factor $\overline{\omega}_f$, which is equivalent to adding $\delta S_f(B) = -i \ln \overline{\omega}_f(B)$ to the action functional (3.3).

In conclusion, we recall that (3.8) is rigorously established in the general theory of the quantization of systems with degenerate Lagrangians.

3.1.3 *Perturbation theory for gauges $nB+c=0$*

To construct the perturbation theory, it is necessary to calculate the functional $\overline{\omega}_f(B)$ and the Gaussian integral (3.7) for a selected gauge. The integral can be done only for simple linear gauges of the form $n_\alpha B^\alpha + c = 0$, where $c(x) = \Sigma_a T_a c_a(x)$ is a given matrix function and n_α is a linear operator on the space of matrix functions. We shall assume that n is either some c-number vector, or a differential operator like ∂_α. The gauge $\partial_\alpha B^\alpha = 0$ is called the *Lorentz* or *transverse* gauge, $\partial_i B^i = 0$ is the *Coulomb* or *radiation* gauge, and $B^0 = 0$ with c-number vector $n = (1, 0, 0, 0)$ is the *axial* gauge.

In the calculation of the integral (3.7) in linear gauges it is easy to isolate the c dependence by making the shift $B^\alpha \to B^\alpha + l^\alpha$ by a longitudinal vector $l^\alpha = \partial^\alpha \varphi$, which should be chosen such that the term c in the argument of the δ function is cancelled. Clearly, it is necessary to take $\varphi = -(n\partial)^{-1}c$. For this shift the factor $\exp iAl$ is isolated from (3.7), and the quadratic term BKB is not changed because K is transverse. The remaining integral can be done using the shift $B \to B + i\Delta A$, with the matrix Δ chosen such that, first, the shift does not change the argument of the δ function, which is equivalent to the requirement $n\Delta A = 0$ for any A, and, second, the shift removes the cross terms (i.e., the terms containing B and A) in the argument of the integrated exponential on the surface $nB = 0$. It is easily verified that these requirements determine the matrix Δ uniquely, if we do not worry about the question of the additions $\pm i0$ in the denominators (this would be taken care of by the correct use of the asymptotic conditions following the general scheme of Sec. 1.6.4). The resulting matrix Δ depends only on n and in momentum space has the form

$$\Delta_n^{a\alpha,\, b\beta} = \delta_{ab} \frac{i}{k^2 + i0} \left[\frac{k^\alpha n^\beta + n^\alpha k^\beta}{nk} - g^{\alpha\beta} - n^2 \frac{k^\alpha k^\beta}{(nk)^2} \right]. \tag{3.9}$$

Therefore, the integral (3.7) in the gauge $nB + c = 0$ is equal to

$$G_f^{(0)}(A) = \exp(-A\Delta_n A/2 + iAl), \tag{3.10}$$

where Δ_n and l are defined above. The matrix Δ_n, which will play the role of the propagator in the graphs of perturbation theory, is symmetric in α and β and even under the substitution $k \to -k$. It should also be noted that this matrix is degenerate, since it is orthogonal to the vector n in each index: $\Delta_n n = n\Delta_n = 0$. Substituting the specific values of n into (3.9), we obtain explicit expressions for the propagator in the selected gauge; in particular, for $n_\alpha = k_\alpha$ we obtain the standard expression for the propagator in the transverse gauge.

Let us now turn to the calculation of the functional $\overline{\omega}_f(B)$ defined by (3.5). This functional enters into (3.8) in combination with $\delta[f(B)]$, so we need to know only its value on the surface $f(B) = 0$. On this surface the argument of the δ function in (3.5) vanishes for $\Omega = 1$, which allows us to restrict ourselves to the neighborhood of unity in the calculation of the integral over Ω. In this neighborhood $\Omega = 1 - i\varepsilon u$, $D\Omega \sim Du$, and $B_u{}^\alpha = B^\alpha + \mathcal{D}^\alpha u$. Taking $f(B) = nB + c = 0$, we obtain $f(B_u) = n_\alpha \mathcal{D}^\alpha u \equiv Mu$ and $\int D\Omega\delta[f(B_\Omega)] \sim \int Du\delta(Mu) = \det M^{-1}$ (we have used the generalization of the well known expression for ordinary integrals). By making the appropriate choice of normalization constant in (3.5), it can be verified that the functional of interest $\overline{\omega}_f(B)$ on the surface $nB + c = 0$ coincides with the functional

$$\omega_f(B) = \det M/\det M_0 = \det[n_\alpha \mathcal{D}^\alpha]/\det[n_\alpha \partial^\alpha], \tag{3.11}$$

which depends on B through the second term in (3.2) and is normalized by the condition $\omega_f(B) = 1$ for $\varepsilon = 0$. For an Abelian group the functional (3.11) becomes unity.

The right-hand side of (3.11) determines the functional ω_f for any B, and the functionals $\overline{\omega}_f$ and ω_f differ off the surface f: the first is gauge-invariant and explicitly depends on the parameter c in the gauge condition, while the second is independent of c, but is not gauge-invariant [it is easy to see that for $c = -n_\alpha B^\alpha$, $\overline{\omega}(B; n, c)$ coincides with $\omega(B; n)$ for all B].

In practice, the functional (3.11) rather than $\overline{\omega}_f$ is always used in (3.8). As already mentioned, the introduction of the factor ω_f is equivalent to the addition $\delta S_f = -i \ln \omega_f$ to the action functional. This addition can be viewed as the result of the interaction of the field B with some auxiliary complex fermionic field ψ, ψ^+, writing the factor $\omega_f \sim \det M$ as a Gaussian integral const $\int D\psi^+ D\psi \exp i\psi^+ M\psi$ [for a bosonic field ψ, ψ^+ we would obtain not $\det M$, but $\det M^{-1}$ from (1.161)]. In the form $\psi^+ M\psi = \psi^+ n_\alpha(\partial^\alpha + i\varepsilon[B^\alpha, \ldots])\psi$ the term $\psi^+ n_\alpha \partial^\alpha \psi$ plays the role of the free action for the field ψ, ψ^+, and the term with the commutator corresponds to a Yukawa interaction (see Sec. 1.4.6) of the scalar fermionic field ψ, ψ^+ with the vector field B. The role of the propagator (the contraction of ψ with ψ^+) for the fermionic field is played by the quantity $i(n\partial)^{-1} \equiv iM_0^{-1}$. Writing $i\varepsilon n_\alpha[B^\alpha, \ldots] \equiv L$, for $\delta S_f = -i \,\mathrm{tr}\, \ln [M_0^{-1}(M_0 + L)]$ we obtain the series

$$-i \,\mathrm{tr}\, \{M_0^{-1}L - \frac{1}{2}M_0^{-1}LM_0^{-1}L + \ldots\}, \tag{3.12}$$

which is graphically represented by the sum, analogous to (1.114), of closed loops of the field ψ, ψ^+ and has the meaning of a polarization addition to the action of the field B owing to its interaction with the field ψ, ψ^+. The symbol tr in (3.12) denotes the trace of the linear operator in the space of matrix functions $u(x) = \Sigma_a T_a u_a(x)$. In the "coordinates" $u_a(x)$, any such operator can be written as a matrix in the isotopic labels a and b. The operators n_α, ∂^α, and M_0 are multiples of the unit matrix in the isotopic indices, and for the operator $[B^\alpha, \ldots]$ the definition

$$[B^\alpha, \ldots]u = [B^\alpha, u] = \sum_{bc} B_c^\alpha u_b [T_c, T_b] = \sum_{abc} B_c^\alpha u_b f_{cba} T_a$$

(where f_{abc} are the structure constants of the group) gives $[B^\alpha, \ldots]_{ab} = \Sigma_c B_c^\alpha f_{cba}$. The trace of this matrix in the isotopic indices is equal to zero owing to the antisymmetry of the structure constants f_{abc}.

We note that in the gauge $nB = 0$ with some c-number vector n (for example, in the axial gauge) the functional (3.11) becomes unity on the surface $nB = 0$, because $n_\alpha[B^\alpha, \ldots] = [n_\alpha B^\alpha, \ldots] = 0$. In such theories there are no additional graphs for δS_f.

Transforming the functional (3.8) according to the rule (3.4) and taking into account the definition (3.7), we obtain

$$G_f(A) = G_f^{(0)}(\delta/\delta iB) \exp i [S_{vf}(B) + AB] \big|_{B=0}, \tag{3.13}$$

where $S_{vf}(B)$ denotes the interaction functional together with the polarization addition $\delta S_f(B)$. Then using the explicit expression (3.10) for $G_f^{(0)}$ in the gauge $nB + c = 0$ and making the shift $\exp(l\delta/\delta B)$, we have

$$G_f(A) = \exp\left[\frac{1}{2}\cdot\frac{\delta}{\delta B}\Delta_n\frac{\delta}{\delta B}\right]\exp\ i[S_{vf}(B+l)+A(B+l)]\,|_{B=0}. \quad (3.14)$$

Up to the factor $\exp iAl$ this expression coincides with the usual expression (1.88) for the Green-function generating functional of the theory with interaction $S_{vf}(B+l)$, which leads to the standard diagrammatic representations of Sec. 1.4 for the functional (3.14). We note that a change of the parameter c in the gauge condition changes the interaction functional, but does not change the lines of the graphs.

3.1.4 *The generalized Feynman gauge*

Averaging the functional (3.8) for the gauge $nB + c = 0$ over the parameter c with weight $\exp icdc$, where d is an arbitrary kernel, we arrive at the functional (the integral over c is done by means of the δ function)

$$G_{nd}(A) = \text{const}\int DB\omega_f(B)\exp\ i[S(B) + Bn^{\mathrm{T}}dnB + AB], \quad (3.15)$$

representing the Green functions in the *generalized Feynman* gauge. The quadratic addition appearing in the argument of the exponential can be added to the free action, which is equivalent to the replacement of its kernel $K \to K' \equiv K + 2n^{\mathrm{T}}dn$. In contrast to K, the kernel K' is nondegenerate, which allows ordinary perturbation theory with the propagator $\Delta' = iK'^{-1}$ to be used to calculate the integral (3.15). In the special case $n_\alpha = \partial_\alpha$, $d = -1/2\rho$ (ρ is a number), the propagator Δ' in momentum space becomes

$$\delta_{ab}\frac{i}{k^2+i0}\left[(1-\rho)\frac{k^\alpha k^\beta}{k^2} - g^{\alpha\beta}\right]. \quad (3.16)$$

For $\rho = 1$ it coincides with the usual expression for the propagator of the electromagnetic field in the Feynman gauge, so these gauges are also referred to as generalized Feynman gauges.

Of course, the functional (3.15) does not coincide with any of the functionals (3.8), but if the on-shell S matrix for the functional (3.8) is independent of c (as it in fact is), it will be exactly the same as for the functional (3.15) for any choice of d.

3.1.5 *The S-matrix generating functional*

Comparing Eq. (3.13) with the usual representations (1.84) and (1.88) for the S matrix and the Green functions and using the fact that the differential reduction operator entering into these representations is none other than $G^{(0)}(\delta/\delta i\varphi)$, it is natural to define the (off-shell) S-matrix generating functional corresponding to the Green functions (3.8) as [29]

$$R_f(A) = G_f^{(0)}(\delta/\delta iA)\exp iS_{vf}(A). \qquad (3.17)$$

Repeating word for word the derivation of (1.93), we easily verify that it remains valid also in this case:

$$G_f(A) = G_f^{(0)}(A)R_f(i\Delta_n A). \qquad (3.18)$$

However, the inverse relation (1.94) now loses meaning, since the propagator (3.9) is degenerate and the inverse matrix Δ_n^{-1} does not exist. There are no such problems for the functional (3.15) with nondegenerate propagator (3.16), and all the usual relations of Chap. 1 remain valid.

The degeneracy of the propagator in the gauge $nB + c = 0$ leads to a specific effect: the relation between the off-shell S matrix and the Green functions is not unique. We see from (3.18) that the functional R_f determines G_f uniquely, but the converse is not true: the same functional G_f can correspond to different functionals R_f. In fact, from (3.18) we see that the graphs of G_f are obtained from those of R_f by multiplying each argument A ("each external line") by $i\Delta_n$, and if we add to R_f a term proportional to n in the vector index for even a single external line, this addition will not contribute to G_f owing to the equation $n\Delta_n = 0$.

Representing the differential operator in (3.17) by the functional integral (3.7) and performing the shift $\exp(B\delta/\delta A)$, we obtain

$$R_f(A) = \text{const}\int DB\ \delta[f(B)]\exp i\left[\frac{1}{2}BKB + S_{vf}(B+A)\right]. \qquad (3.19)$$

Now making the change of integration variable $B \to B - A$, we find

$$R_f(A) = \exp\left[\frac{i}{2}AKA\right]\text{const}\int DB\ \delta[f(B-A)] \times \\ \times\exp i[S_f(B) - AKB]. \qquad (3.20)$$

This equation looks analogous to (1.94), but it has been obtained by correct (at our level of rigor) transformations without the use of the meaningless symbol Δ_n^{-1}.

It should be noted that in defining the S-matrix functional the problem arises of choosing the factor ω_f generating the additional graphs. We have in mind the following: the factor ω_f entered into the integral (3.8) in combination with $\delta[f(B)]$ and, accordingly, we could replace ω_f by any other functional coinciding with ω_f on the surface $f(B) = 0$ (we have already done this once). However, this is no longer so in (3.20), because the factor $\omega_f(B)$ now enters in combination with $\delta[f(B - A)]$, where A is arbitrary. Therefore, different S matrices (3.20) will correspond to different choices of the functional ω_f in (3.8), which is a particular manifestation of the nonuniqueness of the relation between the S matrix and the Green functions mentioned above. When discussing the properties of the S-matrix functional, it is necessary to clearly fix the form of the factor ω_f. We shall take the determinant (3.11) as ω_f, and in this case the additional S-matrix graphs have the same form as for the Green functions.

3.2 Gauge Invariance

3.2.1 The Ward–Slavnov identities

The Ward–Slavnov identities are various differential relations describing the dependence of functionals of the type (3.8), (3.15), and (3.20) on the choice of gauge. They are obtained by means of a suitable change of integration variable $B \to B + \delta B$, which is accompanied, if necessary, by infinitesimal transformations of the parameters A, c, or n. If the variable substitution is an ordinary gauge transformation $B^\alpha \to B^\alpha + \mathcal{D}^\alpha u$, then $DB = DB'$, $\delta S(B) = 0$, but $\delta\omega_f(B) \neq 0$ owing to noninvariance of the functional (3.11). The basic idea of [30] is that it is more convenient to use a substitution with $\delta B^\alpha = \mathcal{D}^\alpha M^{-1}\varphi$, having the form of a gauge transformation with parameter $u = M^{-1}\varphi$ depending functionally on B (we recall that $M \equiv n_\alpha \mathcal{D}^\alpha$). For this substitution the variation of gauge-invariant quantities remains zero in first order in φ, but the Jacobian of the transformation DB'/DB is now not equal to unity, and it turns out that this Jacobian exactly cancels the variation of $\omega_f(B)$, i.e., the product $DB\omega_f(B)$ is invariant under this substitution.

If the substitution $B \to B + \mathcal{D}M^{-1}\varphi$ is made in the integral (3.15), it is necessary to take into account only the variations $\delta(Bn^{\mathrm{T}}dnB) = 2Bn^{\mathrm{T}}dn\mathcal{D}M^{-1}\varphi = 2Bn^{\mathrm{T}}d\varphi$ and $\delta(AB) = A\mathcal{D}M^{-1}\varphi$, which leads to the Slavnov identity [30]:

$$\langle 2Bn^{\mathrm{T}}d\varphi + A\mathcal{D}M^{-1}\varphi \rangle = 0. \tag{3.21}$$

We use the symbol $\langle F \rangle$ to denote the integral (3.15) in which we have introduced the additional factor F. This factor can be taken outside the integral in the usual manner in the form of a differential operator with the substitution $B \to \delta/\delta iA$ (see Sec. 1.7.1).

Upon the substitution $B \to B + \mathcal{D}M^{-1}\varphi$ the argument of the δ function $nB + c$ in the integral (3.8) receives the addition $n\mathcal{D}M^{-1}\varphi = \varphi$, which can be cancelled by a shift $c \to c - \varphi$ of the parameter c in the gauge condition. In the end we obtain an identity expressing the variation of the functional (3.8) when the parameter c is changed in terms of an integral of the type (3.21) with the variation of the single noninvariant quantity AB.

Analogous identities for the S-matrix generating functional will be studied in the following sections.

3.2.2 *Transversality and gauge invariance of the* S *matrix in electrodynamics*

The electromagnetic field is the Yang–Mills field for the one-parameter Abelian group Γ. In this case the transformation (3.1) reduces to the gradient transformation, the functional (3.11) becomes unity, and the gauge invariance of any functional $F(B)$ implies that it is transverse: $F(B + \partial u) = F(B)$.

Let us consider the S-matrix functional (3.20) with arbitrary transverse action functional $S_f \equiv S$ not depending on the choice of gauge, i.e., on n and c. From (3.20) we see that under a gradient transformation $A \to A + \partial u$ only the argument of the δ function changes and its change is equivalent to the shift $c \to c - n\partial u$ of the parameter c: $R(A + \partial u; n, c) = R(A; n, c - n\partial u)$. On the other hand, the argument of the δ function is not changed if the gradient transformation $A \to A + \partial u$ is accompanied by an analogous transformation $B \to B + \partial u$ of the integration variable. The latter does not change the functionals $S(B)$ and AKB owing to their transversality, which proves the transversality (off the mass shell) of the S-matrix functional (3.20) and, consequently, its independence of c and the independence of the S matrix in the generalized Feynman gauge of the choice of averaging method, in particular, of the choice of kernel d.

The change $n \to n'$ of the vector n in the gauge condition can also be reproduced by a gradient transformation of B, setting $B' = B + \partial u$, where $u = (n'\partial)^{-1}(n - n')B$. The Jacobian of this transformation is not unity but some constant, which is cancelled by the corresponding transformation of the normalization constant in front of the integral (3.20) [this constant is the same in (3.7), (3.8), and (3.20)], and we conclude that the normalized functional (3.20) is independent of n. We have thereby proved the transversality and gauge invariance (more precisely, independence of the choice of gauge) of the off-shell S-matrix functional (3.20) for any transverse action $S(B)$.

In real quantum electrodynamics there is a spinor Dirac field ψ, $\overline{\psi}$ in addition to the electromagnetic field B (see Sec. 2.3). The arguments of the

S-matrix functional will now be A, η, and $\bar{\eta}$, where η and $\bar{\eta}$ are anticommuting variables corresponding to the fermionic field ψ, $\bar{\psi}$. The factor in front of the integral on the right-hand side of (3.20) is replaced by $\exp i[AKA/2 + \bar{\eta}K'\eta]$, where $\bar{\eta}K'\eta$ is the free action functional (2.41) for the fermionic field, and the integral (3.20) takes the form

$$\text{const} \int D\bar{\psi} D\psi DB \ \delta[n(B-A)+c] \exp i \times$$
$$\times [S(\bar{\psi}, \psi, B) - AKB - \bar{\psi}K'\eta - \bar{\eta}K'\psi], \tag{3.22}$$

where $S(\bar{\psi}, \psi, B)$ is the full action functional for the electromagnetic and spinor fields. In quantum electrodynamics this functional is invariant under the combined gauge transformation

$$B^{\alpha} \to B^{\alpha} + \partial^{\alpha} u, \quad \psi \to \psi \exp(i\varepsilon u), \quad \bar{\psi} \to \bar{\psi} \exp(-i\varepsilon u). \tag{3.23}$$

The coupling constant ε is interpreted as the charge of the fermionic particle.

For zero values of the arguments $\bar{\eta}$, η everything reduces to the case considered earlier, but instead of a gradient transformation of the field we have the combined gauge transformation (3.23). This proves that the purely vector part of the off-shell S matrix (graphs without external fermionic lines) is transverse and independent of gauge choice. For nonzero η and $\bar{\eta}$ the proof becomes meaningless owing to the noninvariance of the terms $\bar{\psi}K'\eta$ and $\bar{\eta}K'\psi$ in the argument of the exponential in (3.22), which in first order in u acquire the extra terms $-i\varepsilon\bar{\psi}uK'\eta$ and $i\varepsilon\bar{\eta}K'u\psi$, respectively. Making the substitution $A \to A + \partial u$ and with it the gauge transformation (3.23) such that the argument of the δ function in (3.22) is not changed, we obtain the Ward identity for the S-matrix generating functional:

$$\delta_u R(A, \bar{\eta}, \eta) = \exp(\ldots) \langle \varepsilon\bar{\eta}K'u\psi - \varepsilon\bar{\psi}uK'\eta \rangle, \tag{3.24}$$

where $\exp(\ldots)$ denotes the factor in front of the integral given above and the symbol $\langle F \rangle$ denotes the integral (3.22) with the additional functional F inside the integral. In our case F is the variation of first order in u of the argument of the exponential in (3.22).

We now assume that the arguments η, η are on their mass shell, i.e., that they satisfy the free equation $K'\eta = \bar{\eta}K' = 0$. We want to show that in this case the right-hand side of (3.24) vanishes, i.e., the on-shell S-matrix functional for fermions is transverse in the vector argument A. At first glance this appears obvious, since, for example, the integral $\langle \bar{\eta}K'u\psi \rangle$ would seem to be zero identically for the condition $\bar{\eta}K' = 0$. This is true in the end; but it is not so simple—a similar argument for the integral $\langle\bar{\eta}K'\psi\rangle$ would be false. The point is that when the functional integral is done the field ψ from the form $\bar{\eta}K'\psi$ is contracted with some field $\bar{\psi}$ and gives a line, the fermionic propagator $\Delta' = iK'^{-1}$. In the end we obtain an expression $\bar{\eta}K'\Delta'\ldots$ corresponding to an uncertainty of the type $0\cdot\infty$: in momentum space a zero of $\bar{\eta}K'$ is multiplied by a pole of $\Delta' = iK'^{-1}$ and the answer depends on how the uncertainty is resolved.

In our case no such problem arises owing to the presence of the factor $u(x)$ separating the zero of $\bar{\eta}K'$ and the pole of Δ'. In momentum space this factor implies the presence of a momentum transfer between $\bar{\eta}K'$ and Δ', so that the pole of Δ' is a pole in a different (relative to $\bar{\eta}K'$) momentum and cannot cancel the zero of $\bar{\eta}K'$.

These arguments prove the transversality in the argument A of the on-shell S-matrix functional of the fermionic arguments η, $\bar{\eta}$. As usual, the transversality leads to independence of the choice of the parameter c in the gauge condition and, consequently, independence of the S matrix in the generalized Feynman gauge of the choice of kernel d.

It is just as simple to show that a change of the vector n in the gauge condition reduces to a renormalization of the fermionic arguments: on the fermionic mass shell $R(A, \bar{\eta}, \eta; n) = R(A, \bar{\eta}\bar{Z}, Z\eta; \partial)$, where $R(A, \bar{\eta}, \eta; \partial)$ is the S-matrix functional in the Lorentz gauge with $n = \partial$, Z is a linear operator on the set of solutions of the free equation $K'\eta = 0$, and $\bar{Z} = \gamma^0 Z^+ \gamma^0$ is the Dirac conjugate operator. The operator Z plays the role of a renormalization dilatation (see Sec. 1.9) and explicitly depends on the choice of n in the gauge condition. We say "operator" rather than "renormalization constant" because for $n \neq \partial$ the Lorentz invariance is violated and Z is some nontrivial matrix with the spinor indices of η.

Therefore, the renormalized S matrix on the fermionic mass shell is transverse and independent of the parameters n and c in the gauge condition (see the following section regarding the connection between transversality and physical unitarity).

3.2.3 *Transversality and gauge invariance of the on-shell* S *matrix for the Yang–Mills field*

The mass shell for the vector field A is taken, as usual, to be the set of solutions of the free equation $KA = 0$. Owing to the transversality of K, any purely longitudinal vector ∂u is a solution. The general solution is the superposition of a purely longitudinal vector and a vector belonging to the *physical mass shell*. The latter is defined as the set of all solutions of the wave equation $\Box A = 0$ simultaneously satisfying both the Lorentz and the Coulomb gauge conditions: $\partial_\alpha A^\alpha = \partial_i A^i = 0$ (the two real states of the photon with transverse polarization). The full mass shell is invariant under gradient transformations, so that we can speak of transversality of the on-shell S matrix.

The first proof of transversality for the S matrix of the Yang–Mills field in the Lorentz gauge was given in [31]. Here we present the results of [29] without derivation for gauges with arbitrary n and c. They are obtained, as in electrodynamics, by analysis of the Ward–Slavnov identities for the S-matrix functional (3.20). We note that the use of these identities directly for the S-matrix functional rather than for the Green functions significantly simplifies the entire proof.

As in electrodynamics, transversality, c independence, and d independence of the S matrix in the generalized Feynman gauge are equivalent, because from (3.20) we see that a gradient transformation of the argument A is equivalent to a shift of the parameter c.

The Ward–Slavnov identity is obtained in the standard way: the changes δn and δc of the parameters of the gauge condition are cancelled in the argument of the δ function in (3.20) by the transformation $B \to B - \mathcal{D}M^{-1}[\delta n(B - A) + \delta c]$ of the integration variable. The quantity $DB\omega_f(B)$ is invariant under this transformation up to an insignificant numerical factor, which is cancelled by the corresponding transformation of the normalization constant in (3.20), so that inside the integral (3.20) only the form AKB gets an additional term. Analyzing the resulting identity, it can be shown that on the mass shell the functional (3.20) is transverse in the argument A for any n with all the attendant consequences (see above), and that the entire dependence on the choice of n is contained in the renormalization of the wave function: $R(A; n) = R(ZA; \partial)$, where $R(A; \partial)$ is the S-matrix functional in the Lorentz gauge and Z is some n-dependent linear operator on the mass shell of A, i.e., on the set of solutions of the equation $KA = 0$.

The transversality of the on-shell S matrix allows us to replace the simple contraction $n^{\alpha\beta}$ for the vector field B^α in the expression analogous to (1.139) by $[PnP]^{\alpha\beta}$, where P is the projection onto the physical mass shell. This is equivalent to the elimination of unphysical intermediate states (with longitudinal and timelike photons) from the unitarity condition.

It should be mentioned that for the Yang–Mills field all the arguments about the S matrix refer only to the functional formally defined by (3.20). In contrast to electrodynamics, here there is no physically correct definition of the asymptotic states and no S matrix (infrared divergences).

Chapter 4

EUCLIDEAN FIELD THEORY

4.1 The Euclidean Rotation

4.1.1 Definitions

The Euclidean theory can be formally defined as the theory which is obtained from the ordinary (pseudo-Euclidean) theory by the substitution $t \to -it$. This substitution transforms the operator K in Eq. (1.2) into the *Euclidean operator* K_e, and the quantum Schrödinger equation becomes an equation of the heat-conduction type: $i\partial_t\psi = \mathbf{H}\psi \to -\partial_t\psi = \mathbf{H}\psi$ (it is assumed that $\mathbf{H}$ does not depend explicitly on time).

The Euclidean evolution operator $\exp \mathbf{H}(\tau_2 - \tau_1)$ in the Schrödinger picture coincides with the *density matrix* $\rho = \exp(-\beta\mathbf{H})$, where the evolution time determines the temperature T: $\beta \equiv 1/kT = \tau_1 - \tau_2$, where k is the Boltzmann constant. Therefore, from the viewpoint of physics the Euclidean theory is of independent interest as the theory of the density matrix.

The time dependence of the operators in the interaction picture, in particular, the free-field operators, is determined in the Euclidean theory by the expression $a(t) = \exp(t\mathbf{H}_0)a\exp(-t\mathbf{H}_0)$; replacing $\mathbf{H}_0$ by $\mathbf{H}$, we obtain the *Euclidean Heisenberg representation.* We note that the transformation $a \to a(t)$ is not unitary and changes the Hermiticity properties: if the operator a is Hermitian, $a(t)$ possesses the property of *combined Hermiticity* $a(t) = a^+(-t)$ rather than ordinary Hermiticity.

The Euclidean evolution operator in the interaction picture

$$\mathbf{U}_e(\tau_1, \tau_2) = \exp(\tau_1\mathbf{H}_0)\exp \mathbf{H}(\tau_2 - \tau_1)\exp(-\tau_2\mathbf{H}_0) \tag{4.1}$$

satisfies the equation $-\partial\mathbf{U}_e(\tau_1, \tau_2)/\partial\tau_1 = \mathbf{V}(\tau_1)\mathbf{U}_e(\tau_1, \tau_2)$, in which $\mathbf{V}(\tau)$ is the interaction Hamiltonian in the Euclidean interaction picture. The solution of this equation with the condition $\mathbf{U}_e(\tau, \tau) = 1$ can be written as a Dyson T-ordered exponential analogous to (1.57):

$$\mathbf{U}_e(\tau_1, \tau_2) = T_D \exp\left(-\int_{\tau_2}^{\tau_1} dt\mathbf{V}(t)\right). \tag{4.2}$$

The Euclidean Green functions are determined by the usual equation (1.58), but the field operators inside the T-product are now taken in the Euclidean Heisenberg picture.

The various reduction formulas for operator functionals of the free field (the Wick theorem) and the rules for transforming from Dyson T-products to Wick products in the evolution operator and Green functions are valid also for the Euclidean theory. The only difference is in the free-field operator itself, which is obtained from the usual (pseudo-Euclidean) operator by the standard substitution $t \to -it$, which leads to a change of the simple (n) and time-ordered (Δ) contractions:

$$n_{\rm e}(x,x') \equiv n_{\rm e}(t,\mathbf{x};t',\mathbf{x}') = n(-it,\mathbf{x};-it',\mathbf{x}'), \tag{4.3}$$
$$\Delta_{\rm e}(x,x') = \theta(t-t')\,n_{\rm e}(x,x') + \varkappa\theta(t'-t)\,n_{\rm e}(x',x).$$

In spectral representations of the type (2.6) and (2.27) the transformation $\Delta \to \Delta_{\rm e}$ corresponds to the simultaneous replacement $t \to -it$, $E \to iE$, $dE \to idE$.

Equations (1.31) relating the contractions and the kernel of the free action K take the following form in Euclidean space: $K_{\rm e}n_{\rm e} = K_{\rm e}n_{\rm e}^{\rm T} = 0$, $K_{\rm e}\Delta_{\rm e} = -1$.

4.1.2 *The formal Euclidean rotation of the action functional*

In this section we study the Euclidean rotation procedure used to put a given pseudo-Euclidean theory into correspondence with some Euclidean theory, the "Euclidean image" of the original theory. All the constructions will be completely formal, and the correctness of any given operation must be established independently for particular cases.

We shall rotate an arbitrary function $F(x_1 \dots x_n)$ using the following technical trick of Schwinger: for a real positive number $z > 0$ we define the function $F^{(z)}(x_1 \dots x_n) = F(x_1^z \dots x_n^z)$, where $x^z = (zt, \mathbf{x})$. We assume without proof that $F^{(z)}$ as a function of z admits analytic continuation to complex z, and the continuation to the point $z = -i$ will be referred to as the *rotated function*, for which we adopt the special notation $F^{(-i)} \equiv \overline{F}$.

The integrals will be rotated according to the following rule:

$$\int\!\dots\!\int dx_1 \dots dx_n F(x_1 \dots x_n) = \int\!\dots\!\int dx_1^z \dots dx_n^z F(x_1^z \dots x_n^z) = \tag{4.4}$$
$$= z^n \int\!\dots\!\int dx_1 \dots dx_n F^{(z)}(x_1 \dots x_n) = (-i)^n \int\!\dots\!\int dx_1 \dots dx_n \overline{F}(x_1 \dots x_n).$$

For a function of a single variable this rule is valid when the contour for the

t integration can be rotated such that it runs from top to bottom along the imaginary axis and the integration over the semicircle can be discarded.

In momentum space $p \to p^z \equiv (z^{-1}p_0, \mathbf{p})$ and $\int dp_0 \ldots \to i\int dp_0\ldots$, i.e., after the rotation the contour for the p_0 integration runs from bottom to top along the imaginary axis. The form of the Fourier transform is not changed: $px = p^z x^z = p_0 t - \mathbf{px}$.

Taking the action functional to have the general form

$$S(\varphi) = \sum_n \frac{1}{n!} A_n \varphi^n \equiv \sum_n \frac{1}{n!} \int \ldots \int dx_1 \ldots dx_n A_n (x_1 \ldots x_n)\, \varphi(x_1) \ldots \varphi(x_n)$$

and rotating all the integrals according to the rule (4.4), we find

$$iS(\varphi) = \sum_n \frac{1}{n!} A_{ne} \overline{\varphi}^n \equiv S_e(\overline{\varphi}), \tag{4.5}$$

where $\overline{\varphi}$ is the rotated field φ and $A_{ne} = (-i)^{n-1}\overline{A}_n$. The Euclidean theory with action S_e will be referred to as the *Euclidean image of the pseudo-Euclidean theory with action S.*

As an example, let us consider the rotation of Eqs. (1.31). According to the definition, the Euclidean operator K_e is $\overline{K}$, and from (4.3) we see that the Euclidean contraction Δ_e coincides with $\overline{\Delta}$, because the θ function is not changed in the rotation: $\theta(zt) = \theta(t)$ for any $z > 0$. The kernel of the unit operator on the right-hand side of the equation $K\Delta = i$ contains $\delta(t-t')$, which picks up a factor of i in the rotation: $\delta(zt) = z^{-1}\delta(t)$. Therefore, the equation $K\Delta = i$ becomes $K_e\Delta_e = -1$ in the rotation.

4.1.3 Euclidean rotation of the Green functions

It follows from the discussion in Sec. 4.1.1 that in the Euclidean theory the generating functionals for the S matrix and the Green functions are determined by the usual expressions (1.84), (1.88) with the substitution $\Delta \to \Delta_e$ and $iS_v \to S_{ve}$, where S_{ve} is the *Euclidean interaction functional* representing the Sym-form of the quantum operator $-\int dt\,\mathbf{V}(t)$, the argument of the exponential in (4.2).

If the operator $\mathbf{V}(t)$ does not depend explicitly on time and does not contain time derivatives of the field, the operator functional representing its Sym-form will be exactly the same as in the pseudo-Euclidean theory; only its argument, the free-field operator, will change: $\hat{\varphi} \to \hat{\varphi}_e$. In this case the Euclidean interaction functional will coincide with the pseudo-Euclidean one: $S_{ve}(\varphi) = S_v(\varphi)$. We note that this is consistent with the general rotation rule (4.5), according to which $iS_v(\varphi) = S_{ve}(\overline{\varphi})$. In our case the replacement

$t \to -it$ is manifested in only two places: first in $\varphi \to \overline{\varphi}$, and, second, in $i\int dt... \to \int dt...$, from which we see that the functional $iS_v(\varphi)$ rotated according to the rules of the preceding section becomes $S_v(\overline{\varphi})$, as stated. The same arguments demonstrate that the quadratic form $iS'_0(\varphi) = i\varphi K\varphi/2$ containing a single time integration is transformed in the rotation into the Euclidean form $S'_{0e}(\overline{\varphi}) = \overline{\varphi}K_e\overline{\varphi}/2$.

Let us consider an arbitrary graph of the full Green function $G_n(x_1...x_n)$ of the pseudo-Euclidean theory with action $S(\varphi)$. Performing the Euclidean rotation of the arguments $x_1...x_n$ simultaneously with the integration arguments and collecting the resulting factors of $\pm i$, we easily verify that we arrive at the corresponding graph of the Euclidean Green function G_{ne}, in other words, upon rotation the Green functions of the pseudo-Euclidean theory become the Green functions of the Euclidean image of this theory:

$$\overline{G}_n(x_1...x_n) = G_{ne}(x_1...x_n). \tag{4.6}$$

This can be stated compactly in the language of functional integrals: making the substitution $x_i \to x_i^z$ in the representation (1.165), we then go to the point $z = -i$ and, using (4.5), we arrive at the integral const $\int D\varphi\overline{\varphi}(x_1)...\overline{\varphi}(x_n)\exp S_e(\overline{\varphi})$, in which we can replace the integration variable $\varphi \to \overline{\varphi}$. The Jacobian of this substitution is an unimportant constant: for real z the Jacobian $D\varphi/D\varphi^{(z)}$ is the determinant of a linear operator with kernel $\delta\varphi(x)/\delta\varphi^{(z)}(x') = \delta(x-(x')^z)$, which does not depend functionally on φ.

We therefore arrive at the equation

$$\overline{G}_n(x_1...x_n) = \text{const}\int D\overline{\varphi}\ \overline{\varphi}(x_1)...\overline{\varphi}(x_n)\exp S_e(\overline{\varphi}) \tag{4.7}$$

with normalization $\overline{G}_0 = 1$ for the free theory. This equation proves (4.6), since the right-hand side of (4.7) represents (at least, in perturbation theory) the Green function G_{ne} of the Euclidean theory with action S_e (see Sec. 4.2 for details).

Selecting the connected parts of G_n in (4.6), we obtain

$$\overline{W}_n(x_1...x_n) = W_{ne}(x_1...x_n). \tag{4.8}$$

For $n = 0$ the quantity W_0 is a constant (the sum of connected vaccum loops) not containing the arguments x, so $\overline{W}_0 = W_0$ and from (4.8) we find $W_0 = W_{0e}$. In theories which are translationally invariant in time the latter equation is certainly incorrect. In fact, according to (1.74) in the pseudo-Euclidean theory the quantity W_0 corresponding to the logarithm of the vacuum expectation value of the S matrix determines the shift of the ground state energy when the interaction is switched on: $W_0 = -i\Delta E\int dt$. The Euclidean shift ΔE_e is defined analogously: $W_{0e} = -\Delta E_e\int dt$. It will be shown below that ΔE_e, like ΔE, is real. Therefore, W_0 is purely imaginary, and W_{0e} is purely real, and they cannot be equal.

The point is that in obtaining (4.6) and (4.8), all the integration times in the graphs were rotated, in particular, the integral $\int dt$ appearing as an isolated factor in each connected vacuum loop was rotated. In rotating this factor according to the rule (4.4) we essentially postulated the equality $\int dt = z\int dt = -i\int dt$, which is, of course, incorrect. Clearly, it is this substitution which transforms the purely imaginary quantity W_0 into the purely real quantity W_{0e}.

We actually should have rotated the coefficient of $\int dt$, i.e., ΔE, and compared the result with the corresponding coefficient of $\int dt$ in the Euclidean graphs. In the end we would have arrived at the equation $\Delta E = \Delta E_e$, which is the correct substitute for the equation $W_0 = W_{0e}$.

The case $n = 0$ in (4.8) is special. For $n > 0$ Eq. (4.8) does not contain such problems.

We have studied the Euclidean rotation procedure formally, but with certain assumptions it can be justified quite rigorously (see, for example, [32]).

4.1.4 *Properties of the field $\bar{\varphi}$ and the action $S_e(\bar{\varphi})$*

The reality properties of the rotated fields $\bar{\varphi}$ in the integral (4.7) differ from those of the original fields φ. For example, if φ is real, then $\bar{\varphi}$, being the analytic continuation of φ to the imaginary t axis, possesses the property of *combined reality* $\bar{\varphi}(x) = \bar{\varphi}^*(-t, \mathbf{x}) \equiv \bar{\varphi}^+(x)$, the conjugation of the fields $\varphi_2 = \varphi_1{}^*$ becomes *combined conjugation* $\bar{\varphi}_2(x) = \bar{\varphi}_1{}^*(-t, \mathbf{x}) \equiv \bar{\varphi}_1{}^+(x)$ after the rotation, and so on. The properties of combined reality exactly correspond to the properties of combined Hermiticity of the field operators in the Euclidean interaction picture.

Let us now turn to the functional $S_e(\bar{\varphi})$ and show that for an action represented by the integral over time of a real Lagrangian the corresponding *Euclidean action* $S_e(\bar{\varphi})$ is also real.

In fact, rotating the integral $iS(\varphi) = i\int dt\mathscr{L}(t)$ according to the rule (4.4), we arrive at the integral $\int dt\bar{\mathscr{L}}(t)$ which, according to (4.5), represents the Euclidean action $S_e(\bar{\varphi})$. Reality of $\mathscr{L}(t)$ leads to combined reality of $\bar{\mathscr{L}}(t)$; the action $S_e(\bar{\varphi})$, being the integral over a symmetric interval of a function possessing the property of combined reality, is real.

The simultaneous reality of $S_e(\bar{\varphi})$ and $S(\varphi)$ obviously contradicts (4.5), which indicates the incorrectness of the formal Euclidean rotation of the action [nevertheless, Eq. (4.8) derived from (4.5) remains valid for $n \neq 0$].

It should be emphasized that reality of the Euclidean action is guaranteed only when S_e is written as a functional of the rotated fields $\overline{\varphi}$. In particular, in the Euclidean version $\overline{\varphi}_2 K_e \overline{\varphi}_1 = \int dx \overline{\varphi}_2(-\partial_t - \mathscr{E})\overline{\varphi}_1$ of the nonrelativistic form (2.20) the fields $\overline{\varphi}_1$ and $\overline{\varphi}_2$ should be assumed to be not simply complex conjugates (as in the pseudo-Euclidean theory), but combined conjugates. In this case the form $\int dx \overline{\varphi}_2 \cdot \partial_t \overline{\varphi}_1$ will be real, while for complex-conjugate $\overline{\varphi}_2$ and $\overline{\varphi}_1$ this form is purely imaginary. The form $\int dx \overline{\varphi}_2 \mathscr{E} \overline{\varphi}_1$, just like the functionals (2.1) and (2.33) for which the operator K contains only second-order time derivatives, is real for both the original and the rotated fields.

In conclusion, let us note yet another property of $S_e(\overline{\varphi})$: the quadratic forms of the free action do not have definite sign. As an example, we consider the Euclidean version of the quantum-mechanical oscillator (2.1), for which $S'_{0e}(\overline{\varphi}) = \overline{\varphi} K_e \overline{\varphi}/2$, where $K_e = m(\partial_t^2 - \omega^2)$. If the field $\overline{\varphi}$ were real, this form would be negative-definite [in momentum space $K_e = -m(E^2 + \omega^2)$]. However, for the rotated field $\overline{\varphi}$ possessing the property of combined reality, this is no longer true. The equation $\overline{\varphi}(t) = \overline{\varphi}^*(-t)$ shows that the complex function $\overline{\varphi}(t)$ can be assumed to be an independent variable only on a single time semiaxis, for example, the semiaxis $t \geq 0$. Expressing the action in terms of the independent variables $\theta(t)\overline{\varphi}$ and $\theta(t)\overline{\varphi}^*$ taking into account the parity of K_e under the replacement $t \to -t$, we obtain $2S'_{0e}(\overline{\varphi}) = \int dt \theta(t)[\overline{\varphi} K_e \overline{\varphi} + \overline{\varphi}^* K_e \overline{\varphi}^*]$. Introducing the two-component field $\Phi = (\overline{\varphi}, \overline{\varphi}^*)$, we arrive at the form

$$S_{0e}'(\Phi) = \frac{1}{2}\int dt\, \theta(t)\, \Phi^* \begin{pmatrix} 0 & K_e \\ K_e & 0 \end{pmatrix} \Phi. \tag{4.9}$$

The Hermiticity of the two-by-two matrix appearing here ensures that the form is real, but at the same time we see that it does not have definite sign: the eigenvalues of the two-by-two matrix are equal to $\pm|K_e|$, where $|K_e|$ are the absolute values of the eigenvalues of K_e. It can be stated that the form (4.9) is "maximally indefinite": its eigenvalues are symmetrically located about zero.

It is easily verified that this is true for any Euclidean form $S'_{0e}(\overline{\varphi})$.

4.1.5 Rotation of the Lorentz group into O_4

In this section we shall show that any Lorentz-invariant action $S(\varphi)$ is transformed into the O_4-invariant (or Euclidean-invariant) action $S_e(\overline{\varphi})$ in a Euclidean rotation.

We use **R** to denote the space of all real four-dimensional vectors $x = t, \mathbf{x}$; **C** is the space of complex x, $\mathbf{C}_e \subset \mathbf{C}$ is the subset of vectors of the form ex, where $x \in \mathbf{R}$, and e is the diagonal matrix with elements $e_{00} = i$, $e_{11} = e_{22} = e_{33} = 1$.

A Lorentz transformation is described by a 4×4 matrix Λ satisfying the condition $\Lambda^{\mathrm{T}} g \Lambda = g$, where $g = -e^2$ is the metric tensor. We use L to denote the group of real Lorentz transformations with $\det \Lambda = 1$ and $\Lambda_{00} > 0$; L_{c} is the group of complex transformations with $\det \Lambda = 1$; L_{e} is the subgroup of L_{c} consisting of all complex matrices $\Lambda \in L_{\mathrm{c}}$ taking the $\mathbf{C}_{\mathrm{e}}$ into themselves. Clearly, L_{e} consists of those $\Lambda \in L_{\mathrm{c}}$ for which the elements Λ_{0i} and Λ_{i0} are purely imaginary and the other elements are real. Such Λ are unitary and vice versa: each unitary matrix $\Lambda \in L_{\mathrm{c}}$ is contained in L_{e}.

Let us associate a matrix $\Lambda_{\mathrm{e}} = e \Lambda e^{-1}$ with each $\Lambda \in L_{\mathrm{e}}$. The matrices Λ_{e} are real, orthogonal

$$(\Lambda_{\mathrm{e}}^{\mathrm{T}} \Lambda_{\mathrm{e}} = e^{-1} \Lambda^{\mathrm{T}} e^2 \Lambda e^{-1} = -e^{-1} \Lambda^{\mathrm{T}} g \Lambda e^{-1} = 1),$$

and have unit determinant, so that the correspondence $\Lambda \rightarrow \Lambda_{\mathrm{e}}$ is an isomorphism between L_{e} and O_4. Therefore, each representation of L_{c} contains a representation of $L_{\mathrm{e}} \sim \mathrm{O}_4$ as a subrepresentation.

We use $\mathcal{N}_R(\mathcal{O})$ and $\mathcal{N}_c(\mathcal{O})$ to respectively denote the spaces of real and complex fields specified on the region $\mathcal{O}$ (i.e., $x \in \mathcal{O}$). Let $\varphi(x)$, $x \in \mathbf{R}$ be some field (a multicomponent field would have an index) on which a linear representation of the group L is specified:

$$\varphi_{\Lambda}(x) = Q(\Lambda) \varphi(\Lambda^{-1} x). \tag{4.10}$$

Let us extend the original space of fields $\varphi(x)$ to $\mathcal{N}_{\mathrm{c}}(\mathbf{C})$. If the field is complex, only the region of definition is extended, and extension from $\mathbf{R}$ to $\mathbf{C}$ is understood as analytic continuation of $\varphi(x)$. If the original field is real, the region of values of φ is also extended: $\mathcal{N}_R \rightarrow \mathcal{N}_c$.

In the space $\mathcal{N}_c(\mathbf{C})$ the representation (4.10) is naturally extended to a representation of the complex group L_{c}; then, narrowing the region of definition of $\varphi(x)$ to $\mathbf{C}_{\mathrm{e}}$ and imposing the restriction $\Lambda \in L_{\mathrm{e}}$, we obtain a representation of L_{e} on the space $\mathcal{N}_{\mathrm{c}}(\mathbf{C}_{\mathrm{e}})$.

Now we associate with each $\varphi \in \mathcal{N}_{\mathrm{c}}(\mathbf{C}_{\mathrm{e}})$ a function $\overline{\varphi} \in \mathcal{N}_{\mathrm{c}}(\mathbf{R})$ (the "Euclidean image") defined as

$$\overline{\varphi}(x) = \varphi(-it, \mathbf{x}) = \varphi(e^{-1} x), \quad x \in \mathbf{R}. \tag{4.11}$$

A representation of L_{e} on $\mathcal{N}_{\mathrm{c}}(\mathbf{C}_{\mathrm{e}})$ is isomorphic to a representation of O_4 on $\mathcal{N}_{\mathrm{c}}(\mathbf{R})$. In fact, from (4.10) we have $\varphi_{\Lambda}(x) = Q(\Lambda)\varphi(\Lambda^{-1}x)$, $x \in \mathbf{C}_{\mathrm{e}}$, $\Lambda \in L_{\mathrm{e}}$.

According to the definition $\overline{\varphi}(ex) = \varphi(x)$ and $\overline{\varphi_\Lambda}(ex) \equiv \overline{(\varphi_\Lambda)}(ex) = \varphi_\Lambda(x)$, from which we find

$$\overline{\varphi_\Lambda}(ex) = Q(\Lambda)\varphi(\Lambda^{-1}x) = Q(\Lambda)\overline{\varphi}(e\Lambda^{-1}x) =$$

$$= Q(\Lambda)\overline{\varphi}(e\Lambda^{-1}e^{-1}ex) .$$

Introducing instead of $x \in \mathbf{C}_e$ the real independent variable $y = ex \in \mathbf{R}$, we rewrite the preceding expression in the form $\overline{\varphi_\Lambda}(y) = Q(\Lambda)\overline{\varphi}(e\Lambda^{-1}e^{-1}y)$, $\Lambda \in L_e$, $y \in \mathbf{R}$. The isomorphism between L_e and O_4 allows us to view this equation as a representation of O_4 and to write for any $\Lambda_e \in O_4$, $y \in \mathbf{R}$: $(\overline{\varphi})_{\Lambda_e}(y) = Q(e^{-1}\Lambda_e e)\overline{\varphi}(\Lambda_e^{-1}y) \equiv \overline{Q}(\Lambda_e)\overline{\varphi}(\Lambda_e^{-1}y)$.

All these arguments can be succinctly summarized by the expression $\overline{\varphi_\Lambda} = (\overline{\varphi})_{\Lambda_e}$, defining a representation of O_4 on the rotated field $\overline{\varphi}$ in terms of a representation of L_e on the original field φ. According to the rule (4.5), each Lorentz-invariant functional $S(\varphi)$ transforms under the rotation into the O_4-invariant functional $S_e(\overline{\varphi})$, because the equation $S(\varphi) = S(\varphi_\Lambda)$, $\Lambda \in L_c$, after the rotation takes the form $S_e(\overline{\varphi}) = S_e(\overline{\varphi_\Lambda}) = S_e((\overline{\varphi})_{\Lambda_e})$.

Care must be taken in the extension of various expressions, including the operation of complex conjugation. For example, the equation $\varphi(x) = \varphi^*(x)$ must first be rewritten as $\varphi(x) = \varphi^*(x^*)$, after which it can be extended to all $x \in \mathbf{C}$ (in the transformation to $\overline{\varphi}$ this equation becomes the condition for combined reality). The reality property of the original representation $Q(\Lambda)$ is extended to all $\Lambda \in L_c$ in the form $Q(\Lambda) = Q^*(\Lambda^*)$ in the same way. For $\Lambda \in L_e$ we have $\Lambda^* = T\Lambda T^{-1}$, where $T = -g$ is the time reflection matrix, and the reality condition for $Q(\Lambda)$ takes the form $\overline{Q}(\Lambda_e) = \overline{Q}^*(T\Lambda_e T^{-1})$.

In the complex case φ and φ^* should be treated as two independent fields $\varphi \equiv \varphi_1$, $\varphi^* \equiv \varphi_2$ on which the representations $Q_1(\Lambda)$ and $Q_2(\Lambda)$ act; each of them is extended independently by the method described above. The equation $\varphi_2 = \varphi_1^*$ is extended in the form $\varphi_2(x) = \varphi_1^*(x^*)$ and becomes the combined conjugation of the rotated fields into each other. The relation $(\varphi_\Lambda)^* = (\varphi^*)_\Lambda$ or, equivalently, $Q_2(\Lambda) = Q_1^*(\Lambda)$ is extended in the form $Q_2(\Lambda) = Q_1^*(\Lambda^*)$, and for the rotated fields takes the form $\overline{Q}_2(\Lambda_e) = \overline{Q}_1^*(T\Lambda_e T^{-1})$. If $\varphi_2(x) = \gamma\varphi_1^*(x)$, where γ is a fixed matrix (for example, γ^0 for Dirac spinors), in a similar way we obtain $\overline{Q}_2(\Lambda_e) = \gamma\overline{Q}_1^*(T\Lambda_e T^{-1})\gamma^{-1}$.

4.1.6 *Examples*

Let us make the general scheme described above specific for the case of scalar, vector, and spinor fields (see Sec. 2.3).

For a scalar field $Q(\Lambda) = 1$, and the kernel of the Euclidean quadratic form $\overline{\varphi}K_e\overline{\varphi}/2$ of the free action is the operator $K_e = \partial_\alpha\partial_\alpha - m^2 \equiv \Box_e - m^2$, which is negative-definite on real fields.

For a contravariant vector field $Q(\Lambda) = \Lambda$, and for a covariant one $Q(\Lambda) = \Lambda^{-1T} = g\Lambda g$. The extension to L_c is trivial, and for the rotated contravariant field we obtain $\overline{Q}(\Lambda_e) = e^{-1}\Lambda_e e$. The complex field transforms like the real field.

It is possible to perform a linear transformation to go from the rotated field $\overline{B}(x)$ (contravariant) to the field $\mathcal{B}(x)= e\overline{B}(x)$, which transforms like an O_4 vector: $\mathcal{B}_{\Lambda_e}(x)= \Lambda_e\mathcal{B}(\Lambda_e^{-1}x)$. The invariant quadratic form then becomes Euclidean: $\overline{B}g\overline{B} = -\mathcal{B}\mathcal{B}$.

The kernel $K^e_{\alpha\beta}$ of the Euclidean quadratic form of the free action is obtained from (2.38) by the standard replacement $t \to -it$, i.e., $\partial \to e\partial$. In the replacement $\overline{B} \to \mathcal{B}$ we obtain $\overline{B}^\alpha K^e_{\alpha\beta}\,\overline{B}^\beta = \mathcal{B}^\alpha q_{\alpha\beta}\mathcal{B}^\beta$, where

$$q_{\alpha\beta} = (e^{-1}K^e e^{-1})_{\alpha\beta} = (\Box_e - m^2)\,\delta_{\alpha\beta} - \partial_\alpha\partial_\beta \tag{4.12}$$

is a real and negative-definite kernel on real fields [in momentum space $q_{\alpha\beta} = -(p^2 + m^2)\delta_{\alpha\beta} + p_\alpha p_\beta$, where $\delta_{\alpha\beta}$ is the Kronecker delta and $p^2 \equiv p_0^2 + \mathbf{p}^2$ is the Euclidean squared length of the vector].

Let us now turn to the four-component Dirac spinor field $\psi_1(x) \equiv \psi(x)$, $\psi_2(x) \equiv \gamma^0\psi^*(x)$ (using the representation of the γ matrices in which γ^0 is symmetric). The rotated fields $\overline{\psi}_1$ and $\overline{\psi}_2$ (in this chapter the notation $\overline{\psi}$ is used only for the rotated field and should not be confused with Dirac conjugation) are related to each other as $\overline{\psi}_2(x) = \gamma^0\overline{\psi}_1{}^*(-t, \mathbf{x}) = \gamma^0\overline{\psi}_1{}^+(x)$.

Let us first give the necessary preliminary information about the parametrization of the complex Lorentz group. Let x be an arbitrary vector from $\mathbf{C}$ and σ_α be a set of four Hermitian 2×2 matrices: $\sigma_0 = 1$ and σ_i are the Pauli matrices. Since the determinant of the matrix $\hat{x} \equiv x^\alpha\sigma_\alpha$ coincides, as easily verified, with the invariant form $xgx = x^\alpha g_{\alpha\beta}x^\beta$, each transformation $\hat{x} \to \hat{x}' = a\hat{x}b^+$ with arbitrary unimodular (i.e., having unit determinant) matrices a and b is a Lorentz transformation $x \to x' = \Lambda x$ of the vector x. Using the equation $2x^\alpha = \mathrm{tr}(\hat{x}\sigma_\alpha)$, we easily find the matrix Λ:

$$\Lambda_{\alpha\beta}(a, b) = \frac{1}{2}\mathrm{tr}\,(\sigma_\alpha a\sigma_\beta b^+) \tag{4.13}$$

(the index α is actually contravariant). It can be shown that the set of matrices $\Lambda(a, b)$ exhausts the entire group L_c, and (4.13) can be used to verify the correctness of the equations

$$\Lambda^* (a,b) = \Lambda(b,a), \quad \Lambda^{\rm T}(a,b) = \Lambda(b^+,a^+),$$
$$\Lambda^{-1}(a,b) = \Lambda(a^{-1},b^{-1}). \tag{4.14}$$

Real transformations $\Lambda \in L$ correspond to $a = b$.

Let us now turn to the case of a spinor field. The representation (4.10) of the complex group L_c on spinors $\psi \equiv \psi_1$ can be specified by a 4×4 matrix $Q_1(\Lambda) = Q_1(a, b) = \begin{pmatrix} a & 0 \\ 0 & b^{-1+} \end{pmatrix}$, setting $\gamma^0 = \begin{pmatrix} 0 & 1 \\ 1 & 0 \end{pmatrix}$. For the field $\psi_2 = \psi_1{}^*\gamma^0$ the matrix Q is by convention written on the right: $(\psi_2)_\Lambda(x) = \psi_2(\Lambda^{-1}x)Q_2(\Lambda)$. For real x and Λ we have $Q_2(\Lambda) = \gamma^0 Q_1{}^{*\rm T}(\Lambda)\gamma^0$, which is extended to complex $\Lambda \in L_c$ in the form $Q_2(\Lambda) = \gamma^0 Q_1{}^{*\rm T}(\Lambda^*)\gamma^0$. It is clear from (4.14) that the replacement $\Lambda \to \Lambda^*$ is equivalent to interchange of a and b, from which it follows that $Q_2(\Lambda) = Q_2(a, b) = \gamma^0 Q_1^+(b, a)\gamma^0 =$

$$= \begin{pmatrix} a^{-1} & 0 \\ 0 & b^+ \end{pmatrix} = Q_1^{-1}(\Lambda).$$

Bilinear forms invariant under the real group L are automatically invariant under L_c. For example, $(\psi_2)_\Lambda(\psi_1)_\Lambda = \psi_2 Q_2(\Lambda)Q_1(\Lambda)\psi_1 = \psi_2\psi_1$. The relations expressing the tensor properties of the γ matrices are also automatically extended to L_c. From this it is clear that the free action (2.41) is L_c-invariant.

As indicated in the preceding section, the subgroup $L_e \subset L_c$ consists of all unitary matrices $\Lambda \in L_c$. We see from (4.14) that unitarity of $\Lambda(a, b)$ is equivalent to unitarity of the matrices a and b, so that the representations $Q_{1,2}(\Lambda) \equiv \bar{Q}_{1,2}(\Lambda_e)$, $\Lambda \in L_e$, $\Lambda_e \in O_4$ are unitary.

The free Euclidean action is obtained from (2.41) by the replacement $t \to -it$, which is equivalent to transforming to rotated fields $\overline{\psi}_{1,2}$ and making the replacement $\gamma^\alpha\partial_\alpha \to \gamma^\alpha(e\partial)_\alpha$ in the operator K. The derivative ∂ on rotated fields transforms as a Euclidean vector: $\partial \to \Lambda_e\partial$. Euclidean "$\gamma$ matrices" $\gamma_e^\alpha \equiv (e\gamma)^\alpha$ can be introduced, i.e., $\gamma_e^0 = i\gamma^0$, $\gamma_e^k = \gamma^k$, $k = 1, 2, 3$; these also transform as a Euclidean vector. The form $\gamma^\alpha(e\partial)_\alpha = \gamma_e^\alpha\partial_\alpha$ is explicitly O_4-invariant on rotated fields.

4.2 Functional Integral Representations

Let us consider a Gaussian integral with Euclidean quadratic form

$$G_e^{(0)}(A) = \text{const}\int D\varphi \exp\left[\frac{1}{2}\varphi K_e\varphi + \varphi A\right], \tag{4.15}$$

normalized by the condition $G_e^{(0)}(0) = 1$. The integral is done formally by means of the shift $\varphi \to \varphi - K_e^{-1}A$, but, to give it a precise meaning, we should take the general rules of Sec. 1.6.4 as a guide and specify the space of functions over which the integration in (4.15) runs, which is equivalent to supplementing the definition of the operator K_e^{-1}. Of course, we want this definition to lead us to the *a priori* known Euclidean propagator Δ_e, which, as seen from the equation $K_e\Delta_e = -1$, is one of the Green functions $-K_e^{-1}$ for the operator K_e.

The general scheme of Sec. 1.6.4 is fully applicable also to this case, and from it we learn that the desired integration space $E(\Delta_e)$ can be defined as the set of all functions of the form $\Delta_e\varphi'$, where φ' runs over the space from which the argument A in (4.15) is taken.

In Sec. 1.6.4 we assumed that A belongs to the space $\bar{E}$ of functions well behaved at infinity and possessing the same reality properties as the field φ. In the Euclidean theory it is natural to consider another possibility: that A belongs to the space $\bar{\bar{E}}$ of functions well behaved at infinity and possessing the same reality properties as the rotated fields $\bar{\varphi}$. Both versions are allowed by the general scheme of Sec. 1.6.4, and both give for $G_e^{(0)}(A)$ the expression $\exp[\varkappa A\Delta_e A/2]$ coinciding with the generating functional of the Green functions of the free Euclidean theory (as always, $\varkappa = \pm 1$, depending on the statistics). The standard arguments of Sec. 1.6.5 then allow us to write the following integral representation for the Green-function generating functional of a Euclidean theory with arbitrary interaction:

$$G_e(A) = \text{const}\int D\varphi \exp\left[S_e(\varphi) + \varphi A\right]. \tag{4.16}$$

Here $S_e = S'_{0e} + S_{ve}$ is the full Euclidean action, $S'_{0e}(\varphi) = \varphi K_e\varphi/2$ is the quadratic form of the free action, and $S_{ve}(\varphi)$ is the Euclidean interaction functional. There are two versions of the representation (4.16): one in terms of the *original fields*, when $A \in E$ and the integration runs over the space $E(\Delta_e) = \Delta_e E$, and one in terms of the *rotated fields*, where $A \in \bar{E}$ and the integration runs over the space $\bar{E}(\Delta_e) = \Delta_e\bar{E}$.

Let us discuss these spaces in more detail. An important difference between the Euclidean and the pseudo-Euclidean theories is the fact that for all the systems studied in Chap. 2 except that of a quantum-mechanical free particle (Sec. 2.1.2), the Euclidean contraction $\Delta_e(x, x')$ is a function of the time difference $t - t'$ which falls sufficiently rapidly (exponentially) at infinity.

Because of this, the linear operator with kernel Δ_e does not take us outside the class of functions which fall off rapidly at infinity, so that the functions from the spaces $E(\Delta_e)$ and $\bar{E}(\Delta_e)$ are well behaved. For the nonrelativistic fermionic contraction (2.26) this is true when the levels are split into groups 1 and 2 according to the sign of the energy: levels with positive energy are free, and levels with negative energy are filled. This will occur if the Green functions are defined by ground-state expectation values analogous to (1.58).

The quantum-mechanical free particle (Sec. 2.1.2) is an exception. In this case the Euclidean and pseudo-Euclidean contractions differ by only a factor of i, and the functions from the spaces $E(\Delta_e)$ and $\bar{E}(\Delta_e)$ have nonzero asymptotes $\varphi_{(\pm)}$ for $t \to \pm\infty$. It is easily shown that these are $\varphi_{(+)} = c$, $\varphi_{(-)} = c't$, where for the space $E(\Delta_e)$ both constants c and c' are real, and for the space $\bar{E}(\Delta_e)$ the constant c' will be real and c will be purely imaginary.

As far as the reality properties of the functions from $E(\Delta_e)$ and $\bar{E}(\Delta_e)$ are concerned, it can briefly be stated that for all systems, except again for the free theory of Sec. 2.1.2, when the spaces E and $\bar{E}$ are defined suitably (see below) the integration variables in (4.16) possess the "correct" reality properties, i.e., the same ones as the original $[E(\Delta_e)]$ or rotated $[\bar{E}(\Delta_e)]$ fields. In the special case of the free particle of Sec. 2.1.2 the functions from $E(\Delta_e)$ are real, but for $\bar{E}(\Delta_e)$ the correct reality (in this case, combined reality) is guaranteed only for those functions which have zero asymptotes for $t \to \pm\infty$. It is useful to recall that in the pseudo-Euclidean theory the correct reality properties were violated not for the free particle, but for the oscillator: functions from the space $E(\Delta) = i\Delta E$ (real E) with nonzero asymptotes $\varphi_{(\pm)}$ were not real. This is a curious phenomenon arising because the problem is ill-posed: the S-matrix formulation of the pseudo-Euclidean theory is natural for a free particle, but unnatural for the oscillator, which has a discrete spectrum, while the Euclidean theory using the density matrix and partition function (see Chap. 5 for more details) is natural for the oscillator and unnatural for a free particle. When a problem is formulated for a given system in an unnatural way, the technique which is used responds by distorting the correct reality properties of the fields over which the functional integration runs.

Let us explain the meaning of the above remark about "suitable choice" of the spaces E and $\bar{E}$ for the example of the nonrelativistic theory of Sec. 2.2. In this theory the original fields $\psi \equiv \varphi_1$, $\psi^* \equiv \varphi_2$ are complex conjugates, and for the rotated fields $\bar{\varphi}_{1,2}$ the term "correct reality" implies combined conjugation. It is easily verified that the vector $\varphi = \Delta_e A$, where $A \equiv (A_1, A_2)$ and Δ_e is the Euclidean version of the contraction matrix (2.26), will possess the property $\varphi_2^+ = \varphi_1$ when and only when $A_2 = \varkappa A_1^+$, which in this case is the definition of the "suitable" space $\bar{E}$. Actually, this definition is completely natural if we write the linear form $\varphi A = \varphi_1 A_1 + \varphi_2 A_2$ in (4.16) in the form usual for a complex field: $\psi^+ a + a^+\psi = \psi^+ a + \psi \varkappa a^+$.

For a relativistic vector theory it is more convenient to deal not with the field $\bar{B}^\alpha(x)$, but with the field $\mathscr{B}(x) = e\bar{B}(x)$ (see the discussion of the vector field in Sec. 4.1.6). Instead of the functional $S_e(B)$ we must write $\tilde{S}_e(\mathscr{B})$, and the form $BA = B^\alpha g_{\alpha\beta} A^\beta$ becomes the Euclidean scalar product $-\mathscr{B}\mathscr{A}$. The Euclidean contraction $\tilde{\Delta}_e = e\Delta_e e$ corresponding to the field $\mathscr{B}$, where Δ_e is the Euclidean version of (2.39), is real and negative-definite on real fields.

Let us conclude by comparing the two variants of the representation (4.16), excluding the special case of the quantum-mechanical free particle. The advantage of the integral *over rotated fields* is the simplicity of the properties of the action functional which, as shown above, will always be real and O_4-invariant if the original pseudo-Euclidean action was Lorentz-invariant. The disadvantage of such integrals is the fact that they always "diverge," and this is true even for the free theory because the Euclidean quadratic form of the free action for the rotated fields does not have definite sign (see Sec. 4.1.4). The divergence of the integrals means that the factor $D\varphi \exp S_e(\varphi)$ cannot be treated as a positive measure on the space of fields, in spite of its formal positivity following from the reality of the action. The most important property of a measure is the positivity of the average value of a positive quantity, and, as is easily verified, this condition is not satisfied for a Gaussian integral formally defined by means of a shift with a quadratic form which is nondegenerate but not of definite sign.

Let us now turn to the integrals *over the original fields*, where for the vector theory we shall by convention take the field $\mathscr{B}(x)$ as the real integration variable. The functional $S_e(\varphi)$ with nonrotated field φ need not be real, but in many cases it is so nonetheless. For example, the Euclidean interaction functional, as a rule, coincides with the pseudo-Euclidean functional (see Sec. 4.1.3), which guarantees that it is real. The quadratic forms of the free action for the quantum-mechanical theories of Sec. 2.1, and also for the relativistic scalar and vector ($\mathscr{B}$) fields, turn out not only to be real, but also to have definite sign* (negative), and in these cases the expontial in the Gaussian integral (4.15) provides a cutoff. The addition of an interaction $S_{ve}(\varphi)$ to the argument of the exponential can spoil the "convergence" of the integral. If this does not occur and the factor $\exp S_e(\varphi)$ provides a cutoff, we shall refer to the corresponding theory as *quasiprobabilistic* (see Sec. 4.3).

In the nonrelativistic theories of Sec. 2.2 the free action $\varphi_2 K_e \varphi_1$ with complex-conjugate $\varphi_2 = \varphi_1{}^*$ is not real; the time derivative in the operator

---

*For the free theory of Sec. 2.1.2 the property of having definite sign is not absolute.

$K_e = -\partial_t - \mathscr{E}$ gives a purely imaginary contribution to the action functional and the single-particle Hamiltonian $\mathscr{E}$ gives a real contribution. This is also the case in relativistic spinor theory.

The most attractive feature of integrals over nonrotated fields is the presence of a clear correlation between the convergence of the integral for a bosonic theory and the stability (or the fact that the Hamiltonian is bounded below) of the corresponding physical system. This is obvious for free oscillator theories: the convergence of the integral and the fact that the quantum-mechanical Hamiltonian is bounded below are ensured simultaneously by the positivity of the squared frequency. In the nonrelativistic bosonic theory, in which the convergence is determined by the real part of the action, the requirement of positivity of the single-particle Hamiltonian $\mathscr{E}$ plays an analogous role. For the anharmonic oscillator with interaction potential $\mathscr{V} = \lambda\varphi^n$ convergence of the integral requires that n be even and λ be positive. These are precisely the conditions which ensure that the quantum-mechanical Hamiltonian is bounded below.

It is likely that these observations are general and that in bosonic Euclidean theories the convergence of the integral (4.16) over nonrotated fields is a criterion for the stability of the corresponding physical system. The stability of fermionic systems for any sign of the single-particle energies is consistent with the fact that Gaussian integrals on a Grassmann algebra always "converge."

4.3 Convexity Properties

4.3.1 Quasiprobabilistic theories

In the preceding section we used the term "quasiprobabilistic" to describe bosonic Euclidean theories in which the action functional $S_e(\varphi)$ for nonrotated fields is real, and the factor $\exp S_e(\varphi)$ is a cutoff factor, so that one can speak of "convergence" of the integral (4.16).

Such theories can be understood as theories of a classical random field which takes different values ("configurations") with different probabilities. The quantity $D\varphi \exp S_e(\varphi)$ is the positive unnormalized measure on the set of configurations, and the factor $\exp S_e(\varphi)$ is proportional to the probability density for the configuration φ. The integrals $\text{const}\int D\varphi F(\varphi) \exp S_e(\varphi) \equiv \langle F \rangle$ normalized by the condition $\langle 1 \rangle = 1$ can then be understood as ordinary expectation values, and the Euclidean Green functions without vacuum loops acquire the meaning of correlation functions of the classical random field

$\varphi(x)$. Unfortunately, a rigorous mathematical theory of the random field, a *generalized random process* in the terminology of [33], exists only for a very restricted class of "good measures," and expressions of the type $D\varphi \exp S_e(\varphi)$ are not, as a rule, among them. Nevertheless, we shall assume without rigor that the symbol $D\varphi \exp S_e(\varphi)$ possesses the simplest properties of a positive measure, namely, the integral $\int D\varphi \exp S_e(\varphi)$ is positive, and the expectation value of any nonnegative functional is nonnegative. Positivity of the expectation value of the square leads to positivity of the dispersion $\langle |F|^2\rangle - |\langle F\rangle|^2$ for any functional $F(\varphi)$.

The condition of positivity of the measure leads to various consequences regarding the convexity properties (i.e., whether or not the second variation has definite sign) of the Green-function generating functionals. In fact, assuming that the linear form φA in (4.16) is real (for a complex field φA is understood as $\varphi^+ A + A^+\varphi$) and constructing the second variation with respect to A of the functional (4.16), we see that it is proportional to the expectation value of the square of $\delta(\varphi A)$ and is therefore nonnegative; the second variation of the functional $W_e(A) = \ln G_e(A)$ reduces to the dispersion of $\delta(\varphi A)$ and therefore is also nonnegative. This proves that $G_e(A)$ and $W_e(A)$ are functionals of A which are convex downwards: $\delta^2 G_e \geq 0$, $\delta^2 W_e \geq 0$.

A more general statement can be made. Let

$$S_e(\varphi) = \sum_n \frac{1}{n!} A_n \varphi^n \equiv$$

$$\equiv \sum_n \frac{1}{n!} \int \ldots \int dx_1 \ldots dx_n A_n(x_1 \ldots x_n)\, \varphi(x_1) \ldots \varphi(x_n) \tag{4.17}$$

be a real functional with potentials $A \equiv \{A_0, A_1, \ldots\}$ such that the symbol $D\varphi \exp S_e(\varphi)$ can be assumed to be a positive measure. Then the integral $G_e(A) = \text{const} \int D\varphi \exp S_e(\varphi)$, like its logarithm $W_e(A)$, has nonnegative second variation with respect to any variation of the potentials A which does not spoil the reality of the action (4.17). It can be stated that G_e and W_e are functionals which are convex downwards of any variables which enter linearly into the action functional.

It is useful to note that in theories with Lagrangians quadratically dependent on velocities (first time derivatives of the fields), the argument of the weight factor $\exp S_e(\varphi)$ has a simple physical meaning: in going from the pseudo-Euclidean to the Euclidean theory only the sign of squared velocities changes, so that the Euclidean Lagrangian turns out to be equal to the classical pseudo-Euclidean Hamiltonian with a minus sign. In such theories $S_e(\varphi) = -\int dt\, \mathcal{H}(t)$, where $\mathcal{H}$ is the classical Hamiltonian.

4.3.2 *Convexity and spectral representations*

The results of the preceding section are valid only for purely bosonic quasiprobabilistic theories and cannot be generalized to theories with fermions or to the nonrelativistic bosonic theory of Sec. 2.2 with nonreal action. However, there are statements about convexity which are valid for all theories.

Let $S'(\varphi)$ be the action functional of some pseudo-Euclidean theory which is translationally invariant in time and $S''(\varphi) = \int dt\,\mathscr{L}(t)$, where

$$\mathscr{L}(t) = \sum_n \int \ldots \int d\mathbf{x}_1 \ldots d\mathbf{x}_n A_n(\mathbf{x}_1 \ldots \mathbf{x}_n)\,\varphi(t, \mathbf{x}_1) \ldots \varphi(t, \mathbf{x}_n) \tag{4.18}$$

is a real functional with potentials A_n which are independent of time. After rotation, the pseudo-Euclidean theory with action $S = S' + S''$ becomes the Euclidean theory with action $S_e = S'_e + S''$, since the form (4.18) is preserved in the rotation. According to (1.74), the integral const $\int D\varphi \exp iS(\varphi) \equiv G(A)$ normalized by the condition $G(0) = 1$ determines the shift of the ground-state energy owing to the addition of the interaction (4.18): $\ln G(A) = -i\varepsilon(A)\int dt$. In the Euclidean version $G_e(A) = \text{const}\int D\varphi \exp S_e(\varphi) = \exp[-\varepsilon(A)\int dt]$, where $\varepsilon(A)$ is the same functional (energy shift) as in the pseudo-Euclidean theory (see Sec. 4.1.3). If the Euclidean theory with action S_e is quasiprobabilistic, the convexity downwards in the potentials A of the functionals $G_e(A)$ and $W_e(A) = \ln G_e(A)$ follows directly from the results of the preceding section. However, now it is possible, while remaining within the framework of the pseudo-Euclidean theory, to give an independent proof of the convexity of the functional $\varepsilon(A)$. The proof we shall give is based on the use of spectral representations for the Green functions and is valid for all theories, not only quasiprobabilistic ones. Here we shall use the convenient terms ***expectation value*** and ***dispersion*** also for pseudo-Euclidean functional integrals with weight $\exp iS(\varphi)$.

It is clear from these definitions that twice the second variation with respect to the potentials A of the pseudo-Euclidean functional $W(A) = \ln G(A)$ is equal to the dispersion of the functional $i\delta S''(\varphi) = i\int dt\,\delta\mathscr{L}(t)$. On the other hand, it follows from the results of Secs. 1.3.3 and 1.6.5 that the functional expectation value of the product $\delta\mathscr{L}(t)\delta\mathscr{L}(t')$ coincides with the vacuum expectation value of the Dyson T-product of the corresponding operators in the Heisenberg picture, from which

$$\delta^{(2)} W(A) = -\frac{1}{2}\int\int dt dt' \,(0|\, T_D [\delta\hat{\mathscr{L}}(t)\,\delta\hat{\mathscr{L}}(t')]\,|\,0)|_c, \tag{4.19}$$

where $|\,0)$ is the exact ground state for the theory with action $S = S' + S''$, and the subscript "c" stands for "connected part," i.e., the squared expectation value of $\delta\hat{\mathscr{L}}(t)$ is subtracted from the matrix element in (4.19).

If the variations of the potentials A do not spoil the reality of $\mathscr{L}(t)$, which is what we assume, the operator $\delta\hat{\mathscr{L}}(t)$, which differs from the corresponding classical functional only by replacement of the argument $\varphi(x)$ by the field operator in the Heisenberg picture, will be Hermitian.

Let us now write down the standard spectral representation [1]:

$$(0|\,T_{\rm D}[\,\delta\hat{\mathscr{L}}(t)\,\delta\hat{\mathscr{L}}(t')\,]|\,0)|_{\rm c} \;=\; \frac{i}{2\pi}\sum_{n\neq 0}\int dE\,\frac{2\varepsilon_n\rho_n \exp iE(t'-t)}{E^2-\varepsilon_n^2+i0}\,. \qquad (4.20)$$

Here $\rho_n \equiv |\,(0\,|\,\delta\hat{\mathscr{L}}|n\,)\,|^2$, $\delta\hat{\mathscr{L}}$ is the operator in the Schrödinger picture, $|n\,)$ is the orthonormalized set of eigenstates of the full Hamiltonian, and $\varepsilon_n = E_n - E_0$ is the excitation energy. The numbers ε_n are nonnegative, since $|0)$ is the ground state. The condition $n \neq 0$ in (4.20) eliminates the vacuum intermediate state, which is equivalent to selecting the connected part, i.e., subtracting the squared expectation value.

By definition, $W(A) = -i\varepsilon(A)\int dt$, so that $\delta^{(2)}W = -i\delta^{(2)}\varepsilon\int dt$. The integral over t and t' of Eq. (4.20) is also proportional to $\int dt$. This infinite factor cancels in (4.19) and we obtain

$$\delta^{(2)}\varepsilon(A) \;=\; -\sum_{n\neq 0}(\rho_n/\varepsilon_n) \leq 0, \qquad (4.21)$$

which proves the convexity upwards of the functional $\varepsilon(A)$. This result is weaker than the statement about convexity in the preceding section, since here we are speaking of convexity with respect to a narrower class of potentials which are independent of time, but on the other hand it is valid for all stable theories (theories having a ground state).* We also note that a

* For spatially uniform systems an obvious consequence of (4.21) is the convexity upwards of the ground-state energy density ε/V as a function of the coupling constant λ in the interaction. If the ground-state energy of the free theory is taken to be zero (which is usually the case in field theory), and the contribution to the energy of first order in λ is also equal to zero (which is also usual), from the convexity it follows that $\varepsilon(\lambda)$ has a maximum for $\lambda = 0$ and is negative for all $\lambda \neq 0$. In other words, the interaction lowers the ground-state energy from zero to a negative value. From the viewpoint of statistical physics (see Chap. 5), negativity of ε implies positivity of the pressure generated by the system at zero temperature. Extending the arguments of Feynman about the role of the ground-state energy in field theory [13], the positivity of the pressure can be related to the expansion of the universe. Of course, we cannot dwell on this here (if only because there is no certainty that a theory created for describing microscopic processes is applicable at macroscopic scales). We note that in [13] it is not the interaction energy but the free energy of the vacuum oscillations which is discussed; the latter is positive and corresponds to negative pressure. It is usually simply discarded, and the Hamiltonian is redefined accordingly.

specific "Goldstone infinity" appears on the right-hand side of (4.21) if the ground state is not separated by an energy gap from all the other states, i.e., if it does not correspond to an isolated, nondegenerate level of the full Hamiltonian.

Chapter 5

STATISTICAL PHYSICS

This chapter is devoted exclusively to systems in equilibrium. First we consider the quantum statistics of systems like the Bose or Fermi gas, then we consider lattice spin systems and the nonideal classical gas.

5.1 The Quantum Statistics of Field Systems

5.1.1 Definitions

In quantum statistics the equilibrium state of a system at temperature T is described by the *density matrix* $\rho = \exp[-\beta \mathbf{H}]$, where $\beta \equiv 1/kT$, k is the Boltzmann constant, and $\mathbf{H}$ is the *generalized Hamiltonian*. For the canonical ensemble the generalized Hamiltonian is the same as the ordinary Hamiltonian, and for the grand canonical ensemble $\mathbf{H} = \mathbf{H}' - \mu\mathbf{N}$, where $\mathbf{H}'$ is the ordinary Hamiltonian (the energy operator), $\mathbf{N}$ is the particle number operator, and μ is a constant called the chemical potential. Henceforth we shall call $\mathbf{H}$ simply the Hamiltonian.

By definition, the partition function Z is the trace of the density matrix: $Z = \mathrm{tr}\,\rho$. The average value $\langle\langle \mathbf{a} \rangle\rangle$ of some operator $\mathbf{a}$ is given by $\langle\langle \mathbf{a} \rangle\rangle = Z^{-1}\mathrm{tr}(\rho\, \mathbf{a})$.

The thermodynamics of a system is completely determined by the partition function or the equivalent functions $\ln Z \equiv W = -\beta\Omega$; the function Ω is called the thermodynamic potential.

In translationally invariant systems $\ln Z$ is proportional to the total (infinite) volume $\int d\mathbf{x}$, so it is necessary to use specific quantities, i.e., quantities divided by the volume. The equation $\ln Z = \beta p \int d\mathbf{x}$ determines the pressure p for a spatially uniform gas or liquid.

Important objects of study for field systems are the *temperature Green functions* of a field, defined as averages of the Dyson T-product [34–36]:

$$H_{n\beta}(x_1 \ldots x_n) = \langle\langle T_D[\hat{\varphi}_H(x_1) \ldots \hat{\varphi}_H(x_n)] \rangle\rangle , \tag{5.1}$$

where $\hat{\varphi}_H(x)$ are the field operators in the Euclidean Heisenberg picture. We should note here that (5.1) defines the Green functions only when the time arguments of all the field operators lie inside the interval [0, β] (see Sec. 5.1.5 for extensions of them).

At zero temperature only the ground state contributes to the average, and the temperature functions (5.1) become the Euclidean Green functions without vacuum loops, which were studied in Chap. 4.

The ratio of ρ matrices $\rho_0^{-1}\rho = \exp[\beta \mathbf{H}_0]\exp[-\beta \mathbf{H}] \equiv \mathbf{U}_e(\beta, 0)$ coincides with the Euclidean evolution operator (4.1) for the time interval [0, β]. Writing it as an ordinary *T*-exponential (4.2), we obtain

$$\rho = \rho_0 \mathbf{U}_e(\beta, 0) = \rho_0 T_D \exp\left[-\int_0^\beta dt\ \mathbf{V}(t)\right]. \tag{5.2}$$

Equations (5.1) and (5.2) can be transformed to the interaction picture using (1.60) and (1.70), which are valid also for Euclidean field operators [see the remark following (1.70)]. Taking into account the fact that by assumption the field times in (5.1) lie inside the interval [0, β], we can set $\tau_1 = \beta$ and $\tau_2 = 0$ in (1.70). We obtain

$$T_D[\hat{\varphi}_H(x_1)\ldots\hat{\varphi}_H(x_n)] = $$
$$= \mathbf{U}_e^{-1}(\beta, 0)\, T[\hat{\varphi}(x_1)\ldots\hat{\varphi}(x_n)\exp S_{v\beta}(\hat{\varphi})], \tag{5.3}$$

where $\hat{\varphi}(x)$ are the Euclidean field operators in the interaction picture; $S_{v\beta}$ is the *temperature interaction functional*, the Sym-form of the argument of the exponential in (5.2). The functional $S_{v\beta}$ differs from the Euclidean functional S_{ve} only in that the time integration in it runs over the interval [0, β] and not the entire axis.

Equation (1.60) allows the Dyson *T*-exponential in (5.2) to be replaced by the Wick exponential $T\exp S_{v\beta}(\hat{\varphi})$. In the Green functions the evolution operator from (5.3), when grouped together with the exact ρ matrix in the average in (5.1), transforms the ρ matrix into the free ρ matrix according to (5.2), and we obtain

$$H_{n\beta}(x_1\ldots x_n) = \frac{\langle\langle T[\hat{\varphi}(x_1)\ldots\hat{\varphi}(x_n)\exp S_{v\beta}(\hat{\varphi})]\rangle\rangle}{\langle\langle T\exp S_{v\beta}(\hat{\varphi})\rangle\rangle}. \tag{5.4}$$

The symbol ⟨⟨...⟩⟩ here and below denotes the average using the free ρ matrix $\exp[-\beta \mathbf{H}_0]$.

The numerator on the right-hand side of (5.4) will be referred to as the *full temperature Green function* and denoted as $G_{n\beta}$; the $H_{n\beta}$ themselves are the Green functions without vacuum loops. The denominator in (5.4) is the ratio of partition functions:

$$Z/Z_0 = \langle\langle \mathrm{U_e}(\beta, 0)\rangle\rangle = \langle\langle T \exp S_{v\beta}(\hat{\varphi})\rangle\rangle. \tag{5.5}$$

The Wick T-products in (5.4) and (5.5) can, as usual, be brought to normal-ordered form, and the Euclidean propagator (4.3) will be the corresponding contraction. The field-theoretic equations analogous to (5.4) and (5.5) involved the vacuum expectation value, which for operators in normal-ordered form was very simply calculated using the rule (1.87). Now it is necessary to find the expression replacing (1.87) for calculating the averages of operators in normal-ordered form, which will be done in Sec. 5.1.3.

From the viewpoint of physics it is not the temperature functions (5.1) but rather the *time-dependent Green functions at finite temperature* which are of greatest interest. The latter differ from (5.1) only in that the operators inside the T-product are operators in the ordinary Heisenberg picture, rather than the Euclidean one. Equation (5.3) remains valid also in this case, but all quantities in it become pseudo-Euclidean. The combination of the pseudo-Euclidean operator $\mathrm{U}^{-1}(\beta, 0)$ in (5.3) with the Euclidean ρ matrix which is averaged over in (5.1) does not produce a simple object, so it is not possible to construct a simple diagrammatic technique for calculating the time-dependent functions. In practice, these are found by analytic continuation of the temperature functions (5.1) in all the time arguments, because a diagrammatic technique based on (5.4) can be constructed for the temperature functions.

5.1.2 The free theory

Let us consider the nonrelativistic theory of Sec. 2.2, which is most important for statistics. For simplicity we assume that the single-particle Hamiltonian $\mathscr{E}$ has a purely discrete spectrum.

The free-field operators and their contractions are obtained from the corresponding expressions of Sec. 2.2 by the standard substitution $t \to -it$, and the free Hamiltonian H_0 is given by (2.21). In the present chapter the normal-ordered product will be understood in its simplest form N_a for both bosons and fermions.

For convenience, let us give the explicit expressions for the fields and the contraction $\Delta_\mathrm{e} \equiv (\Delta^a_{12})_\mathrm{e}$:

$$\hat{\psi}(x) = \sum_\alpha a_\alpha \Phi_\alpha(\mathbf{x}) \exp[-\varepsilon_\alpha t], \quad \psi^+(x) = \sum_\alpha a^+_\alpha \Phi^*_\alpha(\mathbf{x}) \exp[\varepsilon_\alpha t]$$

$$\Delta_e(x,x') \equiv (\Delta^a_{12})_e(x,x') = \theta(t-t') \sum_\alpha \Phi_\alpha(\mathbf{x}) \Phi^*_\alpha(\mathbf{x}') \exp \varepsilon_\alpha[t'-t]. \tag{5.6}$$

We shall also need the explicit expressions for the logarithm of the partition function and the average values of the operators $a^+_\alpha a_\alpha$, which for the Hamiltonian (2.21) are easily found to be [37]

$$W_0 \equiv \ln Z_0 = -\varkappa \sum_\alpha \ln[1-\varkappa\xi_\alpha], \quad \xi_\alpha \equiv \exp[-\beta\varepsilon_\alpha], \tag{5.7}$$

$$\langle\langle a^+_\alpha a_\alpha \rangle\rangle \equiv \bar{n}_\alpha = \xi_\alpha/(1-\varkappa\xi_\alpha). \tag{5.8}$$

As always, $\varkappa = \pm 1$ depending on the statistics. Equations (5.7) and (5.8) can be generalized directly to the case of a continuous spectrum.

Let us now consider the simplest Green functions (5.1) for the free field. The average value of a single operator $\hat{\psi}$ or $\hat{\psi}^+$ is obviously zero. For two operators we have

$$g(x,x') \equiv \langle\langle T[\hat{\psi}(x)\hat{\psi}^+(x')]\rangle\rangle = \Delta_e(x,x') + \langle\langle N[\hat{\psi}(x)\hat{\psi}^+(x')]\rangle\rangle, \tag{5.9}$$

where Δ_e is the Euclidean propagator (5.6). In universal notation $\psi \equiv \varphi_1$, $\psi^+ \equiv \varphi_2$ the contraction g is g_{12}, $g_{21} = \varkappa g^T_{12}$, $g_{11} = g_{22} = 0$. The average of the normal-ordered product in (5.9) is expressed in terms of $\bar{n}_\alpha$:

$$\langle\langle N[\hat{\psi}(x)\hat{\psi}^+(x')]\rangle\rangle = \varkappa\langle\langle \hat{\psi}^+(x')\hat{\psi}(x)\rangle\rangle = \\ = \varkappa\sum_\alpha \bar{n}_\alpha \Phi_\alpha(\mathbf{x})\Phi^*_\alpha(\mathbf{x}') \exp \varepsilon_\alpha(t'-t). \tag{5.10}$$

Adding Δ_e to this, we obtain the explicit expression for the temperature function.

5.1.3 *The average of an operator in normal-ordered form*

To calculate the average of an operator in normal-ordered form it is sufficient to calculate the functional $\langle\langle N \exp \hat{\varphi} A\rangle\rangle$ and then use equations of the type

(1.12). For a complex field $\hat{\psi}$, $\hat{\psi}^+$ the symbol $\hat{\varphi}A$ is understood as $\hat{\psi}^+A + A^+\hat{\psi} = \Sigma_\alpha(a_\alpha^+c_\alpha + c_\alpha^+a_\alpha)$, where

$$c_\alpha \equiv \int dx\ A(x)\,\Phi_\alpha^*(\mathbf{x})\exp(\varepsilon_\alpha t)\,,$$

$$c_\alpha^+ \equiv \int dx\ A^+(x)\,\Phi_\alpha(\mathbf{x})\exp(-\varepsilon_\alpha t)\,. \tag{5.11}$$

We shall assume that the time integration in the linear form $\hat{\varphi}A$ and, consequently, in (5.11) runs only over the interval $[0, \beta]$.

For a bosonic field the quantities c_α, c_α^+ are ordinary numbers and for a fermionic field they are anticommuting numbers, i.e., the generators of a Grassmann algebra. The spectrum of the single-particle Hamiltonian $\mathscr{E}$ is again assumed to be purely discrete.

Obviously, the average of $N\exp\hat{\varphi}A$ breaks up into the product of averages for each α, so that it is sufficient to calculate the average for the one-level problem:

$$F(c, c^+) \equiv \langle\langle \exp a^+c \exp c^+a \rangle\rangle =$$

$$= Z^{-1}\ \mathrm{tr}\,[\exp(-\beta\varepsilon a^+a)\exp a^+c \exp c^+a]\,, \tag{5.12}$$

where we have used $Z \equiv \mathrm{tr}\exp(-\beta\varepsilon a^+a)$ and have taken into account the fact that $N\exp(a^+c + c^+a) = \exp a^+c \exp c^+a$. Equation (5.12) is easy to calculate for fermions. Owing to the anticommutativity of the symbols c, c^+ and a, a^+ (inside the N-product) we have $\exp a^+c = 1 + a^+c$ and $\exp c^+a = 1 + c^+a$, from which

$$F(c, c^+) = 1 + cc^+\langle\langle a^+a\rangle\rangle = 1 - \bar{n}c^+c = \exp(-\bar{n}c^+c)\,.$$

Multiplying these exponentials together for all levels α and using the explicit expressions (5.11) for c, c^+, we finally obtain

$$\langle\langle N\exp[\hat{\psi}^+A + A^+\hat{\psi}]\rangle\rangle = \exp A^+dA\,, \tag{5.13}$$

where the kernel of the quadratic form $d(x, x')$ turns out to be exactly equal to the average of the N-product (5.10) for the fermionic field.

In the bosonic case the calculation of the function (5.12) is more complicated, but the final result (5.13) is the same. Let us present this calculation.

The quantity Z on the right-hand side of (5.12) is known:

$$Z = \sum_{n=0}^{\infty} \exp(-\beta\varepsilon n) = (1-\xi)^{-1}, \tag{5.14}$$

where $\xi \equiv \exp(-\beta\varepsilon)$. We still need to calculate the sum

$$\sum_{n=0}^{\infty} \xi^n \langle n | \exp a^+ c \exp c^+ a | n \rangle,$$

which is conveniently rewritten as

$$\sum_{n=0}^{\infty} \frac{1}{n!} \xi^n \left(\frac{\partial^2}{\partial y^+ \partial y} \right)^n \langle 0 | \exp y^+ a \exp a^+ c \exp c^+ a \exp a^+ y | 0 \rangle \Big|_{y = y^+ = 0}, \tag{5.15}$$

using the fact that $|n\rangle = (n!)^{-1/2}(a^+)^n|0\rangle$.

The product of operator exponentials is reduced to normal-ordered form using the rule (1.49), where it is assumed that the contraction of a with a^+ is unity and all other contractions are zero. As a result, for the matrix element in (5.15) we obtain the expression $\exp[y^+y + y^+c + c^+y]$. Substituting this into (5.15) and shifting $y \to y + c$, $y^+ \to y^+ + c^+$, we bring (5.15) to the form

$$\exp(-c^+c) \sum_{n=0}^{\infty} \frac{1}{n!} \xi^n \left(\frac{\partial^2}{\partial y^+ \partial y} \right)^n \exp(y^+ y) \Big|_{y = c,\ y^+ = c^+}.$$

Then, representing the factor $\exp(y^+y)$ as a Gaussian integral

$$\exp(y^+ y) = \text{const} \int dx dx^+ \exp[-x^+ x + y^+ x + x^+ y] ,$$

we can explicitly perform the differentiation with respect to y, y^+ and assemble the series in n. The result is again a Gaussian integral. Calculating it and dividing the result by (5.14), we find $F(c, c^+) = \exp(\bar{n}c^+c)$, where $\bar{n} \equiv \xi/(1-\xi)$ is the average (5.8) for bosons. From this, exactly as for fermions, we arrive at (5.13), where now the kernel d is the average (5.10) for a bosonic field.

Therefore, the result is given by (5.13) for both statistics. In the derivation it was assumed that the spectrum of the single-particle problem is discrete, but the result (more precisely, the kernel d) admits direct generalization to a continuous spectrum. We shall assume it is valid for any spectrum.

We have obtained (5.13) for the particular field (5.6), but it is quite clear that an equation of this type will hold in any theory. In fact, in proving that the average of $N \exp \hat{\varphi}A$ has the form of a Gaussian exponential, we have used only the linearity of the fields in the operators a, a^+ and the properties of the latter, so this part of the proof remains valid for any theory of the oscillator type (for the time being we exclude the free particle of Sec. 2.1.2 from consideration). On the other hand, if an expression of the type (5.13) with some unknown kernel d has already been obtained, the latter must be given by an expression like (5.10), which is clear from comparing the contributions quadratic in A on both sides of the equation. These arguments prove the expression

$$\langle\langle N \exp \hat{\varphi}A \rangle\rangle = \exp(\varkappa A d A/2), \qquad (5.16)$$

in which $d(x, x') = \langle\langle N[\hat{\varphi}(x)\hat{\varphi}(x')]\rangle\rangle$. Equation (5.16) is written in universal notation and is valid for any theory except the quantum-mechanical free particle of Sec. 2.1.2.

From (5.16) using (1.12) and (1.92) we obtain the expression

$$\langle\langle NF(\hat{\varphi})\rangle\rangle = \exp\left[\frac{1}{2}\cdot\frac{\hat{\varphi}}{\delta\varphi}d\frac{\delta}{\delta\varphi}\right]F(\varphi)\big|_{\varphi=0}, \qquad (5.17)$$

which can be interpreted as the Wick theorem for the averages of operators in normal-ordered form. In the fermionic case the two derivatives of the quadratic form are assumed, as usual, to be right ones.

5.1.4 *Diagrammatic representation of the partition function and the Green functions*

Using (5.17), it is easy to calculate the averages of T-products:

$$\langle\langle TF(\hat{\varphi})\rangle\rangle = \langle\langle N \exp\left[\frac{1}{2}\cdot\frac{\delta}{\delta\varphi}\Delta_e\frac{\delta}{\delta\varphi}\right]F(\varphi)\big|_{\varphi=\hat{\varphi}}\rangle\rangle =$$

$$= \exp\left[\frac{1}{2}\cdot\frac{\delta}{\delta\varphi}(\Delta_e + d)\frac{\delta}{\delta\varphi}\right]F(\varphi)\big|_{\varphi=0}. \qquad (5.18)$$

Therefore, two operations—reduction to the N-form (1.46) and averaging of the N-form (5.17)—are combined into a single operation having the form of the Wick theorem with the contraction $g \equiv \Delta_e + d$, where Δ_e is the Euclidean propagator and d is the average of the N-product. We shall refer to g as the *temperature contraction* or the *temperature propagator*. In contrast to the Euclidean propagator, g depends explicitly on temperature via the averages $\bar{n}_\alpha$ entering into d.

Using (5.18), we obtain the following expressions for the partition function:

$$Z/Z_0 = \langle\langle U_e(\beta, 0)\rangle\rangle = \exp\left(\frac{1}{2}\cdot\frac{\delta}{\delta\varphi}g\frac{\delta}{\delta\varphi}\right)\exp S_{v\beta}(\varphi)|_{\varphi=0} \quad (5.19)$$

and for the generating functional of the full Green functions:

$$G_\beta(A) \equiv \langle\langle T\exp[S_{v\beta}(\hat{\varphi}) + \hat{\varphi}A]\rangle\rangle =$$

$$= \exp\left(\frac{1}{2}\cdot\frac{\delta}{\delta\varphi}g\frac{\delta}{\delta\varphi}\right)\exp[S_{v\beta}(\varphi) + \varphi A]|_{\varphi=0}. \quad (5.20)$$

In all the expressions the time integration runs only over the interval $[0, \beta]$. The complete analogy between (5.19), (5.20) and the field-theoretic expressions (1.84), (1.88) means that all the results of Sec. 1.4 regarding diagrammatic representations apply automatically to (5.19) and (5.20). The right-hand side of (5.19) before setting $\varphi = 0$ has the form of the S-matrix generating functional, Z/Z_0 is analogous to the vacuum expectation value of the S matrix, and so on. Of course, all statements of a topological nature, for example, the connectedness of the logarithms of Z/Z_0 and G_β or the cancellation of vacuum loops in the ratios (5.4), remain valid.

5.1.5 *Periodic extensions of the Green functions*

In the definition (5.1) the time for each field satisfies the condition $0 \le t \le \beta$. The right-hand side of (5.1) is meaningful for all times and it could be viewed as the natural extension of $H_{n\beta}$ to any values of $t_1 \ldots t_n$, but then (5.3) and its consequence (5.4) would become invalid. There is another extension which is more convenient, namely, a periodic one, which for fields is defined by the rule [34–36]

$$\varphi(t + n\beta, \mathbf{x}) = \varkappa^n \varphi(t, \mathbf{x}), \quad (5.21)$$

where n is any integer and $\varkappa$ has the usual meaning. The periodic extension (5.21) of an arbitrary function specified on the interval $[0, \beta]$ is given by a Fourier series:

$$f(t) = \sum_n f_n \exp i\omega_n t, \quad f_n = \frac{1}{\beta}\int_0^\beta dt\, f(t) \exp(-i\omega_n t), \tag{5.22}$$

in which $\omega_n = \pi n/\beta$, and the summation runs only over even n for an even extension [$\varkappa = 1$ in (5.21)] and only over odd n for an odd extension ($\varkappa = -1$). Everywhere below the summation over frequencies is understood in the same way.

Functions of many variables are extended by a Fourier series in each argument. In particular, the extension of the temperature contraction $g(x, x')$ is given by a double series with coefficients

$$g_{nm}(\mathbf{x}, \mathbf{x}') = \frac{1}{\beta^2}\int_0^\beta\int_0^\beta dt dt' g(x, x') \exp[-i\omega_n t - i\omega_m t']. \tag{5.23}$$

Using the explicit expressions (5.6) and (5.10), we can calculate the coefficients (5.23), which turn out to be nonzero only for $m = -n$. Omitting the simple calculations, we give only the result. The extension of g has the form

$$g(x, x') = \frac{1}{\beta}\sum_{n\alpha} \frac{\Phi_\alpha(\mathbf{x})\Phi^*_\alpha(\mathbf{x}')}{\varepsilon_\alpha - i\omega_n} \exp i\omega_n(t' - t) \tag{5.24}$$

and is formally the same for both statistics, the difference between which is manifested only in the conditions on the frequencies ω_n: they are even for bosons and odd for fermions. It is easy to see that (5.24) is obtained by simple replacement of the integration by summation over the frequencies ω_n in the spectral representation of the Euclidean propagator (5.6):

$$1/2\pi \cdot \int dE \ldots \to 1/\beta \cdot \sum_{\omega_n} \ldots. \tag{5.25}$$

We recall that, according to the general rules of Sec. 4.1.1, the spectral representation of the Euclidean propagator is obtained from the pseudo-Euclidean representation (2.27) by the standard substitutions $t \to -it$, $E \to iE$, $dE \to idE$.

The right-hand side of (5.25) is the Riemann sum for the integral on the left-hand side. Actually, the interval between adjacent frequencies of identical parity is $2\pi/\beta = \Delta\omega$, so that for $\beta \to \infty$

$$\frac{1}{\beta}\sum_{\omega_n}\ldots = \frac{1}{2\pi}\sum_{\omega_n}\Delta\omega\ldots \rightarrow \frac{1}{2\pi}\int dE\ldots. \tag{5.26}$$

Let us now determine the relation between the temperature propagator g and the linear operator K_e, the kernel of the Euclidean free action. For the theory we are considering $K_e = -\partial_t - \mathscr{E}$, where $\mathscr{E}$ is the single-particle Hamiltonian. Acting with the operator K_e on the first argument of the contraction (5.24) and taking into account the equation $\mathscr{E}\Phi_\alpha = \varepsilon_\alpha \Phi_\alpha$, we obtain

$$-\frac{1}{\beta}\sum_{n\alpha}\Phi_\alpha(\mathbf{x})\,\Phi^*_\alpha(\mathbf{x}')\exp i\omega_n(t'-t) =$$

$$= -\delta(\mathbf{x}-\mathbf{x}')\,\frac{1}{\beta}\sum_n \exp\, i\omega_n(t'-t)\,. \tag{5.27}$$

The sum over α gives $\delta(\mathbf{x}-\mathbf{x}')$, because the Φ_α form a complete orthonormal set. Now we note that the sum over frequencies divided by β on the right-hand side of (5.27) is just the kernel of the unit operator on the space of periodic functions of t, because this quantity coincides, as is easily verified, with the standard periodic extension of $\delta(t-t')$. We have thereby proved the validity of the equation $K_e g = -1_p$, where 1_p should be understood as the unit operator on the space of periodic functions of time. We recall that in Euclidean field theory we had $K_e\Delta_e = -1$, but there 1 was the unit operator on the space of all functions of time without any periodicity requirements.

The equation $K_e g = -1_p$ and the rule (5.25) turn out to be valid for all theories. As yet another example, let us consider the quantum-mechanical oscillator of Sec. 2.1.1. The Euclidean contraction $\Delta_e(t, t')$ is obtained from (2.6) by discarding the factor of i and making the substitution $\omega^2 \rightarrow -\omega^2$. The kernel $d(t, t') \equiv \langle\langle N[\hat{\varphi}(t)\hat{\varphi}(t')]\rangle\rangle$ is easily calculated using the explicit expression (2.3) for the field operator, in which we must make the substitution $t \rightarrow -it$. As a result, we obtain $d(t, t') = (\bar{n}/\omega\, m)\cosh\omega(t-t')$, where $\bar{n} \equiv \langle\langle a^+ a\rangle\rangle = \exp(-\beta\omega)/[1-\exp(-\beta\omega)]$. Representing the temperature propagator as a double Fourier series and calculating the series coefficients in terms of the known functions Δ_e and d, we find

$$g(t, t') = \frac{1}{m\beta}\sum_n \frac{1}{\omega^2 + \omega_n^2}\exp\, i\omega_n(t'-t) \tag{5.28}$$

(the sum runs over even frequencies). The series on the right-hand side of (5.28) is obtained from the Euclidean propagator Δ_e by the same substitution

(5.25). The equation $K_e g = -1_p$, where $K_e = m[\partial_t^2 - \omega^2]$, is also satisfied.

Let us conclude by indicating the changes which must be made in (5.19) and (5.20) to generalize them to the extended theory. We recall that the representations (5.19) and (5.20) were derived for the theory on a finite time interval [0, β], and all the time integrations ran only over this interval. The expressions will remain valid for the extended theory if the argument A is taken to be an arbitrary function given on the entire time axis, and if the field operator $\hat{\varphi}$, its classical analog φ, and the contraction g are understood as the periodic extensions from the interval [0, β]. The time integrations in the forms $\hat{\varphi}A$, φA and in the quadratic form of the derivatives should run over the entire axis.

From the mathematical point of view it is more elegant and natural to take the interval [0, β] to be a ring rather than a line segment, so that the fields and other quantities are understood as functions on the ring which are single-valued for bosons and double-valued for fermions. This language is equivalent to the language of periodic extensions.

5.1.6 *Representations by functional integrals*

The relation $K_e g = -1_p$ established in the preceding section can be used to transfer the results of Sec. 4.2 on integral representations of various functionals in Euclidean theory to the present case. The only difference will be that now the functional integration will be carried out over the space of fields with the required (in accordance with the statistics) periodicity properties, and the time integration in the quadratic form of the free action $S_{0\beta}(\varphi) = \varphi K_e \varphi/2$ and in the interaction functional $S_{v\beta}(\varphi)$ will run over the finite interval [0, β]. For the free theory

$$G_\beta^{(0)}(A) = \exp\left[\frac{\varkappa}{2}AgA\right] = \text{const}\int D\varphi \exp\left[\frac{1}{2}\varphi K_e \varphi + \varphi A\right], \quad (5.29)$$

and in the general case

$$G_\beta(A) = \text{const}\int D\varphi \exp\left[S_\beta(\varphi) + \varphi A\right], \quad (5.30)$$

where $S_\beta = S_{0\beta} + S_{v\beta}$ is the full (Euclidean) action functional for the interval [0, β]. The periodicity condition allows the surface terms to be dropped in the integration by parts in the free action functional, so that K_e can be treated as a symmetric operator on the space of periodic functions of time; this condition ensures the uniqueness of the definition of the inverse operator $-K_e^{-1} = g$ arising in the calculation of the Gaussian integral (5.29) using a shift.

The integral (5.30) is normalized by the usual condition $G_\beta(0) = 1$ for the free theory, and for a theory with an interaction $G_\beta(0)$ is equal to the ratio of partition functions Z/Z_0, as seen from (5.19) and (5.20).

For the extended theory the parameter A in (5.29) and (5.30) should be treated as an arbitrary function given on the entire time axis, and the integration over time in the linear form φA runs along the entire axis.

From the viewpoint of the reality properties of the fields over which the functional integration runs, there are two possible variants of the representations (5.29) and (5.30): one in terms of the original fields and one in terms of the rotated fields, as discussed in detail in Sec. 4.2.

We can summarize briefly as follows: quantum statistics is Euclidean field theory on a finite time interval with the additional condition that the fields be periodic.

5.1.7 *The zero-temperature limit*

We see from (5.26) that in the zero-temperature limit the temperature propagator becomes the Euclidean propagator: for $\beta \to \infty$

$$g(x, x') \to \Delta_e(x, x'). \tag{5.31}$$

The convergence is pointwise, i.e., the statement (5.31) is true for any fixed x, x'.

The expressions of the preceding sections involved the action functional for the interval $[0, \beta]$, and so at first glance it appears that for $\beta \to \infty$ quantum statistics must go into Euclidean field theory on the positive time semiaxis. On the other hand, assuming the action is a functional of periodic fields and using the fact that the Euclidean Lagrangian, being a bosonic quantity, has even extension, we can rewrite the action functional as an integral over a symmetric interval:

$$S_\beta = \int_0^\beta dt \mathcal{L}_e(t) = \int_{-\beta/2}^{\beta/2} dt \mathcal{L}_e(t). \tag{5.32}$$

In the last form, the formal limit $\beta \to \infty$ leads us to the usual Euclidean theory on the entire time axis.

In spite of the fact that (5.32) is actually true, the two formal limits—the theory on the time semiaxis and the theory on the entire axis—are essentially different. This shows only that the formal limit is, in general, incorrect. The reason for this is understood: the graphs of perturbation theory correspond to multiple integrals of products of propagators, and the pointwise convergence of (5.31) is not sufficient to ensure the possibility of taking the limit inside the integral.

More detailed analysis shows that one of the formal limits, the theory on the entire time axis, is nevertheless correct. Without dwelling on the detailed proof, we only give a simple example illustrating this statement.

Let $F(t)$ be an arbitrary function specified on the entire axis, and $f(\omega)$ be its Fourier transform: $F(t) = (1/2\pi)\int d\omega f(\omega)\exp i\omega t$. We define the periodic function $F_\beta(t)$:

$$F_\beta(t) = \frac{1}{\beta}\sum_n f(\omega_n)\exp i\omega_n t \tag{5.33}$$

(the sum runs over even frequencies), which becomes $F(t)$ in the limit $\beta \to \infty$. Owing to the periodicity, the integrals of $F_\beta(t)$ over the intervals $[0, \beta]$ and $[-\beta/2, \beta/2]$ coincide. We see from (5.33) that each is equal to $f(0)$ and is independent of β. On the other hand, the formal limits of the corresponding integrals are integrals of the limit function $F(t)$ over the semiaxis and the entire axis, respectively. Since $F(t)$ is arbitrary, these limits are essentially different, and the second one (over the entire axis), equal to $f(0)$, is correct, i.e., it coincides with the limit of the integral over the finite interval.

5.1.8 The Feynman–Kac formula

Repeating word for word the discussion of Sec. 2.4, we can derive the Euclidean analog of (2.56). The result, as usual, is obtained from the pseudo-Euclidean expression by the standard substitutions $t \to -it$, $iS \to S_e$, and we quote it without derivation:

$$\langle \mathbf{x} | \exp \mathbf{H}(\tau' - \tau) | \mathbf{x}' \rangle = \text{const} \int_{(\mathbf{x}\mathbf{x}')} D\varphi \exp S_e(\varphi) .$$

The functional $S_e(\varphi)$ on the right-hand side is the Euclidean action on the interval $\tau' \le t \le \tau$, and the integration runs over trajectories with given boundary values $\mathbf{x}, \mathbf{x}'$ at the ends of the interval.

If we set $\tau = \beta$ and $\tau' = 0$, the left-hand side can be interpreted as the coordinate representation of the density matrix $\rho = \exp(-\beta \mathbf{H})$:

$$\rho(\mathbf{x}, \mathbf{x}') = \text{const} \int_{(\mathbf{x}\mathbf{x}')} D\varphi \exp S_\beta(\varphi) . \tag{5.34}$$

Calculating the trace of the density matrix, we obtain the partition function Z:

$$Z = \text{tr}\,\rho = \text{const}\int d\mathbf{x} \int_{(\mathbf{x},\mathbf{x})} D\varphi \exp S_\beta(\varphi)\,. \tag{5.35}$$

This result is valid, in particular, also for the quantum-mechanical free theory of Sec. 2.1.2, which up to now has been excluded from consideration. We note that the right-hand side of (5.35) is, like (5.30), an integral over all periodic fields: the periodicity condition (5.21) for the bosonic field $\varphi(t)$ implies $\varphi(\beta) = \varphi(0)$, and integration over fields satisfying the condition $\varphi(\beta) = \varphi(0) = \mathbf{x}$ followed by integration over $\mathbf{x}$ is obviously equivalent to integration over all periodic fields (i.e., over closed trajectories).

In contrast to most such expressions, (5.34) admits a mathematically correct interpretation [38], namely, for the free theory of Sec. 2.1.2 in which the action $S_{0\beta}$ is simply the time integral of the kinetic energy with a minus sign, the symbol

$$\text{const} \int_{(\mathbf{x}_1\mathbf{x}_2)} D\varphi \exp S_{0\beta}(\varphi) \ldots \tag{5.36}$$

actually defines the Wiener measure $\mu(\varphi;\, \beta, \mathbf{x}_1, \mathbf{x}_2)$ on a space of continuous but not necessarily differentiable functions with given boundary conditions. To stress this fact, (5.34) is rewritten as

$$\rho(\mathbf{x}_1, \mathbf{x}_2) = \int D\mu(\varphi;\beta, \mathbf{x}_1, \mathbf{x}_2) \exp S_{\nu\beta}(\varphi) \tag{5.37}$$

and called the Feynman–Kac formula. Here the measure μ turns out to be concentrated on nondifferentiable functions, so the notation in (5.36) is purely symbolic.* The mathematical correctness of the representation (5.37) for sufficiently well behaved $S_{\nu\beta}$ allows it to be used to obtain various rigorous results concerning, for example, the existence of the thermodynamical limit in quantum statistics [39].

5.1.9 Convexity properties

Functional integrals in statistics differ from the integrals of Euclidean field theory only by the finiteness of the range of integration over time in the action

* We recall that $S_{0\beta}(\varphi)$ contains a time derivative of φ.

functional and the periodicity requirements on the fields. The arguments of Sec. 4.3.1 remain in force, so it can be stated that in quasiprobabilistic theories $G_\beta(A)$ and its logarithm are convex-downwards functionals of A and the other variables entering linearly into the action.

Actually, the only quasiprobabilistic theory of practical importance for statistics is quantum mechanics with a well behaved potential. Relativistic quasiprobabilistic theories of scalar and vector fields are apparently only of academic interest, and the nonrelativistic theory of Sec. 2.2 of greatest practical importance is not quasiprobabilistic.

As in field theory, in statistics there exists the universal property of convexity for time-independent potentials—the analog of (4.21). The proof, to which we now turn, is nearly word for word identical to the arguments of Sec. 4.3.2. Let $S_\beta = S'_\beta + S''_\beta$, where S''_β is the integral over time from 0 to β of the real (for unrotated fields) functional (4.18), and let $G_\beta(A) = \text{const}\int D\varphi \exp S_\beta(\varphi)$ be a functional normalized by the condition $G_\beta(0) = 1$. We want to show that $G_\beta(A)$ and its logarithm $W_\beta(A)$ are convex-downwards functionals of the potentials A_n in (4.18). First of all, we note that the functional G_β itself is positive, since it is equal to the ratio of the partition functions of the theories with action S_β and S'_β, and each partition function is nonnegative because it is the trace of a positive operator $\exp[-\beta\mathbf{H}]$.

Furthermore, it is obvious that the second variation of G_β is proportional to the functional average of the squared variation of the action $\delta S''_\beta$ with weight $\exp S_\beta(\varphi)$, and the second variation of the logarithm of G_β is proportional to the corresponding dispersion. The positivity of the dispersion is an automatic consequence of the positivity of the average of the square.

The variation of the action $\delta S''_\beta$ is the time integral of the variation of the Lagrangian $\delta\mathscr{L}(t)$. Therefore, the functional average of the square of $\delta S''_\beta$ is expressed as a double integral over t and t' of the functional average of the product $\delta\mathscr{L}(t)\delta\mathscr{L}(t')$. The latter can be expressed in terms of the statistical average of the Dyson T-product of the corresponding operators in the Euclidean Heisenberg picture using the relation between functional integrals and Wick T-products and the rules for transforming to the interaction picture in the Green functions [the analog of (5.3)]:

$$\langle \delta\mathscr{L}(t)\,\delta\mathscr{L}(t')\rangle = (\,(T_{\mathrm{D}}[\delta\hat{\mathscr{L}}(t)\,\delta\hat{\mathscr{L}}(t')])\,)\,. \tag{5.38}$$

The symbol $\langle\ldots\rangle$ denotes the functional average with weight $\exp S_\beta(\varphi)$, and $\delta\hat{\mathscr{L}}(t)$ is the operator in the Euclidean Heisenberg picture corresponding to the functional $\delta\mathscr{L}(t)$. If the variation of the potentials does not take $\mathscr{L}(t)$ outside the class of real (in the pseudo-Euclidean version) Lagrangians, as we shall assume, the Schrödinger operator $\delta\hat{\mathscr{L}} = \delta\hat{\mathscr{L}}(0)$ is Hermitian, while $\delta\hat{\mathscr{L}}(t)$ possesses the property of combined Hermiticity. We shall show that in this case the average on the right-hand side of (5.38) is nonnegative; this is more than sufficient for proving the convexity properties.

In fact, assuming that the Lagrangian $\mathscr{L}(t)$ and its variation $\delta\mathscr{L}$ are bosonic, we have $T_D[\delta\hat{\mathscr{L}}(t)\,\delta\hat{\mathscr{L}}(t')] = \theta(t-t')\,\delta\hat{\mathscr{L}}(t)\,\delta\hat{\mathscr{L}}(t') + (t \rightleftarrows t')$. Let us consider the average of the product $\delta\hat{\mathscr{L}}(t)\,\delta\hat{\mathscr{L}}(t')$, which is proportional to the trace of the operator $\delta\hat{\mathscr{L}}(t)\,\delta\hat{\mathscr{L}}(t')\exp[-\beta\mathbf{H}]$, where $\mathbf{H}$ is the full Hamiltonian of the theory with action S_β. The desired trace is written as a double series

$$\sum_{nm} |(n|\delta\hat{\mathscr{L}}|m)|^2 \exp[-\beta E_n + (E_n - E_m)(t-t')], \tag{5.39}$$

where $|n)$ is a complete set of eigenstates of $\mathbf{H}$, E_n are the corresponding eigenvalues, and $\delta\hat{\mathscr{L}}$ is the operator in the Schrödinger picture. All the terms of the series (5.39) are positive; assuming that the series converges, we conclude that its sum is positive. This provides the desired proof of the statement, because the second term of the T-product differs only by the interchange of the arguments t and t'.

We have therefore proved the nonnegativity of the second variations of the functionals $G_\beta(A)$ and $W_\beta(A) = \ln G_\beta(A)$ with respect to the potentials. The functional $\Omega_\beta(A)$, defined by $W_\beta(A) = -\beta\Omega_\beta(A)$, has the meaning of the shift of the thermodynamic potential due to the addition of the interaction S''_β and is analogous to the shift of the ground-state energy in field theory. It follows from this discussion that $\Omega_\beta(A)$ is, like the energy shift $\varepsilon(A)$, a convex-upwards functional of the potentials, i.e., its second variation is negative.

As in Sec. 4.3.2, in the proof it is implicitly assumed that the system in question is stable, i.e., that the Hamiltonian is bounded below. Actually, the series (5.39) determines the Green function in the region $0 \le t' \le t \le \beta$. The coefficients of the energies E_n and E_m in the argument of the exponential in (5.39) are negative at all points in this region. From this we see that growth ($E_n \to \infty$) of the eigenvalues favors the convergence of the series, but if the Hamiltonian $\mathbf{H}$ is unbounded below and there exists a sequence $E_n \to -\infty$, the series certainly diverges and the proof becomes meaningless.

5.1.10 Convexity of the logarithm of the partition function

Let us discuss an important special case of the general statement proved in the preceding section: the convexity properties of $W = \ln Z$ with respect to numerical parameters entering linearly into the exponent of the ρ matrix. Let

$\mathbf{a} \equiv -\beta\mathbf{H} = \Sigma_i x_i \mathbf{a}_i$, where the x_i are real numerical coefficients and the $\mathbf{a}_i$ are certain Hermitian operators. The parameters x_i will be treated as independent variables. As one of them we can take the coefficient β itself; the other x_i can have different meanings: the chemical potential, a uniform external field, and so on. For example, for the grand canonical ensemble $-\beta\mathbf{H} = -\beta\mathbf{H}' + \beta\mu\mathbf{N}$, where $\mathbf{H}'$ is the ordinary Hamiltonian and $\mathbf{N}$ is the particle number operator. In this case we can take $x_1 = -\beta$, $x_2 = \beta\mu$, $\mathbf{a}_1 = \mathbf{H}'$, and $\mathbf{a}_2 = \mathbf{N}$.

The logarithmic derivatives of the partition function $\alpha_i \equiv \partial W/\partial x_i = ((\mathbf{a}_i))$ define the variables α_i *thermodynamically conjugate* to the x_i. In the example given above the average energy is conjugate to the variable x_1, and the average particle number is conjugate to x_2.

The convexity downwards of W as a function of the variables x follows formally from the results of the preceding section, but we shall give another version of the proof. We have

$$\delta W = \sum_i \frac{\partial W}{\partial x_i}\delta x_i = \sum_i ((\mathbf{a}_i))\,\delta x_i \equiv ((\delta\mathbf{a})) = Z^{-1}\,\mathrm{tr}\,[\delta\mathbf{a}\exp\mathbf{a}]\,.$$

Using the well known equation

$$\delta\exp\mathbf{a} = \int_0^1 dt\exp[\mathbf{a}(1-t)]\,\delta\mathbf{a}\exp[t\mathbf{a}]\,, \tag{5.40}$$

we easily obtain the following expression for the second variation:

$$\delta^{(2)} W = \frac{1}{2\beta}\int_0^\beta dt\,[((\delta\mathbf{a}(t)\,\delta\mathbf{a}(0))) - ((\delta\mathbf{a}))^2]\,, \tag{5.41}$$

in which $\delta\mathbf{a}(t)$ is the operator $\delta\mathbf{a}$ in the Euclidean Heisenberg picture and $\delta\mathbf{a}(0) = \delta\mathbf{a}$. The right-hand side of (5.41) has the form of a dispersion, and the positivity of the corresponding average of the square, i.e., the first term on the right-hand side of (5.41), directly follows from the positivity, proved in the preceding section, of the function $((\delta\mathbf{a}(t)\delta\mathbf{a}(0)))$ for the operator $\delta\mathbf{a}$, which is Hermitian at $t = 0$. As already mentioned several times, the positivity of the dispersion is a formal consequence of the positivity of the average of the square.

We see from (5.39) that the average of the square can vanish only when the operator $\delta\hat{\mathcal{L}}$ is equal to zero. From this it follows that the dispersion (5.41) can vanish only if $\delta\mathbf{a}(t) - ((\delta\mathbf{a})) = 0$, i.e., when and only when the operator $\delta\mathbf{a}$ is a multiple of the unit operator.

The requirement that the function $W(x)$ be convex downwards contains all the usual conditions of positivity in thermodynamics such as positivity of the specific heat, the susceptibility, and so on. The proof of positivity of the second variation (5.41) given above will be completely rigorous only when the ρ matrix is actually a well defined operator with finite trace and all quantities are mathematically correct. Such properties are possessed by systems in a finite volume, with certain assumptions about the nature of the interaction [40]. The convexity requirement automatically extends to systems of infinite extent if the quantity $W(x)$ is determined by rigorously taking the limit from finite volume.

5.1.11 Representation of the partition function of the free theory by a functional integral

We shall show that when the free action is correctly defined, the expression

$$Z = \text{const}\int D\varphi \exp S_\beta(\varphi) \tag{5.42}$$

is valid, where the normalization constant is independent both of the type of interaction and of the single-particle energies. This differs from the expressions of Sec. 5.1.6 in that (5.42) defines the partition function itself, not only the ratio Z/Z_0.

Since we know that the ratio of the partition functions is equal to the ratio of the integrals entering into (5.42), it is sufficient to verify that (5.42) is valid for free theories.

We begin with the one-dimensional quantum-mechanical oscillator, for which

$$Z = \sum_{n=0}^{\infty} \exp\left[-\beta\omega\left(n+\frac{1}{2}\right)\right] = \left[2\sinh\frac{\beta\omega}{2}\right]^{-1}. \tag{5.43}$$

On the other hand, for the oscillator $2S_\beta = \varphi K_e \varphi$, where $K_e = m[\partial_t^2 - \omega^2]$. Calculating the Gaussian integral on the right-hand side of (5.42) using the rule (1.145), we obtain $\text{const}\cdot\det K_e^{-1/2}$. Here and below const denotes any constant independent of the single-particle energies, in the present case, of ω.

The operator K_e acts on the space of periodic functions (5.22) with even harmonics. The latter are eigenfunctions of K_e with eigenvalues $-m[\omega_n^2 + \omega^2]$, where $\omega_n = 2\pi n/\beta$. Calculating the determinant as the product of eigenvalues, we obtain

$$\det K_e^{-1/2} = \text{const}\left[\omega \prod_{n=1}^{\infty}\left(1+\frac{\omega^2\beta^2}{4\pi^2 n^2}\right)\right]^{-1} \tag{5.44}$$

We have isolated the factor with $n = 0$ and have taken into account the doubling in the sign of n.

The infinite product in the square brackets together with the factor of ω converges to $2\beta^{-1}\sinh[\beta\omega/2]$, so that the Gaussian integral on the right-hand side of (5.42) is proportional to (5.43), as we wanted to prove.

Let us now turn to the nonrelativistic theory of Sec. 2.2, for which $S_\beta = \psi^+ K_e \psi$, where $K_e = -\partial_t - \mathscr{E}$. Calculating the Gaussian integral in (5.42) using (1.146) and (1.160), we obtain $\det K_e^{-\varkappa}$ up to an unimportant coefficient. As usual, $\varkappa = \pm 1$ depending on the statistics.

We assume that the single-particle Hamiltonian $\mathscr{E}$ has a purely discrete spectrum ε_α. Then $\det K_e$ obviously breaks up into a product over all levels α. We therefore need only to solve the one-level problem, i.e., to calculate the determinant of the operator $K = -\partial_t - \varepsilon$, acting on the space of complex periodic functions, up to a factor independent of ε.

However this, at first glance, elementary problem reveals a serious difficulty: for an operator K containing a first-order (in contrast to second-order) time derivative, $\det K$ has no unambiguous, correct definition even after division by an infinite factor independent of ε. In fact, after dividing out this "trivial infinity" we obtain an infinite product of the type $\Pi_n(1 + a_n)$, the criterion for the convergence of which is the absolute summability of the series for the a_n. In the case of an operator of second order in ∂_t (the oscillator) we had $a_n \sim 1/n^2$, so the convergence condition was satisfied. For a first-order operator we have $a_n \sim 1/n$, so the series $\Sigma_n |a_n|$ diverges logarithmically and the expression $\Pi_n(1+ a_n)$ is undefined. It is, of course, possible to reduce the second case to the first by grouping the factors with $\pm n$ in pairs, but this will be only one of the possible ways of defining an infinite product which does not exist in a strict sense.

Therefore, for operators of first order in ∂_t the quantity $\det K$ needs to be defined if we do not want to completely give up general expressions of the type (1.146) for this case. The problem can thus be stated as follows: is there any way to define $\det K$ for operators of first order in ∂_t such that all the general expressions, including the representation (5.42), remain valid?

We shall now show that this is possible if not only $\det K$ but also the action functional itself is defined consistently with the definition of the T-product for coincident times (Sec. 1.2.2). For $\det K$ with $K = -\partial_t - \varepsilon$ such consistency is ensured by the use of the expression following from (1.147),

$$\partial \ln \det K/\partial\varepsilon = -\mathrm{tr}\, K^{-1} = \mathrm{tr}\, g = \int_0^\beta dt\; g(t,t)\,. \tag{5.45}$$

as the definition of det K up to a factor independent of ε. In (5.45) the quantity $g = -K^{-1}$ is the temperature propagator for the one-level problem, known from Eqs. (5.6), (5.9), and (5.10): $g(t,t') = [\theta(t-t') + \varkappa n]\exp\varepsilon(t'-t)$, where $n = [\exp(\beta\varepsilon) - \varkappa]^{-1}$ is the occupation number known from (5.8). The quantity $g(t,t)$ on the right-hand side of (5.45) is defined by the rule $\theta(0) = 1/2$, which follows from the way we have defined the T-product for coincident times (Sec. 1.2.2). Then (5.45) takes the form $\partial \ln\det K/\partial\varepsilon = \beta[\varkappa n + 1/2]$, from which by integrating over ε we find $\ln\det K^{-\varkappa} = \mathrm{const} - \varkappa\ln[1 - \varkappa\exp(-\beta\varepsilon)] - \varkappa\beta\varepsilon/2$ (the same result obtained by pairwise grouping of the contributions with $\pm n$ in the infinite product).

This result does not coincide with the logarithm of the free partition function for the one-level problem. The difference is the terms $-\varkappa\beta\varepsilon/2$, which do not appear in (5.7). However, this discrepancy is due not to use of an incorrectly defined det$K$, but to the incorrectness the free action functional represented as a simple quadratic form $\psi^+K\psi$, as has been used up to now. Actually, the correct expression for the free action is the functional $S_{0\beta} = \psi^+K\psi + \varkappa\beta\varepsilon/2$, with an added constant which in (5.42) cancels the unwanted contributions $-\varkappa\beta\varepsilon/2$ in the expression obtained above for $\ln\det K^{-\varkappa}$. We note that this constant is important only in the calculation of the free partition function Z_0; in all expressions of the type (5.30) the contribution of the constant cancels in the ratio Z/Z_0, and then the simple form $\psi^+K\psi$ can, as usual, be used as the free action $S_{0\beta}$.

Let us explain in more detail the origin of this constant. When studying any specific quantum problem, the primary objects are the quantum operators. Classical functionals of the type $S(\varphi)$ are secondary and arise in the process of interpreting the original quantum problem into the classical language of functional integrals. The rule for the correspondence between a quantum operator and its "classical image" in functional integrals itself needs to be defined so as to be consistent with the definition of the T-product for coincident times. This problem has been discussed above in Sec. 1.3.2 in determining the form of the classical functional $S_V(\varphi)$ corresponding to a given quantum interaction operator **V**. There it was shown that the definition of the T-product for coincident times as the Sym-product necessarily leads to $-\int dt\, \mathbf{V}(t) = \mathrm{Sym}\,\mathrm{S}_V(\hat\varphi) = \mathrm{S}_V(\hat\varphi)$ as the definition of the symmetric (the second equality) classical interaction functional $\mathrm{S}_V(\varphi)$. In other words, the quantum interaction operator must be represented by its Sym-form when going to the classical language of functional integrals (Sec. 1.2.4).

This rule should actually be considered universal and extended to the free Hamiltonian $\mathbf{H}_0$ if we want to use functional integrals to calculate not only ratios like Z/Z_0, but the partition functions Z and Z_0 themselves. In the problem we are considering now $\mathbf{H}_0 = \varepsilon a^+ a$, where a^+ and a are quantum creation and annihilation operators. Since $a^+a = (a^+a + \varkappa aa^+)/2 + (a^+a - \varkappa aa^+)/2 = \mathrm{Sym}[a^+a - \varkappa/2]$, the quantum Hamiltonian $\mathbf{H}_0 = \varepsilon a^+a$ in the functional integrals should correspond to the classical expression $\varepsilon[a^+a - \varkappa/2]$ with the addition of a constant $-\varepsilon\varkappa/2$, equivalent to the addition of $\beta\varepsilon\varkappa/2$ to the free action functional. We note that for the oscillator such a problem did not arise, because as its quantum Hamiltonian we used the operator $\varepsilon(a^+a + 1/2) = \mathrm{Sym}[\varepsilon a^+a]$ with the addition $\varepsilon/2$ from vacuum oscillations.

Therefore, when all the needed quantities (the T-product for coincident times, the θ functions at points of discontinuity, $\det K$, and the "classical images" of quantum operators) are accurately and consistently defined, the general expression (5.42) remains valid also for problems with first-order time derivatives in the free action. The starting point is the definition of the T-product for coincident times, and everything else must be consistent with this. The definition of this T-product as the Sym-product used in this book (Sec. 1.2.2) is not, of course, the only one possible: one could, for example, define it not as the Sym-product, but as the N-product, and then define all other quantities accordingly. General expressions of the type (5.42) remain valid for any choice of definition, as long as it ensures their mutual consistency. In our opinion, the definition of the T-product using the Sym-product is the most natural in the general case, especially for problems with second-order time derivatives, which correspond to fields φ (of the oscillator type) which commute on the surface $t = \mathrm{const}$. But, anyway, the choice of definition is essentially one of convenience; only the mutual consistency of the definitions of all the quantities involved is important. As a technical point, we might add that a great deal about the nature of functional integrals can be learned by using (5.42) to calculate the partition function of the one-site Hubbard model with Hamiltonian $\mathbf{H} = \varepsilon(a_1^+a_1 + a_2^+a_2) + g a_1^+ a_2^+ a_2 a_1$, in which a_i^+, a_i are pairs of fermionic operators with $i = 1,2$ (up and down spin) and the fourth term is their Coulomb interaction. The partition function is known: $Z = \mathrm{tr}\exp[-\beta\mathbf{H}] = 1 + 2\exp(-\beta\varepsilon) + \exp(-2\beta\varepsilon - \beta g)$, but obtaining this result using (5.42) is a nontrivial and very instructive problem.

At zero temperature, i.e., in field theory, expressions of the type (5.42) determine the ground-state energy up to an additive constant independent of the single-particle energies. The expression analogous to (5.45) for the fermionic theory of Sec. 2.2 involves the propagator (2.26) of the one-level problem rather than the temperature propagator g: $\Delta(t, t') = \theta(t - t')\exp i\varepsilon(t' - t)$ if the level is free and $\Delta(t, t') = -\theta(t' - t)\exp i\varepsilon(t' - t)$ if the level is filled.

For $t' = t$ we respectively obtain $\Delta(t,t) = \pm 1/2$, which gives ε for the energy difference, as required.

5.2 Lattice Spin Systems

5.2.1 The Ising model

In this model a spin operator s_i having only two eigenvalues ± 1 (spin up and spin down) is associated with each site i of a spatial lattice. The operators s_i at different sites commute with each other, so that there are no noncommuting operators at all in this system and it can be considered classical.

The Hamiltonian is given by [41]

$$\mathrm{H} = -\frac{1}{2}\sum_{ik} s_i \mathcal{V}_{ik} s_k - \sum_i s_i h_i \equiv -\frac{1}{2} s \mathcal{V} s - hs\,, \tag{5.46}$$

the first term of which represents the exchange interaction of the spins, while the second is the interaction with a nonuniform external field h. The interaction is given by a real symmetric matrix $\mathcal{V}$ having zeros along the diagonal. It is usual to consider translationally invariant systems, for which $\mathcal{V}_{ik}$ depends only on the difference of the coordinates of sites i and k, and the external field h_i is uniform, i.e., independent of i. However, from the technical point of view it is more convenient to first take $\mathcal{V}$ and h to be arbitrary parameters, the specific values of which are assigned only in the final expressions. To make the notation concise we define $A \equiv \{A_i = \beta h_i\}$ and $\Delta \equiv \{\Delta_{ik} = \beta \mathcal{V}_{ik}\}$. Then

$$Z = \mathrm{tr}\exp\left[\frac{1}{2} s\Delta s + As\right] = \exp\left[\frac{1}{2}\cdot\frac{\delta}{\delta A}\Delta\frac{\delta}{\delta A}\right] Z^{(0)}(A)\,, \tag{5.47}$$

where $Z^{(0)}(A) = \mathrm{tr}\exp As$. In obtaining the second equation we have used the fact that differentiation with respect to A_i is equivalent to multiplication by s_i inside the trace owing to the commutativity of all the s_i. The functional $Z^{(0)}$ breaks up into a product over sites and is easily calculated: it is equal to the product of factors $2\cosh A_i$. From this we find

$$\mathcal{M}(A) \equiv \ln Z^{(0)}(A) = \sum_i \ln 2\cosh A_i \equiv \ln 2\cosh A\,. \tag{5.48}$$

The right-hand side of (5.47) has the form of the S-matrix generating functional (1.84), so for the partition function of the Ising model we can use all the results of Sec. 1.4 concerning diagrammatic representations of the S matrix. The partition function Z is the sum of unity and all graphs with line Δ and generating vertex (5.48), and its logarithm $W = \ln Z$ is the sum of all connected graphs. In this theory a line rather than a vertex is an interaction. The line Δ contains the factor β, so that by classifying graphs according to the number of lines we arrive at high-temperature expansions in powers of β.

The generating vertex (5.48) is represented as a sum over sites, so the vertex factors (1.97) are local: the derivatives $\mathcal{M}_n(i_1...i_n) \equiv \partial^n\mathcal{M}/\partial A_{i_1}...\partial A_{i_n}$ are nonzero only when all the indices $i_1...i_n$ coincide. For example, $\mathcal{M}_1(i) = \tanh A_i$, $\mathcal{M}_2(ik) = \delta_{ik}(1-\tanh^2 A_i)$, and so on.

In addition to the partition function Z and its logarithm W, the derivatives of these functionals with respect to the variable A are also of physical interest. Of special importance are the derivatives $\partial W/\partial A_i = ((s_i))$ determining the average value of the spin s_i, i.e., the *magnetization* of the site i, and the second derivatives $\partial^2 W/\partial A_i\partial A_k = ((s_i s_k)) - ((s_i))((s_k))$ representing the *spin correlation matrix*. The diagrammatic representations of these quantities are obtained automatically from that of W.

In the practical calculation of the coefficients of high-temperature expansions in zero external field it is usual to use not the Feynman diagrammatic technique discussed above, but the technique of lattice graphs, which is based on the specific form of the exchange interaction (i.e., the matrix $\mathcal{V}$) and is weakly related to the standard formalism of field theory. We shall not dwell on it here, but refer the interested reader to the books [42,43].

Let us give the integral representation for the partition function obtained in the usual way from (5.47) and completely analogous to the S-matrix representation (1.168):

$$Z = \text{const}\int D\varphi \exp\left[-\frac{1}{2}\varphi\Delta^{-1}\varphi + \ln 2\cosh(\varphi + A)\right]. \tag{5.49}$$

Here $\varphi \equiv \{\varphi_i\}$ is a real field on the lattice, $D\varphi = \Pi_i d\varphi_i$, and the normalization constant is defined such that the right-hand side of (5.49) becomes unity when $\ln 2\cosh(\varphi + A)$ is discarded.

We note that the stationary-phase approximation for the integral (5.49) leads to the well known equation of the self-consistent Weiss field for the magnetization. The latter, being the logarithmic derivative of Z with respect to A, is the functional average of $\tanh(\varphi + A)$.

The definition (5.47) can be understood as yet another integral representation of the partition function:

$$Z = \text{const}\int D\varphi\ \delta(\varphi^2-1)\exp\left[\frac{1}{2}\varphi\Delta\varphi + A\varphi\right],$$

where $\delta(\varphi^2 - 1) = \Pi_i\delta(\varphi_i^2 - 1)$. Representing this δ function as a Fourier functional integral $\sim\int D\psi \exp i\psi(\varphi^2 - 1)$, for Z we obtain a representation in the form of an integral over the pair of fields φ, ψ.

In conclusion, let us give the differential equations for Z. The ordinary Schwinger equation (Sec. 1.7.1) is ineffective for the integral (5.49) owing to the fact that the interaction $\ln 2\cosh\varphi$ is not a polynomial, but the desired equations are very simply obtained directly from the definition (5.47):

$$2\partial Z/\partial\Delta_{ik} = \partial^2 Z/\partial A_i\partial A_k,\ \ i\neq k;\ \ \partial^2 Z/\partial A_i\partial A_i = Z. \tag{5.50}$$

The second of these expresses the condition $s_i^2 = 1$. Iterating these equations in powers of Δ with the known zeroth-order approximation $Z^{(0)}(A) = \exp\ln 2\cosh A$, we arrive at the standard diagrammatic representations for Z.

Let us again recall the classical Heisenberg model, which differs from the Ising model in that with each lattice site is associated a classical d-dimensional spin vector $\mathbf{s}$ of unit length, and the partition function is defined as the integral over the directions of all vectors $\mathbf{s}_i$; in the special case $d = 1$ we return to the Ising model. The diagrammatic representations of the partition function will be exactly the same as for the Ising model; the only difference is in the appearance of additional spin labels and in the explicit form of the generating vertex $\mathcal{M}(\mathrm{A})$ analogous to (5.48).

5.2.2 *The Heisenberg quantum ferromagnet*

In this model, each site i is associated with a quantum-mechanical spin operator $s_{i\alpha}$, $\alpha = 1, 2, 3$, with a given value of s, where s is any positive integer or half-integer. The operators s_α are understood as matrices acting in the finite-dimensional space of states of a single site of dimensionality $2s + 1$. The operators at different sites commute with each other, and the operators at the same site satisfy the commutation relations of the rotation group: $[s_\alpha, s_\beta] = i\varepsilon_{\alpha\beta\lambda}s_\lambda$.

The Hamiltonian is again written as a quadratic form [41]:

$$\mathbf{H} = -\frac{1}{2}\sum_{i\alpha,\, k\beta} s_{i\alpha}\mathcal{V}_{i\alpha,\, k\beta}s_{k\beta} - \sum_{i\alpha} h_{i\alpha}s_{i\alpha} \equiv -\frac{1}{2}s\mathcal{V}s - hs\,, \tag{5.51}$$

where it is assumed that the matrix elements $V_{i\alpha,k\beta}$ vanish for $i = k$, i.e., only the spins at different sites interact. In this case no problem arises with the ordering of the factors $s_{i\alpha}s_{k\beta}$.

By definition,

$$Z = \operatorname{tr}\exp[-\beta\mathbf{H}] = \operatorname{tr}\exp\left[\frac{\beta}{2}sVs + \beta hs\right]. \qquad (5.52)$$

The difference from the classical spin systems discussed above is that now the factor $\exp[\beta sVs/2]$ cannot be taken outside the trace as a differential operator owing to the noncommutativity of the spin operators. Therefore, to construct the perturbation theory it is necessary to go to the interaction picture. The starting point is (5.2) for the density matrix $\rho = \exp[-\beta\mathbf{H}]$, from which

$$Z/Z_0 = \langle\langle T_D \exp\left[-\int_0^\beta dt\, \mathbf{V}(t)\right]\rangle\rangle. \qquad (5.53)$$

The second term in (5.51) is taken as the free Hamiltonian, and the first is the interaction; we recall that the symbol $\langle\langle\ldots\rangle\rangle$ denotes the average using the free ρ matrix and $\mathbf{V}(t)$ is the interaction Hamiltonian in the Euclidean interaction picture. The symbol T_D in (5.53) denotes the ordinary Dyson T-product of the interaction operators, but if desired it can be understood as the time-ordered T-product of the spin operators themselves $s(x) \equiv s_{i\alpha}(t)$ in the interaction picture:

$$T[s(x_1)\ldots s(x_n)] = \sum_P P[\theta(t_1\ldots t_n)\, s(x_1)\ldots s(x_n)]. \qquad (5.54)$$

When all or some of the time arguments $t_1\ldots t_n$ coincide, the T-product is defined as the Sym-product (see Sec. 1.2.2). Replacement of T_D in (5.53) by T does not lead to any confusion, because spin operators with identical t in (5.53) commute (only spins at different sites interact).

Let us introduce the auxiliary variable $A(x) \equiv A_{i\alpha}(t)$ and determine the functional

$$Z(A)/Z_0 = \langle\langle T\exp\left[As - \int_0^\beta dt\mathbf{V}(t)\right]\rangle\rangle, \qquad (5.55)$$

in which $As \equiv \int dxA(x)\,s(x) \equiv \sum_{i\alpha}\int_0^\beta dtA_{i\alpha}(t)\,s_{i\alpha}(t)$.

The spin operators inside the T-product behave like classical commuting objects, and this makes it possible to extract the term containing $\mathbf{V}(t)$, which is quadratic in the spin, in the argument of the exponential in (5.55) in the form of a differential operator:

$$Z(A)/Z_0 = \exp\left[\frac{1}{2}\cdot\frac{\delta}{\delta A}g\frac{\delta}{\delta A}\right]\langle\langle T\exp As\rangle\rangle, \tag{5.56}$$

where, as usual, $\frac{\delta}{\delta A}g\frac{\delta}{\delta A} \equiv \iint dxdx'\frac{\delta}{\delta A(x)}g(x,x')\frac{\delta}{\delta A(x')}$, and the "contraction" $g(x, x') = \delta(t - t')\mathcal{V}_{i\alpha,i'\alpha'}$ is determined by the exchange-interaction matrix in the Hamiltonian (5.51).

The right-hand side of (5.56) has the form of the S-matrix generating functional (1.84) for the theory with propagator g and generating vertex

$$\mathcal{M}(A) = \ln\langle\langle T\exp As\rangle\rangle \tag{5.57}$$

and therefore admits the standard diagrammatic representations of Sec. 1.4. In order to obtain Z/Z_0, which is analogous to the vacuum expectation value of the S matrix in field theory, it is necessary to set $A = 0$ in (5.56).

From the viewpoint of ordinary field theory, (5.56) is similar to the S-matrix generating functional, but it is actually the generating functional of the full temperature Green functions of the spin operators $s(x)$. Replacing, as above, T_D by T, it is easy to show that

$$((T_D[s_H(x_1)\dots s_H(x_n)])) =$$

$$= \text{const}\,\langle\langle T\left[s(x_1)\dots s(x_n)\exp\left(-\int_0^\beta dt\mathbf{V}(t)\right)\right]\rangle\rangle. \tag{5.58}$$

This expression is analogous to (5.4); $s_H(x)$ are the spin operators in the Euclidean Heisenberg picture, the symbol $((\dots))$ denotes the average with the exact ρ matrix, and the normalization constant on the right-hand side of (5.58) is determined by the condition $((1)) = 1$ and is numerically equal to Z_0/Z.

The diagrammatic representations of the Green functions of the spin operators are uniquely determined by the known diagrammatic representation of the generating functional (5.55). In practical calculations it is most difficult to calculate the vertex factors, i.e., multiple derivatives of the functional (5.57) for $A = 0$; comparing (5.58) and (5.57), we see that these factors have the meaning of connected Green functions of the spin operators for the free theory. The average $T\exp As$ obviously breaks up into a product over all sites, so that the generating vertex (5.57) breaks up into a sum over sites; therefore, the vertex factors $\mathcal{M}_n(x_1 \ldots x_n; A)$ are "local in the sites," i.e., they are nonzero only when all the indices $i_1 \ldots i_n$ coincide (we recall that $x \equiv t, i, \alpha$). For zero A the coefficient of the Kronecker δ for the sites is independent of the number of the lattice site (it is the same for all), and so the problem is to calculate the functional (5.57) and the coefficients of its expansion in A for a single spin operator. A special diagrammatic technique (the "Wick theorem for spin operators") has been developed to calculate these coefficients; it is described in [44]. The diagrammatic technique we have described above was first formulated in [45]. In conclusion, we note that finding the functional (5.57) in closed form for a single site is equivalent to exactly solving the problem of a spin in an arbitrary, time-dependent external field.

5.3 The Nonideal Classical Gas

5.3.1 A gas with two-body forces

The partition function Z_N of a system of N classical point particles in d-dimensional space is, as is well known (see, for example, [7,8]), defined by the integral

$$Z_N = \frac{1}{N!\,(2\pi)^{Nd}} \int \ldots \int d\mathbf{p}_1 \ldots d\mathbf{p}_N d\mathbf{x}_1 \ldots d\mathbf{x}_N \exp\left(-\beta \mathcal{H}_N\right) \qquad (5.59)$$

(as always, $\hbar = 1$), in which $\mathcal{H}_N$ is the full N-particle Hamiltonian:

$$\mathcal{H}_N(\mathbf{p}_1 \ldots \mathbf{p}_N \mathbf{x}_1 \ldots \mathbf{x}_N) = \sum_i \left(\frac{\mathbf{p}_i^2}{2m} + \mathcal{V}_1(\mathbf{x}_i)\right) + \sum_{i<k} \mathcal{V}_2(\mathbf{x}_i, \mathbf{x}_k) .$$

The function $\mathcal{V}_1(\mathbf{x})$ is the external-field potential, and the symmetric function $\mathcal{V}_2(\mathbf{x}, \mathbf{x}')$ is the two-body interaction potential of the particles; the summation over i, k runs from 1 to N, i.e., over all the particles, and the integration over $\mathbf{x}$ runs over the given volume of the system, which in a translationally invariant theory is all space.

Performing the Gaussian integral over the momenta in (5.59), we obtain

$$Z_N = (N!\lambda^{Nd})^{-1}\int\ldots\int d\mathbf{x}_1\ldots d\mathbf{x}_N\times$$
$$\times\exp\left[\sum_i A_1(\mathbf{x}_i) + \sum_{i<k} A_2(\mathbf{x}_i, \mathbf{x}_k)\right], \tag{5.60}$$

where $A_i \equiv -\beta\mathcal{V}_i$, and $\lambda \equiv (2\pi\beta/m)^{1/2}$ is the thermal wavelength.

The partition function Z of the grand canonical ensemble is defined by the series

$$Z = \sum_{N=0}^{\infty} Z_N \exp(\beta\mu N), \tag{5.61}$$

in which $Z_0 = 1$ and we have introduced the additional parameter μ, the chemical potential.

The coefficient of the integral in the general term of the series (5.61) is equal to $a^N/N!$, where $a = \lambda^{-d}\exp\beta\mu$ is a parameter called the activity. The factor a^N can be incorporated as an addition $\ln a$ to the potential A_1, so we shall drop it, treating A_1 and A_2 as independent functional variables whose specific values will be assigned only in the final expressions. We therefore set

$$Z = \sum_{N=0}^{\infty}\frac{1}{N!}\int\ldots\int d\mathbf{x}_1\ldots d\mathbf{x}_N \exp\{\sum_i A_1(\mathbf{x}_i) + \sum_{i<k} A_2(\mathbf{x}_i, \mathbf{x}_k)\}. \tag{5.62}$$

The absence of an external field now corresponds to $A_1(\mathbf{x}) = \ln a$ and not to $A_1 = 0$.

Differentiation of the general term of the series (5.62) with respect to $A_1(\mathbf{x})$ is equivalent to multiplication of the integrand by $\Sigma\delta(\mathbf{x}-\mathbf{x}_i)$; the observable (a random quantity) $n(\mathbf{x})$, represented in each N-particle sector as the operation of multiplication by $\Sigma\delta(\mathbf{x}-\mathbf{x}_i)$, is interpreted as the particle number density at the point $\mathbf{x}$. This explains the meaning of the multiple variational derivatives of the functional Z and its logarithm W with respect to A_1: $\delta^n Z/\delta A_1(\mathbf{x}_1)\ldots\delta A_1(\mathbf{x}_n) = Z((n(\mathbf{x}_1)\ldots n(\mathbf{x}_n)))$, where $((\ldots))$ denotes the average over the exact distribution. The derivatives of Z are the analogs of the full Green functions in field theory. Dividing them by Z, we obtain the analogs of the Green functions without vacuum loops. The multiple derivatives of the functional $W = \ln Z$ are the analogs of the connected Green functions. The first two derivatives of W are of greatest interest: $\delta W/\delta A_1(\mathbf{x}) = ((n(\mathbf{x})))$ is the average particle number density at the point $\mathbf{x}$ [it is usually referred to as simply the density, but should not be confused with the random quantity $n(\mathbf{x})$], and $\delta^2 W/\delta A_1(\mathbf{x})\delta A_1(\mathbf{x}') = \langle\langle n(\mathbf{x})n(\mathbf{x}')\rangle\rangle - \langle\langle n(\mathbf{x})\rangle\rangle\langle\langle n(\mathbf{x}')\rangle\rangle$ is the correlation function of the density fluctuations.

Comparing the derivative of Z with respect to $A_2(\mathbf{x}, \mathbf{x}')$ to the second derivative with respect to A_1, we easily obtain the following equation:

$$\frac{\delta Z}{\delta A_2(\mathbf{x}, \mathbf{x}')} = \frac{1}{2}\left[\frac{\delta^2 Z}{\delta A_1(\mathbf{x})\,\delta A_1(\mathbf{x}')} - \delta(\mathbf{x}-\mathbf{x}')\,\frac{\delta Z}{\delta A_1(\mathbf{x})}\right]. \tag{5.63}$$

The second term on the right-hand side appears because the summation of the paired contributions $A_2(\mathbf{x}_i, \mathbf{x}_k)$ in (5.62) runs over $i \neq k$. Equation (5.63) will be needed in the next chapter.

Let us now turn to the diagrammatic representations. Introducing the Mayer line (or the superpropagator) $g_{ik} \equiv -1 + \exp A_2(\mathbf{x}_i, \mathbf{x}_k)$, we rewrite (5.62) as

$$Z = \sum_{N=0}^{\infty} \frac{1}{N!}\int\ldots\int d\mathbf{x}_1\ldots d\mathbf{x}_N \prod_i \exp A_1(\mathbf{x}_i) \prod_{i<k} (1+g_{ik}). \tag{5.64}$$

The general term of this series is written as the sum of all possible labeled graphs (see Sec. 1.4.2) with N vertices; each vertex is associated with a factor $\exp A_1(\mathbf{x})$, and a line joining vertices i and k is associated with the superpropagator g_{ik}. All the graphs are Mayer graphs, i.e., any pair of vertices is connected by no more than a single line, and there are no self-contracted lines.

Comparison with the expressions of Sec. 1.4.5 shows that we are dealing with the ordinary graphs of field theory with an exponential interaction in N-form. The analogy becomes obvious if (5.62) is rewritten as

$$Z = \exp\left[\frac{1}{2}\cdot\frac{\delta}{\delta A_1}A_2\frac{\delta}{\delta A_1}\right]\sum_{N=0}^{\infty}\frac{1}{N!}\int\ldots\int d\mathbf{x}_1\ldots d\mathbf{x}_N \times$$
$$\times \exp\sum_i\left[A_1(\mathbf{x}_i) - \frac{1}{2}A_2(\mathbf{x}_i, \mathbf{x}_i)\right]. \tag{5.65}$$

The differential operator acting on $\exp\Sigma A_1(\mathbf{x}_i)$ produces the factor $\exp[\frac{1}{2}\Sigma A_2(\mathbf{x}_i, \mathbf{x}_k)]$, and the superfluous diagonal terms of the form are cancelled by the additional terms $A_2(\mathbf{x}_i, \mathbf{x}_i)$ in (5.65).

Summing the series over N in (5.65), we arrive at an expression like (1.84):

$$Z = \exp\left[\frac{1}{2}\cdot\frac{\delta}{\delta A_1}A_2\frac{\delta}{\delta A_1}\right]\exp \mathcal{M} \tag{5.66}$$

with generating vertex

$$\mathcal{M} = \int d\mathbf{x}\exp\left[A_1(\mathbf{x}) - \frac{1}{2}A_2(\mathbf{x},\mathbf{x})\right]. \tag{5.67}$$

Therefore, the partition function Z coincides with the S-matrix generating functional of field theory with the line A_2 and generating vertex (5.67). The exponential nature of the interaction allows the ordinary graphs with the line A_2 to be resummed to give Mayer graphs with the line g, as explained in detail in Sec. 1.4.5. It is also clear that the substitution $A_1(\mathbf{x}) \to A_1(\mathbf{x}) - A_2(\mathbf{x},\mathbf{x})/2$ in the generating vertex eliminates graphs with self-contracted lines. In the operator language of field theory the vertex (5.67) corresponds to the exponential interaction $\int d\mathbf{x}\exp A_1(\mathbf{x})$ with normal ordering, and the functional (5.67) represents its Sym-form. The corresponding reduced vertex (1.99) is a simple exponential:

$$\mathcal{M}_{\text{red}} = \exp\left[\frac{1}{2}\cdot\frac{\delta}{\delta A_1}A_2\frac{\delta}{\delta A_1}\right]\mathcal{M} = \int d\mathbf{x}\exp A_1(\mathbf{x}). \tag{5.68}$$

The topological concepts of graph theory and symmetry coefficients discussed in Chap. 1 were first used in the statistics of the nonideal gas. The theorem of the connectedness of the logarithm (the first Mayer theorem) was also first proved for this problem. This technique appeared somewhat later in quantum field theory in connection with Feynman diagrams, and then it was borrowed from there for use in quantum statistics. It is a rather curious history.

5.3.2 A gas with many-body forces

This problem is interesting to us mainly because it leads to a diagrammatic technique which lies outside ordinary graph theory. In this respect a classical gas with many-body forces is more complicated than any quantum field theory.

Let us introduce many-body potentials into the exponent in (5.62):

$$Z = \sum_{N=0}^{\infty}\frac{1}{N!}\int\ldots\int d\mathbf{x}_1\ldots d\mathbf{x}_N \exp\,\{\sum_i A_1(\mathbf{x}_i) +$$

$$+ \sum_{i<k} A_2(\mathbf{x}_i, \mathbf{x}_k) + \sum_{i<k<m} A_3(\mathbf{x}_i, \mathbf{x}_k, \mathbf{x}_m) + \ldots \} . \tag{5.69}$$

The number of potentials A_n is not bounded, but naturally an integral with a given N contains only A_n with $n \leq N$. For a given A_n the summation in (5.69) runs over all possible different groups of n pairwise differing indices; the set $i_1 < i_2 < \ldots < i_n$ uniquely parametrizes this group. Introducing the Mayer potentials $g_{i_1 \ldots i_n} \equiv -1 + \exp A_n(\mathbf{x}_{i_1} \ldots \mathbf{x}_{i_n})$, we rewrite (5.69) as

$$Z = \sum_{N=0}^{\infty} \frac{1}{N!} \int \ldots \int d\mathbf{x}_1 \ldots d\mathbf{x}_N \times$$

$$\times \prod_i \exp A_1(\mathbf{x}_i) \prod_{i<k} (1 + g_{ik}) \prod_{i<k<m} (1 + g_{ikm}) \ldots . \tag{5.70}$$

Graphs corresponding to individual terms of the product (5.70) will be called supergraphs [46]. To specify a labeled supergraph with N vertices it is necessary to indicate which pairs of vertices are connected by a potential (line) g_{ik}, which triplets of vertices are connected by a potential g_{ikm}, which quadruplets of vertices are connected by a potential g_{ikms}, and so on. In ordinary graphs the vertices are connected only by two-body potentials, i.e., lines, and each labeled graph is uniquely specified by the adjacency matrix π (see Sec. 1.4.2): by definition, $\pi_{ik} = 1$ if the vertices i and k are connected by a line, and $\pi_{ik} = 0$ otherwise. To specify a labeled supergraph it is necessary to introduce a more complicated object, the *adjacency exponent* [46]: $\pi \equiv \{\pi_{ik}, \pi_{ikm}, \ldots\}$. By definition, the symbol $\pi_{i_1 \ldots i_n}$ is symmetric under any permutation of the indices, and for a given set of indices it takes the value 1 or 0 depending on whether or not the corresponding group of vertices of the supergraph is connected by the potential $g_{i_1 \ldots i_n}$. In particular, π_{ik} is the usual adjacency matrix. The symbols $\pi_{i_1 \ldots i_n}$ are tensors under the group of vertex number permutations: $\pi'_{i_1 \ldots i_n} = P_{i_1 k_1} \ldots P_{i_n k_n} \pi_{k_1 \ldots k_n}$, where P is the permutation matrix (see Sec. 1.4.2).

The concepts of equality, equivalence, and symmetry group of labeled graphs defined in Sec. 1.4.2 can be directly generalized to supergraphs if the term "adjacency matrix" is everywhere replaced by "adjacency exponent." The number of equivalent but different labeled supergraphs is then determined by the usual expression $N!/s$, where s is the symmetry number, the order of the symmetry group of the supergraph. The product (5.70) obviously contains all possible labeled supergraphs, none of which are identical, so that the coefficient of the free graph (supergraph) is determined by the usual expression $(1/N!)(N!/s) = 1/s$. Because of this, the proofs of the two Mayer theorems (the first is the statement that $\ln Z$ is connected, and the second concerns the virial expansion and will be discussed in detail in the next chapter) generalize directly to systems with many-body forces [46].

Chapter 6

VARIATIONAL METHODS AND FUNCTIONAL LEGENDRE TRANSFORMS

6.1 Phase Transitions

6.1.1 Introduction

The theory of phase transitions is certainly one of the most interesting areas of statistical physics. To give a general definition of a phase transition in the formalism we are using, it is necessary to use the concepts, introduced in Sec. 1.10, of normal and anomalous solutions and spontaneous symmetry breaking, which carry over to Euclean field theory and statistical physics without change.

In this language a phase transition can briefly be defined as a change of solution. In practice this looks as follows. For a physical system studied in a wide range of temperatures T, the normal solution for the Green functions is realized in the region $T > T_c$, and the anomalous solution is realized for $T < T_c$. At the critical point $T = T_c$ the solution changes from the normal to the anomalous one. This transition is accompanied, as a rule, by the spontaneous breakdown of some symmetry, which is expressed in the appearance of nonzero Green functions or averages which under normal conditions should be equal to zero owing to a symmetry of the theory. Let us list a few of the best known phase transitions.

1. ***Superfluidity***. The system is a quantum Bose gas with suitable two-body interaction; the anomalous averages which are nonzero for $T < T_c$ are the first Green functions $\langle\langle\hat{\psi}(x)\rangle\rangle$ and $\langle\langle\hat{\psi}^+(x)\rangle\rangle$; the broken symmetry is the gauge group $\psi(x) \to \psi(x)\exp i\theta$, $\psi^+(x) \to \psi^+(x)\exp(-i\theta)$.

2. ***Superconductivity***. The system is a quantum Fermi gas of electrons with suitable two-body interaction or interacting with phonons; the anomalous averages are the Green functions $\langle\langle T_D[\hat{\psi}(x)\hat{\psi}(x')]\rangle\rangle$ and $\langle\langle T_D[\hat{\psi}^+(x)\hat{\psi}^+(x')]\rangle\rangle$; the broken symmetry is the same as above.

3. *Ferromagnetism.* The system is a classical or quantum spin system with isotropic interaction in zero external field; the anomalous average is the magnetization, i.e., the average value of the angular momentum (spin) operator s; the broken symmetry is the rotation group for an isotropic Heisenberg ferromagnet and the reflection of s for the Ising model.

4. *Liquid–vapor condensation.* The system is a nonideal classical gas with suitable interaction. Here there is no spontaneous breakdown of any symmetry: for $T \leq T_c$ the solution splits into two branches corresponding to a liquid and a gas. The nonuniqueness of the solution implies that it is anomalous in the sense of the definition in Sec. 1.10.

It is clear from these examples that the problem of phase transitions is a special case of the general problem of constructing anomalous solutions, which was discussed in Sec. 1.10. There it was noted that it is natural to solve such problems using the variational method, to which this entire chapter is essentially devoted. The general idea behind this method is best understood from a simple example from ordinary thermodynamics, which will be discussed in the following sections.

6.1.2 Transformation to the variational problem in thermodynamics

We shall take thermodynamics to mean the study of the partition function Z or quantities equivalent to it as functions of various numerical parameters: the temperature, an external field, the chemical potential, and so on. In quantum theory $Z = \mathrm{tr}\exp[-\beta\mathbf{H}]$, and we shall assume, as in Sec. 5.1.10, that the numerical parameters $x \equiv \{x_i\}$ are coefficients in the linear expansion of the argument of the exponential, i.e., $-\beta\mathbf{H} = \Sigma_i x_i \mathbf{a}_i$, where $\mathbf{a}_i$ are operators. Without loss of generality, the operators $\mathbf{a}_i$ can be assumed to be independent in the following sense: no linear combination of the $\mathbf{a}_i$ is a multiple of the unit operator (otherwise, one of the operators $\mathbf{a}_i$ could be eliminated). In this section we shall assume that the system is enclosed in a finite volume V (for a lattice the number of sites plays the role of V) and that its partition function Z, like the averages of the operators $\mathbf{a}_i$, is correctly defined. The last assumption is obviously satisfied for lattice spin systems with a finite number of sites, and in other cases it should be considered as a physically reasonable restriction on the form of the interaction [40].

We set $W(x) = V^{-1}\ln Z$ and define the variables conjugate to x:

$$\alpha_i = \partial W/\partial x_i = V^{-1}\partial \ln Z/\partial x_i = V^{-1}(\langle \mathbf{a}_i \rangle), \tag{6.1}$$

which have the meaning of the specific averages of the operators whose coefficients are the variables x_i.

The problem of the theory is to calculate the averages α and W from given values of the variables x. This can be formulated as a variational problem if we consider the Legendre transformation $\Gamma(\alpha)$ of the function $W(x)$: $\Gamma(\alpha) = W(x) - \Sigma_i x_i \partial W/\partial x_i \equiv W - x\alpha$. Here the variables x on the right-hand side of the equation are expressed in terms of α using Eqs. (6.1), which implicitly determine the x_i as functions of α. Differentiating $\Gamma(\alpha)$ with respect to the variables α_i, which are here assumed to be independent, we obtain the equations

$$\frac{\partial\Gamma}{\partial\alpha_i}=\sum_k\frac{\partial W}{\partial x_k}\frac{\partial x_k}{\partial\alpha_i}-\sum_k\left(\frac{\partial x_k}{\partial\alpha_i}\alpha_k+x_k\frac{\partial\alpha_k}{\partial\alpha_i}\right)=-x_i, \tag{6.2}$$

giving explicit expressions for x in terms of α for a known function $\Gamma(\alpha)$. Equations (6.2) can also be used to determine unknown α from given x, namely: the desired values $\alpha(x)$ are the points in the space of the variables α at which the first derivatives $\partial\Gamma/\partial\alpha_i$ take the given values $-x_i$. Introducing the function $\Phi(\alpha;\, x) = \Gamma(\alpha) + \alpha x$, it can be said that the desired $\alpha(x)$ are stationarity points of $\Phi(\alpha;\, x)$ with respect to variations of α for fixed x; from the definitions of Γ and Φ it is clear that the value of Φ at the stationarity point coincides with $W(x)$.

Therefore, in the language of the Legendre transform Γ the problem of determining the averages α and $W(x)$ is a variational problem: it is necessary to find the stationarity point of the function $\Phi(\alpha;\, x)$ and the value of the function at this point. The transformation to a variational problem allows the natural introduction of such concepts as degeneracy (nonuniqueness) of a solution and spontaneous symmetry breaking. Degeneracy implies nonuniqueness of the solution of the variational problem, i.e., nonuniqueness of the stationarity point, and spontaneous symmetry breaking corresponds to the situation where the function $\Phi(\alpha;\, x)$ invariant under some group of transformations of the variables α has noninvariant stationarity points. Spontaneous symmetry breaking automatically implies degeneracy, because a stationarity point "shifted by the group" will also be a stationarity point; it is also clear that the value of Φ itself at a stationarity point is the same for all points connected by a group shift owing to the assumed invariance of Φ.

It should be noted immediately that degeneracy and spontaneous symmetry breaking can arise only after taking the limit to infinite volume. For a system in a finite volume satisfying the requirements listed at the beginning of this section, the function $W(x)$ is rigorously convex downwards (see the proof in Sec. 5.1.10), i.e., the matrix of second derivatives $\partial^2 W/\partial x_i\partial x_k$ is rigorously positive-definite. This ensures that Eqs. (6.1) can be solved for x and that there is a one-to-one correspondence between the variables x and α, so that degeneracy is excluded.

Using (6.1) and (6.2), we obtain

$$\delta_{ik} = \frac{\partial x_i}{\partial x_k} = \sum_s \frac{\partial x_i}{\partial \alpha_s}\frac{\partial \alpha_s}{\partial x_k} = -\sum_s \frac{\partial^2 \Gamma}{\partial \alpha_i \partial \alpha_s}\frac{\partial^2 W}{\partial x_s \partial x_k}, \tag{6.3}$$

which shows that the matrices of second derivatives of the function $W(x)$ and of its Legendre transform $\Gamma(\alpha)$ are the inverse of each other up to a sign. The property of having definite sign is preserved in going to the inverse matrix, so for a theory in a finite volume the Legendre transform $\Gamma(\alpha)$, like $\Phi(\alpha; x) = \Gamma + x\alpha$, is a rigorously convex-upwards function of the variables α.

We can consider Legendre transforms with respect to some rather than all of the arguments x ("incomplete transforms"). We assume that the x_i are split into two groups x', x'', and that the transformation to the conjugate variables α' is done only for the arguments x': $\Gamma(\alpha', x'') = W(x) - \Sigma_i x'_i \alpha'_i$. Equations (6.2) and (6.3) clearly remain true for the arguments of the first group. Furthermore, assuming that α' and x'' are independent variables and that x' are functions of them, we obtain

$$\frac{\partial \Gamma(\alpha', x'')}{\partial x''_i} = \frac{\partial W(x)}{\partial x''_i} + \sum_k \frac{\partial W(x)}{\partial x'_k}\frac{\partial x'_k}{\partial x''_i} - \sum_k \frac{\partial x'_k}{\partial x''_i}\alpha'_k = \frac{\partial W(x)}{\partial x''_i}. \tag{6.4}$$

The derivatives on the two sides of this equation have different meanings: on the left is the derivative of Γ at fixed α', and on the right is the derivative of W at fixed x'.

For completeness, we also give the equations relating the matrices of second derivatives of Γ and W with respect to their arguments (α' and x'' for Γ and x' and x'' for W): $\Gamma_{11}W_{11} = -1$, $\Gamma_{22} = W_{22} - \Gamma_{21}W_{12}$, $\Gamma_{12} = -\Gamma_{11}W_{12}$. We use the shorthand notation Γ_{11}, Γ_{12}, Γ_{22} and similarly for W to denote the matrices of second derivatives with respect to the arguments of the first and second groups, respectively ($1 = \alpha'$, $2 = x''$ for Γ and $1 = x'$, $2 = x''$ for W).

In conclusion, let us formulate and prove one simple property of convex functions which will be used often in what follows. Let $f(\alpha)$ be a convex function having finite piecewise-continuous second derivatives. The set of stationarity points of this function is convex, i.e., together with any two points $\alpha^{(1)}$ and $\alpha^{(2)}$ it also contains the entire line segment connecting them: $c_1\alpha^{(1)} + c_2\alpha^{(2)}$, $c_1 \geq 0$, $c_2 \geq 0$, $c_1 + c_2 = 1$.

We use f_1 to denote the derivative along the direction of this segment and f_2 to denote the derivative along (any) one of the directions orthogonal to **1**. By assumption, $f_1 = f_2 = 0$ at the ends of the segment.

The increase Δf_1 in going from one of the end points of the segment to any interior point is equal to the integral of f_{11} over the corresponding interval.

Taking into account the convexity of f (the fact that f_{11} has definite sign) and the fact that $\Delta f_1 = 0$ for the entire segment (stationarity of the end points), we conclude that $f_1 = f_{11} = 0$ throughout the entire segment. From this, taking into account the inequality $|f_{12}|^2 \leq f_{11} f_{22}$ following from the convexity of f and the assumed finiteness of the second derivatives f_{22}, it follows that the mixed derivative f_{12} is also zero throughout the entire segment, so that $f_2 = 0$ on the segment, because the increase of this derivative is expressed as an integral of f_{12}. This is true for any direction **2** orthogonal to **1**, so throughout the entire segment all the first derivatives of f are equal to zero, as we wanted to prove. It is clear from the proof that degeneracy (nonuniqueness) of the stationarity point is possible only when the convexity is not rigorous. It is also clear that when a stationarity segment is present, the function f is constant on the segment.

Therefore, the set of all stationarity points of a convex function forms a convex "stationarity region" $\mathcal{O}$, and f is constant on this region. It follows from the continuity of the first derivatives of f that the region $\mathcal{O}$ is also closed, i.e., it contains all its boundary points. For a rigorously convex function $\mathcal{O}$ can consist of only a single point.

Let us carefully define the meaning of the variables x and α for several specific systems. We begin with the Ising model in a uniform external field h, for which, according to (5.46), $-\beta\mathbf{H} = -\beta\mathbf{H}_{ex} + \beta h s$, where $\mathbf{H}_{ex}$ is the Hamiltonian of the exchange interaction of the spins, and $\mathbf{s} \equiv \Sigma s_i$ is the total spin, i.e., the sum of the spins of all the lattice sites. If we take $x_1 = -\beta$ and $x_2 = \beta h$ (the reduced external field), the conjugate variables α will respectively be the average exchange energy and the average spin (magnetization) per site. We choose not to use the traditional variables of temperature and external field because, of course, in these variables the simple and universal property of convexity of W would take an unreasonably complicated form.

The function $W(x)$ is even in x_2, i.e., invariant under change of sign of the field, and its Legendre transform is even in the magnetization α_2. For zero external field the varied function $\Phi(\alpha; x) = \Gamma(\alpha) + \alpha_1 x_1$ is even in α_2, and the spontaneous symmetry breaking is expressed as the appearance of a stationarity point with nonzero value of the magnetization α_2 which is degenerate in the sign of α_2 owing to the parity of Φ.

The classical and quantum Heisenberg ferromagnets differ from the Ising ferromagnet only in that the reduced external field and its conjugate variable, the magnetization, become three-vectors, and the invariance under change of sign of these quantities becomes invariance under rotations of the corresponding vectors (for isotropic exchange interaction of the spins). In the presence of spontaneous symmetry breaking in zero external field the solution is degenerate in all the directions of the magnetization vector.

Let us now consider the nonrelativistic quantum Bose gas (see Sec. 2.2) with Hamiltonian $\mathbf{H}'_0 - \mu\mathbf{N} + \mathbf{V}$, where $\mathbf{H}'_0$, $\mathbf{N}$, and $\mathbf{V}$ are respectively the single-particle energy operator, the particle number operator, and the interaction operator, and the constant μ is the chemical potential [we note that the free Hamiltonian (2.21) is $\mathbf{H}_0 = \mathbf{H}'_0 - \mu\mathbf{N}$]. The Hamiltonian is assumed to be invariant under gauge transformations $\hat{\psi}(\mathbf{x}) \to \hat{\psi}(\mathbf{x}) \exp i\theta$, $\hat{\psi}^+(\mathbf{x}) \to \hat{\psi}^+(\mathbf{x}) \exp(-i\theta)$ of the field operators.

We introduce into the Hamiltonian an additional noninvariant term of the form $h^+ \int d\mathbf{x} \hat{\psi}(\mathbf{x}) + h \int d\mathbf{x} \hat{\psi}^+(\mathbf{x})$, where h is an arbitrary complex number; the field ψ is assumed to be spinless. For a real system $h = h^+ = 0$, but first it is convenient to assume that h is an independent variable.

The natural variables x_i of the form $-\beta\mathbf{H}$ for the Hamiltonian with noninvariant addition are $-\beta$, $\beta\mu$, $-\beta h^+$, and $-\beta h$, which are respectively conjugate to the specific averages of the energy operator $\mathbf{H}'_0 + \mathbf{V}$, the particle number operator $\mathbf{N}$, and the additions $\int d\mathbf{x} \hat{\psi}(\mathbf{x})$ and $\int d\mathbf{x} \hat{\psi}^+(\mathbf{x})$. The last two variables, which we denote as λ, λ^+ since they are obviously conjugates of each other, are the analog of the magnetization, while the variables h, h^+ are the analog of the external field. The function W is invariant under gauge transformations of h, h^+, and its Legendre transform is invariant under gauge transformations of λ, λ^+. For a real problem with $h = h^+ = 0$ the varied function Φ is also invariant, and when there is spontaneous symmetry breaking the solution is degenerate in the phase of λ and the stationarity points form a circle in the plane of the complex variable λ (and if Φ is convex, it is a complete circle).

Superconductivity can be studied in a completely analogous manner, only now it is necessary to add to the Hamiltonian noninvariant terms of the type $h^+ \int d\mathbf{x} \hat{\psi}(\mathbf{x})\hat{\psi}(\mathbf{x}) + \text{h. c.}$

As a final example, let us consider a nonideal classical gas in zero external field. According to the expressions of Sec. 5.3, the partition function Z of the grand canonical ensemble has the form

$$Z = \sum_{N=0}^{\infty} \frac{1}{N!} \int \ldots \int \prod_i \frac{d\mathbf{p}_i d\mathbf{x}_i}{(2\pi)^d} \exp\left[-\beta\mathcal{H}_N + \beta\mu N\right],$$

where $\mathcal{H}_N$ is the classical Hamiltonian for N particles and μ is the chemical potential. If we take $x_1 = -\beta$ and $x_2 = \beta\mu$, the conjugate variables will be α_1, the specific average energy, and α_2, the specific average particle number, i.e., the density. The meaning of the phase transition is most easily understood in the language of the incomplete Legendre transform $\Gamma(x_1, \alpha_2)$, which is a function of temperature and density. For given values of the temperature and chemical potential the equilibrium value of the density is found by solving the equation $\partial\Gamma/\partial\alpha_2 = -x_2$ for α_2. It turns out that at high temperatures the equation has a unique solution, and at some critical temperature the solution breaks up into branches corresponding to a liquid and a gas. In contrast to all the preceding examples, the degeneracy of the solution is not related to the spontaneous breakdown of any symmetry.

6.1.3 The infinite-volume limit

In this section we shall discuss, with no attempt at mathematical rigor, the infinite-volume limit or, as it is often called, the thermodynamical limit.

As mentioned above, for reasonable assumptions the functions W, Γ, and Φ for a theory in a finite volume are rigorously convex, and there is a one-to-one correspondence between the conjugate variables x and α. This makes spontaneous symmetry breakdown impossible, so that the averages α for a theory in a finite volume always satisfy all the requirements following from the symmetry of the problem. For example, the magnetization of a finite lattice of spins in zero external field will always be zero. We remind the reader that here we are dealing with equilibrium statistics: real finite magnets exist in states which are not equilibrium states in the strict sense of the definition.

If the limits for $V \to \infty$ of the specific logarithm of the partition function W and the averages α exist, the appearance of anomalous solutions in the $V \to \infty$ limit can be explained as follows: the limit function $W(x)$ is finite, but is not everywhere smooth as a function of the parameters x. At certain singular points the first derivatives of W with respect to x become multivalued, i.e., the limits of the first derivatives are different when the singular point is approached from different directions. The anomalous averages α will be exactly these nonunique values of the derivatives of the limit function $W(x)$ which do not coincide with the single-valued limits of the derivatives (6.1) of the function $W(x)$ for finite volume.

For example, for the Ising ferromagnet such a singular point will be the value $\beta h = 0$ (zero field) at temperatures below the critical value; the derivatives of the limit function W with respect to the reduced external field βh will tend to two different (differing by a sign) finite limits as the point $\beta h = 0$ is approached from $h > 0$ and from $h < 0$. These two different limits correspond to the two possible values of the spontaneous magnetization in zero field, while the limit of the magnetization of the finite lattice in zero field is obviously equal to zero (as the limit of zero).

Let us now discuss the mechanism by which anomalous solutions appear in the infinite-volume limit in the language of the variational problem. For a theory in a finite volume the functions Γ and Φ are rigorously convex, so their limits for $V \to \infty$ will also be convex. However, since the convexity of the limit functions is not necessarily rigorous, anomalous solutions can appear in the limit theory: the desired values of the averages $\alpha(x)$ are defined as stationarity points of the limit function $\Phi(\alpha; x)$, and a nonrigorously convex function can have more than one stationarity point. As shown earlier, the set of all stationarity points of a convex (even a nonrigorously convex) function is convex, and the function itself is constant on this set.

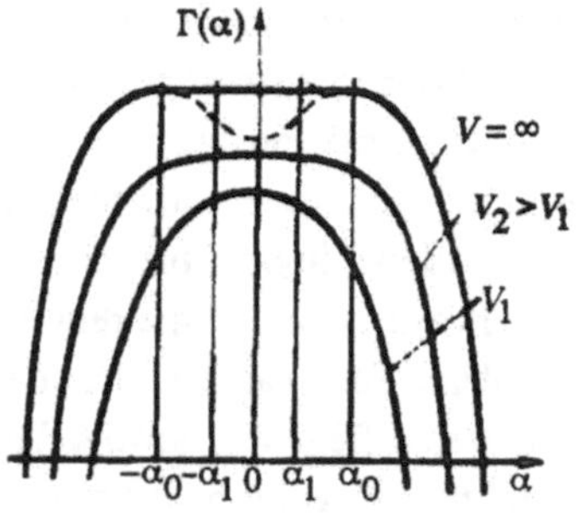

The change of the function Γ as $V \to \infty$ is shown schematically in the figure for the case of an even function of a single variable α. An example of such a function is the Legendre transform with respect to the reduced external field in the Ising model, viewed as a function of the magnetization α at a fixed temperature below the critical value. The same picture is also applicable to the Heisenberg model if α is interpreted as the length of the magnetization vector and its values are restricted to the semiaxis $\alpha > 0$.

Therefore, in the language of the variational problem, degeneracy of a solution in the infinite-volume limit becomes possible because the Legendre transform ceases to be rigorously convex, and flat segments, plateaus, appear on the surface Γ. At the corresponding values of the parameters x (in the figure, for $d\Gamma/d\alpha = -x = 0$) we obtain not a stationarity point $\alpha(x)$, but an entire convex region of stationarity $\mathcal{O}_x$. In the figure the segment $|\alpha| \leq \alpha_0$ is such a region. Its end points correspond to "pure states" with spontaneous magnetization $\pm\alpha_0$, and its interior points can be associated with statistical mixtures of the two pure states with definite direction of the spontaneous magnetization (see the following sections for more detail).

Looking ahead, let us compare the picture described above of the behavior of the function Γ obtained from general considerations with the results of an approximate calculation of this function, for example, using the self-consistent field method. The approximate calculations will be discussed in detail below; here we wish only to note that, as a rule, all simple approximations give for the function Γ a well rather than a plateau, as shown by the dashed line in the figure. The well edges (the intervals $\alpha_1 \leq |\alpha| \leq \alpha_0$ in the figure), inside which the approximate function still has the correct convexity, are referred to as *regions of metastability*, and the interior of the well, where the approximate function has the incorrect convexity, is referred to as the *region of absolute instability*. In the theory of condensation the metastable segments correspond to regions of supercooling and superheating, and in magnetic systems they correspond to hysteresis loops.

In fact, such behavior of Γ is impossible in an equilibrium theory and is always the result of the approximate nature of the calculations. The inclusion of corrections to the simple self-consistent field approximation will lead to a gradual filling in of the well; in addition, as the accuracy of the approximation increases, the behavior of the approximate function in the vicinity of the well becomes more and more complicated (multiple oscillations arise), so that the actual classification in terms of convexity loses meaning. In practice, after the function Γ is calculated approximately, it should simply be replaced by a convex envelope (which leads, in particular, to the well known Maxwell equal-area rule [8] in condensation theory).

In conclusion, we recall that for liquids and gases the quantity W in the limit $V \to \infty$ determines the pressure of the medium. In fact, the pressure p_V for a system in volume V is given by the well known expression $\beta p_V = \partial \ln Z_V / \partial V$ [37]. If the limit $W \equiv \lim V^{-1} \ln Z_V$ exists and $V^{-1} \ln Z_V = W + O(V^{-1/3})$ (the surface correction), then $W = \beta p$, where p is the pressure of a medium of infinite extent.

6.1.4 Singular and critical points

In this section we shall give a more detailed discussion of the properties of the surface $\Gamma(\alpha)$ in the invinite-volume limit and introduce some concepts useful for describing phase transitions.

For a theory in a finite volume it is natural to assume that the function $W(x)$ is infinitely differentiable with respect to the parameters x, because its multiple derivatives reduce to various Green functions (see Sec. 5.1.10) and in a "good" theory they must be finite. Smoothness and rigorous convexity of W ensure smoothness of its Legendre transform.

When anomalous solutions appear in the infinite-volume limit, the functions W and Γ lose the property of infinite differentiability: according to the viewpoint adopted in the preceding section, the limit function $W(x)$ has discontinuities in the first derivatives at several singular values of x, which corresponds to the appearance of flat segments on the limit surface $\Gamma(\alpha)$. At the point where the plateau is reached the first derivatives of $\Gamma(\alpha)$ remain continuous [because, according to (6.2), these derivatives determine a value of x common to all the anomalous solutions], but several of the second derivatives of Γ must be discontinuous owing to the abrupt change in the curvature of the surface Γ. In the rest of our discussion we shall assume that there are no other violations of the smoothness of Γ, owing to the "smooth as possible" assumption: everything which can remain smooth in the infinite-volume limit does.

Let n be the total number of real variables x (and α), and D_x and D_α be the regions of variation of the variables x and α. For each $x \in D_x$, the corresponding values of α are defined as a stationarity point of the convex function $\Phi(\alpha; x)$. If a given x corresponds not to a single point but to an entire region of stationarity $\mathcal{O}_x$, we shall say that the point x is *singular*; the closure (in the topology $D_x \cap \mathbf{R}_n$) of the set of singular points x will be denoted M_x and called a *singular region* in D_x. Each point x not belonging to M_x will be called *nonsingular*; for any nonsingular x a stationarity point of the function $\Phi(\alpha; x)$ is unique, and we denote it $\alpha(x)$. The nonsingular points form an open set in the topology $D_x \cap \mathbf{R}_n$, i.e., for each nonsingular point it is possible to find a sufficiently small neighborhood wholly lying inside the region of nonsingular points.

Each of the regions $\mathcal{O}_x$ is convex and closed (the latter property follows from the assumed continuity of the first derivatives of Γ). It is also clear that the regions $\mathcal{O}_x$ for diferent x do not intersect: from (6.2) we see that a single point α cannot be a stationarity point of two functions $\Phi(\alpha; x)$ with different x. As the point x moves inside the singular region M_x the corresponding region $\mathcal{O}_x$ also moves, changing its shape and location. Since the regions $\mathcal{O}_x$ for close but different x do not intersect, it is clear that the region $\mathcal{O}_x$ must have fewer dimensions than the space of all variables α (i.e., fewer than n).

We shall call the union of the regions $\mathcal{O}_x$ for all $x \in M_x$ a *singular region* in D_α and denote it M_α. When x runs through M_x, the corresponding region $\mathcal{O}_x$ runs through M_α. If the motion is smooth, as we assume it is, the size (number of dimensions) of the region M_α is equal to the sum of the sizes of the region $\mathcal{O}_x$ and the region M_x, i.e., $\dim M_x + \dim \mathcal{O}_x = \dim M_\alpha \le n$.

For all the systems studied in Sec. 6.1.2 it turns out that $\dim M_\alpha = n$. For the Ising model and the classical gas $n = 2$, and the region of stationarity $\mathcal{O}_x$ is the straight-line segment connecting the two extremal stationarity points. Therefore, $\dim \mathcal{O}_x = 1$ and $\dim M_x = 1$, i.e., the singular points in D_x form a line. For the Heisenberg ferromagnet $n = 4$ (the temperature and the three-vector of the external field), and when there is degeneracy the region $\mathcal{O}_x$ is a three-sphere, so that the singular points in D_x also form a line. For a quantum gas $n = 4$ (the temperature, the chemical potential, and an additional complex parameter playing the role of an external field), and when there is degeneracy the region $\mathcal{O}_x$ is a two-sphere—a circle in the complex plane, so that $\dim M_x = 2$, i.e., the singular points form a two-dimensional manifold in D_x (a region in the temperature—chemical potential plane for zero external field).

Let us now define the concept of a *critical point*. When x moves along a singular region M_x the size of the corresponding region $\mathcal{O}_x$ changes continuously. When some points $x^c \in M_x$ are approached from M_x the size of $\mathcal{O}_x$ can tend to zero, i.e., $\mathcal{O}_x$ can shrink to a point; these points x^c will be called critical points. It usually turns out that the critical points lie on the boundary of M_x. For example, if M_x is a line ending somewhere inside D_x, the endpoint is a critical point. The second end of the singular line usually lies on the boundary of D_x (for example, at zero temperature), and this endpoint will not be a critical point.

6.1.5 Description of phase transitions

We shall assume that the point x moves along a smooth trajectory in D_x and that at some point $\bar{x}$ the trajectory intersects a singular region M_x (we recall that $\dim M_x < \dim D_x$). All the points on the trajectory except $\bar{x}$ are nonsingular, so each of them uniquely corresponds to a stationarity point $\alpha(x)$; the singular point $\bar{x}$ corresponds to a convex closed region of stationarity $\mathcal{O}_{\bar{x}}$. When the singular point $\bar{x}$ is approached from either side, the corresponding trajectories $\alpha(x)$ lying outside the singular region M_α approach two, in general different, points of the boundary of $\mathcal{O}_{\bar{x}}$ (the trajectory cannot fall directly inside $\mathcal{O}_{\bar{x}}$ if the singular region M_α, which is the union of nonintersecting regions $\mathcal{O}_x$, has total dimension n). At the instant of passing through the singular point $\bar{x}$ the corresponding stationarity point "jumps over" the region $\mathcal{O}_{\bar{x}}$. It is said that a *first-order phase transition* occurs at the point $\bar{x}$.

We now assume that the x trajectory from the region of nonsingular points arrives at the critical point x^c and then moves along the region of singular points M_x. Before reaching x^c the corresponding trajectory $\alpha(x)$ is single-valued; at the point x^c the trajectory $\alpha(x)$ branches, because each of the subsequent $x \in M_x$ corresponds to an entire region of stationarity $\mathcal{O}_x$. It is said that a *second-order phase transition* occurs at x^c.

Points which are not interior points for any line segment lying wholly in $\mathcal{O}$ are called the extremal points of a convex region $\mathcal{O}$. All extremal points lie on the boundary of $\mathcal{O}$, but not every point of the boundary is necessarily extremal: for example, if $\mathcal{O}$ is a triangle, only its vertices will be extremal points. We note that a region of stationarity of this type can appear in real physical problems. For example, for an anisotropic Heisenberg ferromagnet with exchange interaction invariant under rotations by angle $2\pi/3$ about the 3 axis, the region of stationarity will be not a sphere, as for an isotropic interaction, but an equilateral triangle.

From the physical point of view, the extremal points of a region of stationarity $\mathcal{O}_x$ correspond to pure states, and the interior points correspond to statistical mixtures of such states. Under the conditions of a specific experiment, for a given $x \in M_x$ a particular state is realized which corresponds to one of the points of the region $\mathcal{O}_x$, but the choice of this point is determined by external conditions (the experimental setup, the preparation procedure) and cannot be made within the framework of the equilibrium theory itself. To explain this, let us briefly discuss the experimental setup for various systems.

In the case of a ferromagnet, the independent variables which can actually vary are the temperature T and the external magnetic field h. The singular points form the line $h = 0$, $0 \leq T \leq T_c$, and the point $h = 0$, $T = T_c$ is the critical point. A first-order phase transition can be observed by smoothly varying h at a fixed temperature $T < T_c$ such that the value of h passes through zero. At the time of the transition there is an abrupt change of the magnetization (if we ignore the phenomenon of hysteresis, which cannot occur in the equilibrium theory).

A second-order phase transition is observed when the temperature is gradually decreased in zero external field. A spontaneous magnetization appears in passing through the critical value T_c (the Curie temperature). In a real sample relatively large segments—domains—are formed with a definite direction of the spontaneous magnetization, but these directions are chaotically distributed for different domains. When it is said that a pure state of definite magnetization is formed in a ferromagnet for $T < T_c$, it is a single domain that is referred to, not the entire sample.

Let us now turn to the nonideal classical gas. From the formal point of view this system is analogous to the Ising ferromagnet (if the absence of the corresponding symmetry is neglected), but there is a very important difference when an actual experiment is carried out. The point is that for a gas, the real independent variable is not the chemical potential—the analog of the external field—but the conjugate variable: the density, which is the analog of the magnetization. The singular points for the gas form a line $\mu_0(T)$, $T \leq T_c$, in the μ–T plane ending at the critical point $T = T_c$, $\mu = \mu_0(T_c)$. If we were able to smoothly vary the chemical potential μ for $T < T_c$, in passing through the value $\mu_0(T)$ we would observe an abrupt change of the density analogous to the abrupt change of the magnetization of the ferromagnet. However, in practice we can change not μ, but the density (by compressing the gas), and the experimental picture of the phase transition changes: by compressing the gas at a fixed temperature $T < T_c$ we smoothly change its density ρ, which corresponds to a smooth change of the chemical potential and pressure. At some value of the density $\rho_1(T)$ the chemical potential reaches the value $\mu_0(T)$, i.e., x reaches the line of singular points M_x, and the trajectory $\alpha(x)$ reaches the boundary of the region of stationarity $\mathcal{O}$ corresponding to the singular point in T, $\mu_0(T)$. The region $\mathcal{O}$ is a line segment joining two extremal points, and in the subsequent smooth variation of the density the stationarity point moves smoothly along this entire segment. During this motion the chemical potential and pressure remain constant [we recall that the pressure is determined by the value of the function $\Phi(\alpha; x)$ at the stationarity point, and the value of $\Phi(\alpha; x)$ is constant on the region of stationarity $\mathcal{O}_x$]. We note that the fact that the pressure is constant in the region $\mathcal{O}$ makes it possible to depict the line of singular points in pressure–temperature variables, as is usually done.

When the value of the density $\rho_2(T)$ corresponding to the second end of $\mathcal{O}$ is reached, the stationarity point moves outside $\mathcal{O}$ to the region of nonsingular points, and the chemical potential and pressure again begin to vary smoothly.

Therefore, in this case the stationarity point does not jump over the region $\mathcal{O}$, but smoothly moves through it. The endpoints correspond to pure states—a liquid and gas, and the interior points correspond to statistical mixtures of these states. Such a mixture is in practice realized as a two-phase system with the ratio of the volumes of the liquid and the gas determined by the density.

Let us now discuss phase transitions in a quantum Bose or Fermi gas. These have a feature in common with the Heisenberg ferromagnet: the degeneracy of the solution in both cases is continuous (the phase of the anomalous average in a superconductor or a superfluid is analogous to the direction of spontaneous magnetization of the ferromagnet). However, also here there is a significant difference between these cases in an actual experiment. For a quantum gas we have introduced four parameters: the temperature, the chemical potential, and the complex "external field," but in fact the only independent variable which can be varied continuously is the temperature. The external field introduced artificially into the Hamiltonian is zero for a real system, and the chemical potential also cannot be considered a real independent variable, because the quantity conjugate to it, the density ρ, is usually fixed in an experiment. The auxiliary condition ρ = const isolates from the two-dimensional manifold of singular points in the μ–T plane a singular line, the endpoint of which is the critical point. In an experiment it is only possible to observe the second-order phase transition analogous to the transition occurring in a ferromagnet in zero external field.

In conclusion, the reader should be warned that the phenomenology of phase transitions described above is not generally accepted. Phase transitions are usually related to singularities in thermodynamical functions, but it seems to us that the language of the variational principle and convex functions is clearer and more universal.

6.1.6 *Critical and Goldstone fluctuations*

As already mentioned several times, the varied function Φ is constant on each region of stationarity $\mathcal{O}_x$. From this it follows that at any interior point $\alpha \in \mathcal{O}_x$ the second (and other) variations of Φ are equal to zero for all directions of variations $\delta\alpha$ which do not lead outside $\mathcal{O}_x$. Owing to the convexity of the region $\mathcal{O}_x$, the set of such directions is identical for all its interior points, and we denote it L_x; L_x is some linear subspace of the same dimension as $\mathcal{O}_x$ showing the orientation of the region $\mathcal{O}_x$ in the space of the variables α.

We shall say that the surface Φ has zero curvature along the directions of L_x at any interior point $\mathcal{O}_x$. The same can be said about the surface Γ, since the functions Φ and Γ differ by a term linear in α.

Let us now consider a point α lying on a smooth segment of the boundary of the region $\mathcal{O}_x$ (for example, not at a vertex if $\mathcal{O}_x$ is a triangle). Among the vectors of L_x it is possible to distinguish one pointing along the normal to the boundary of $\mathcal{O}_x$; all the vectors of L_x orthogonal to it point in directions tangent to the boundary of $\mathcal{O}_x$. It makes sense to speak of tangent vectors only when $\dim\mathcal{O}_x > 1$. For a one-dimensional region $\mathcal{O}_x$ which is a line segment (the Ising model and the classical nonideal gas), L_x contains a single direction which should be assumed to be normal to the boundary.

For all the systems studied in Sec. 6.1.2, the curvature of the surface Γ undergoes a finite jump at the points where the flat segment of $\mathcal{O}_x$ starts, because the second derivatives along the direction of the normal to the boundary undergo a finite jump at the boundary. The size of this jump decreases in approaching the critical point x^c, and at the point x^c itself the jump vanishes. Meanwhile, the second derivatives along the directions tangent to the boundary are continuous at the boundary; from continuity and the fact that these derivatives are zero inside $\mathcal{O}_x$, it follows that the derivatives along the directions in L_x tangent to the boundary tend to zero also in approaching an extremal point of $\mathcal{O}_x$ from the side of the nonsingular points.

Therefore, at extremal points of $\mathcal{O}_x$ the curvature of the surface Γ along the directions of L_x tangent to the boundary vanishes (the "Goldstone zero"); the directions of zero curvature are determined by the choice of region $\mathcal{O}_x$, i.e., by the choice of x, and by the choice of the point α on the boundary of $\mathcal{O}_x$.

As mentioned above, when a singular point x tends to x^c, the jumps of the second derivatives along the direction of the normal to the boundary of $\mathcal{O}_x$ tend to zero for any point on the boundary, so that at the critical point itself these derivatives also vanish. It is clear from this that at the critical point the surface Γ has zero curvature along all directions of L_{x^c} (a "critical zero"). Of course, we assume that for $x \to x^c$ the space L_x characterizing the orientation of the region $\mathcal{O}_x$ turns continuously, tending to some limit L_{x^c}.

Let us now show that the curvature of the surface Γ determines the size of the fluctuations in the system. We use Γ'' to denote the matrix of second derivatives of Γ at some point α^0 and W'' to denote the matrix of second derivatives of W at the corresponding point x^0 (α^0 determines x^0 uniquely, but the reverse is not true in general). Let e be a vector in whose direction the surface Γ at the point α^0 has zero curvature. This means that the average of the matrix Γ'' over the vector e is equal to zero. Since the matrix Γ'' is of fixed sign ($\Gamma'' \leq 0$), it can be stated that e is an eigenvector of it with zero eigenvalue, and the set of all vectors orthogonal to e is an invariant subspace of the matrix Γ'', i.e., the off-diagonal matrix elements between this subspace and e are equal to zero. Furthermore, according to (6.3), the matrices Γ'' and W'' are the inverses of each other up to a sign, so that e will also be an eigenvector of W'' with eigenvalue $+\infty$. It would be more rigorous to state this as follows: when x tends to x^0 from the region of nonsingular points, one of the eigenvectors of the positive-definite matrix W'' tends to e, and the corresponding eigenvalue tends to infinity.

In analyzing the convexity of W in Sec. 5.1.10, we saw that the second variation of W along the direction e is proportional to the "temperature dispersion" (5.41) of the operator $\mathbf{a}_e = \Sigma_i e_i \mathbf{a}_i$, where $\mathbf{a}_i$ are operators whose specific averages are the variables α_i. If all the operators $\mathbf{a}_i$ commute with each other, the temperature dispersion coincides with the ordinary dispersion. Therefore, the infinity of the second variation of W is usually referred to as a *fluctuation infinity*.

From this analysis of the curvature of the surface Γ we can conclude that in those cases where the region of stationarity $\mathcal{O}_{\bar{x}}$ has more than one dimension, infinite *Goldstone fluctuations* arise in the system when the singular point $\bar{x}$ is approached from the region of nonsingular points x. The directions for which the fluctuations become infinite are determined by the point of the boundary of $\mathcal{O}_{\bar{x}}$ at which the trajectory $\alpha(x)$ arrives from the region of nonsingular points. This trajectory is, in turn, determined by the trajectory along which the point x approaches $\bar{x}$.

Goldstone fluctuations are usually related to the spontaneous breakdown of some continuous symmetry group. In this case the boundary of $\mathcal{O}_x$ consists of points obtained by group shifts of any single boundary point, and the directions of the tangents to the boundary coincide with the directions of the group shifts. We note that the vanishing of the curvature of the surface Γ along the directions of the group shifts at the stationarity points of Φ (in particular, on the boundary of $\mathcal{O}_x$) is now an automatic consequence of the symmetry of the function Φ, so that the assumption that the second derivatives are continuous along directions tangent to the boundary of $\mathcal{O}_x$ is certainly valid. However, it should be noted that the spontaneous breakdown of a continuous symmetry is a sufficient but not at all necessary condition for the appearance of infinite Goldstone fluctuations, though we do not have any example of a particular system which manifests Goldstone fluctuations without spontaneous symmetry breaking.

At a critical point x^c the fluctuations in all the directions of L_{x^c} become infinite: these are *critical fluctuations*. For the Ising model and the classical nonideal gas with one-dimensional region $\mathcal{O}$ only critical fluctuations are possible; L_{x^c} contains only a single direction, and the corresponding second derivative of W is interpreted as the susceptibility for a magnet and the compressibility for a gas. These quantities become infinite at the critical point.

The second derivatives of W along directions orthogonal to L_{x^c} could in principle remain finite, but for all real systems it for some reason turns out that at the critical point these derivatives also become infinite (for the Ising model and the gas this corresponds to infinite specific heat). No universal and convincing explanation of the cause of this phenomenon has been found so far.

The phenomenological similarity hypothesis or scaling has become very popular in recent years [42,43,47]. This hypothesis amounts to a general statement about the behavior of thermodynamical functions of the type $W(x)$ near a critical point which is in agreement with a large amount of experimental data (although there are some exceptions; see the lecture by M. Fisher in Ref. 48). The second important idea, also relatively recent, is that of the universality of critical behavior: the thermodynamical functions of systems with the same number of variables, for example, the Ising magnet and the classical gas, behave practically identically near a critical point.

All of these general regularities in critical behavior discovered experimentally still lack a clear and convincing explanation at the level of the microscopic theory.*

---

*This is done in the modern RG-theory of critical behavior [88].

With this we conclude our discussion of phase transitions and the variational formulation in thermodynamics. Of course, we have only touched upon this interesting and broad topic, but its only relevance to us is the fact that it involves all the ideas of the variational principle. When discussing the more complicated functional versions of this principle below, we shall pay particular attention to the purely technical aspects, and assume that the general meaning is now clear.

6.2 Legendre Transformations of the Generating Functional of Connected Green Functions

6.2.1 *Functional formulations of the variational principle*

Let us consider field theory or quantum statistics of the field type, for now without regard to ferromagnets and the classical gas. In the cases we are interested in the dynamics of a system is determined by the action functional $S(\varphi)$, which consists of the free part and the interaction. We shall use universal notation and assume that the unified field $\varphi(x)$ is bosonic. This limitation is not important: to generalize to fermions one needs only to keep track of the type (left, right) and ordering of the derivatives.

Let $S'(\varphi)$ be the action functional of the theory we are interested in and $S''(\varphi)$ be an addition to the action, the time integral of the real Lagrangian (4.18). The symmetric "potentials" $A_n(\mathbf{x}_1\ldots\mathbf{x}_n)$ entering into this Lagrangian will henceforth be treated as independent variables playing the role of the numerical parameters x in thermodynamics. As usual, we shall consider only variations of the potentials A which do not spoil the reality of the Lagrangian (4.18).

In field theory the analog of the thermodynamical function $W(x)$ will be the functional $\varepsilon(A)$ defined in Sec. 4.3.2, the shift of the ground-state energy due to the addition S'' in the action. We recall that in the pseudo-Euclidean theory this functional is determined by the equation $\ln G(A) = -i\varepsilon(A)\int dt$, in which $G(A) = \text{const}\int D\varphi \exp iS(\varphi)$, where $S \equiv S' + S''$ is the total action including the addition, and the normalization constant is fixed by the condition $G(0) = 1$. In the Euclidean version of the theory the same functional $\varepsilon(A)$ is determined by the equation $\ln G_e(A) = -\varepsilon(A)\int dt$, in which $G_e(A) = \text{const}\int D\varphi \exp S_e(\varphi)$ with the usual normalization, and $S_e = S'_e + S''_e$ is the Euclidean action functional ($S''_e = S''$).

For quantum statistics at finite temperature the analog of $\varepsilon(A)$ will be the functional $\Omega(A)$ defined in Sec. 5.1.9, interpreted as the shift of the thermodynamic potential owing to the addition S''. Let us recall its definition: $\ln G_\beta(A) = -\beta\Omega(A)$, where $G_\beta(A) = \text{const}\int D\varphi \exp S_\beta(\varphi)$ with the usual normalization, and $S_\beta = S'_\beta + S''_\beta$ is the total temperature action, i.e., the Euclidean action for the time interval $[0, \beta]$. When the temperature goes to zero, $\beta \to \int dt$ and $\Omega(A) \to \varepsilon(A)$.

As shown in Secs. 4.3.2 and 5.1.9, ε and Ω are convex-upwards functionals of the potentials A.

In changing over to the variational problem one introduces variables α conjugate to the potentials A. In field theory $\alpha_n(\mathbf{x}_1 \ldots \mathbf{x}_n) = \delta\varepsilon(A)/\delta A_n(\mathbf{x}_1 \ldots \mathbf{x}_n)$. Using the general rules for the relation between the functional and operator constructions (Secs. 1.3.3 and 1.6.5), it is easily verified that in field theory

$$\alpha_n(\mathbf{x}_1 \ldots \mathbf{x}_n) = -\frac{1}{n!}(0|\,\hat{\varphi}(t, \mathbf{x}_1)\ldots\hat{\varphi}(t, \mathbf{x}_n)|0), \tag{6.5}$$

where $|0)$ is the ground state of the full Hamiltonian of the theory with action $S = S' + S''$ and $\hat{\varphi}$ is the field operator of this theory. The vacuum expectation value (6.5) is actually independent of the time, so the field operator can be used in any representation. In quantum statistics the average would be obtained rather than the vacuum expectation value (6.5).

All the equations of Sec. 6.1.2 have an obvious generalization to the case under consideration: functions are replaced by functionals and ordinary derivatives are replaced by variational ones. An expression analogous to (6.3) is valid for the Legendre transform

$$\Gamma(\alpha) = \varepsilon(A) - \sum_n \int \ldots \int d\mathbf{x}_1 \ldots d\mathbf{x}_n A_n(\mathbf{x}_1 \ldots \mathbf{x}_n)\, \alpha_n(\mathbf{x}_1 \ldots \mathbf{x}_n), \tag{6.6}$$

which shows that the functional $\Gamma(\alpha)$ is convex downwards. Stationarity equations analogous to (6.2), $\delta\Gamma/\delta\alpha_n = -A_n$, determine the unknown α in terms of the known A. The original theory with action S' corresponds to the point $A = 0$ in the space of potentials. If this point is singular in the sense of the definition in Sec. 6.1.4, the solution of the variational problem is degenerate, and instead of a stationarity point α we obtain an entire convex region of stationarity $\mathcal{O}$ with all the attendant consequences. In particular, each continuous degeneracy of a solution is accompanied by the appearance of infinite Goldstone fluctuations. From the spectral representation (4.21), we see that the second variation of the functional $\varepsilon(A)$ becomes infinite when the Hilbert space of states of the corresponding quantum-mechanical theory contains states which are different from the ground state but have the same or arbitrarily close energy. This means that the ground state cannot correspond to an isolated, nondegenerate level of the Hamiltonian: either there is degeneracy, or the level is not isolated, i.e., there is no gap in the Hamiltonian spectrum separating the ground state from all the other states [49]. This statement is a careful formulation of the Goldstone theorem, well known in the relativistic theory [50]. This theorem states that every spontaneous violation of a continuous symmetry is accompanied by the appearance of massless particles in the theory. Spontaneous violation is necessary only inasmuch as it automatically entails a continuous degeneracy, and a massless particle is the relativistic version of a gapless excitation. The analog of the Goldstone theorem for quantum statistics at finite temperature was obtained in [51].

The essential feature of the variational principle described above was the fact that the potentials and the variables conjugate to them depended only on the coordinates $\mathbf{x}_i$ and not on the time. Now let us turn to the formulation of the variational principle with potentials of the most general form. We introduce the functional

$$A(\varphi) = \sum_{n=1}^{\infty} \frac{1}{n!} A_n \varphi^n \equiv$$

$$\equiv \sum_{n=1}^{\infty} \frac{1}{n!} \int \ldots \int dx_1 \ldots dx_n A_n (x_1 \ldots x_n) \varphi(x_1) \ldots \varphi(x_n) \qquad (6.7)$$

with arbitrary symmetric potentials $A_n(x_1 \ldots x_n)$. In the pseudo-Euclidean theory we shall assume that $A(\varphi) = iS(\varphi)$, where $S(\varphi)$ is the action functional, and in the Euclidean theory and quantum statistics we shall assume that the action functional is $A(\varphi)$ itself, so that in all cases the functional integrals are written with the weight $\exp A(\varphi)$. For the finite-temperature statistics the fields φ and the potentials A_n should, of course, be considered as functions which are periodic in each of the time arguments, and should be integrated only over the interval $[0, \beta]$ in these arguments. We shall consider only bosonic theories. When generalizing to fermions, potentials with even subscript should be considered bosonic quantities, and ones with odd subscript should be considered fermionic.

We define the functionals $G(A)$ and $W(A)$ by the equation

$$G(A) = \exp W(A) = \text{const} \int D\varphi \exp A(\varphi) . \qquad (6.8)$$

The numerical factor in front of the integral fixes the normalization of G.

The quantities $G(A)$ and $W(A)$ are the generating functionals of the full and connected Green functions, respectively, for the theory with action $A(\varphi)$ (Euclidean version) or $-iA(\varphi)$ (pseudo-Euclidean version). These Green functions are multiple derivatives of the corresponding functional with respect to the potential $A_1(x)$.

The functional W will henceforth play the same role as the function $W(x)$ in thermodynamics. The variable conjugate to the potential A_n is

$$\alpha_n(x_1...x_n) = \delta W(A)/\delta A_n(x_1...x_n) = \\ = (1/n!)\langle\varphi(x_1)...\varphi(x_n)\rangle, \tag{6.9}$$

where $\langle...\rangle$ denotes the functional average with weight $\exp A(\varphi)$. In field theory this average corresponds to the full Green function without vacuum loops (the function H_n of Secs. 1.3.1 and 1.4.9), and in quantum statistics it is the temperature Green function without vacuum loops (5.1).

The transformation with respect to the first m potentials $A' \equiv \{A_1... A_m\} \to \alpha \equiv \{\alpha_1...\alpha_m\}$ will be called the Legendre transform of order m:

$$\Gamma(\alpha;A'') = W(A) - \sum_{k=1}^{m} A_k\delta W/\delta A_k = W(A) - \sum_{k=1}^{m} A_k\alpha_k, \tag{6.10}$$

where A'' is the set of "higher" potentials not involved in the transformation, and the product $A_k\alpha_k$ is, of course, understood as contraction in all the arguments x. Equations (6.2) take the form

$$\delta\Gamma/\delta\alpha_k = -A_k,\ k = 1, 2, ..., m, \tag{6.11}$$

and their solutions $\alpha(A)$ are the stationarity points of the functional

$$\Phi(\alpha;A) = \Gamma(\alpha;A'') + \sum_{k=1}^{m} A_k\alpha_k, \tag{6.12}$$

taking the value $W(A)$ at the stationarity point.

The other expressions of Sec. 6.1.2 are generalized just as easily. In what follows we shall need the analogs of (6.3) and (6.4):

$$\sum_{s=1}^{m} \frac{\delta^2\Gamma}{\delta\alpha_i\delta\alpha_s}\cdot\frac{\delta^2 W}{\delta A_s\delta A_k} = -\delta_{ik};\ \frac{\delta\Gamma}{\delta A_n} = \frac{\delta W}{\delta A_n} \text{ for } n > m. \tag{6.13}$$

The product of second derivatives is now understood as a contraction. As before, the matrices of second derivatives of Γ and W with respect to the arguments α and A' are respectively the inverses of each other, but now they must be understood as linear integral operators on the space of "columns" of symmetric functions with increasing number of arguments x, and the unity on the right-hand side of the first of Eqs. (6.13) should be understood as the unit operator on this space.

In contrast to the earlier formulations of the variational principle, these functionals in general are no longer convex. Convexity is ensured only in quasiprobabilistic Euclidean theories (Secs. 4.3.1 and 5.1.9), in other words, the second variation of $W(A)$ can be assumed positive-definite at those points of the space of potentials A for which $D\varphi \exp A(\varphi)$ can be treated as positive measure on the manifold of fields φ and integrals with weight $\exp A(\varphi)$ "converge."

The absence of convexity is compensated for by one very important advantage of the Legendre transform (6.10): the relative simplicity of constructing iteration diagrammatic expansions for these functionals. This is very important, because from the practical point of view Legendre transforms are needed in using the variational principle to find anomalous, i.e., noniteration, solutions for $W(A)$, and this presupposes the ability to construct expressions for Γ which, even if approximate, are explicit. Of course, we do not assume that the functional Γ can be determined exactly, but this is not necessary: after constructing Γ in some approximation, we can still hope to find an anomalous solution for the Green functions α (stationarity points). However, for this the stationarity equations must be solved exactly, not iteratively. In practice, this procedure is quite efficient: all the presently known anomalous solutions have been or can be obtained exactly in this manner, and in all cases it is necessary to use only the lowest-order approximation of the varied functional. We note that we have already discussed this procedure in detail for the special case of the first Legendre transform, i.e., for the transform with respect to the single variable A_1, which can be used in the variational method to find the Green function α_1, the average value of the field. The diagrammatic representation for this functional was constructed in Sec. 1.8.3; in all specific models like the Goldstone model [20] the anomalous solution for α_1 is found by taking the functional Γ in first-order perturbation theory, which corresponds to the loopless approximation (1.230).

The relative simplicity of the diagrammatic technique for the functionals (6.10) is explained by the fact that the original variables, the potentials A, enter into the diagrammatic representations of the functional W in a simple way: up to an additive normalization constant, W is the sum of $1/2 \cdot \mathrm{tr} \ln \Delta$ and all connected graphs with the line $\Delta = -A_2^{-1}$ (the bare propagator) and vertices A_n, $n \neq 2$ (see Sec. 1.7.4). The conjugate "dressed" variables α_k are represented just as simply by diagrammatic series with bare lines and vertices. Solving these equations iteratively for the first m bare variables $A' \equiv \{A_1 \ldots A_m\}$, we can write them as infinite series of skeleton graphs with dressed lines and vertices with up to m legs and bare vertices A_n, $n > m$.* Substitution of the diagrammatic series obtained for the first potentials A' into the right-hand side of (6.10) leads to the desired diagrammatic representation of Γ. De Dominicis and Martin used this direct method to analyze the diagrammatic representations of the first, second, third, and fourth Legendre transforms [52].

In this text we shall construct the diagrammatic representations of Γ in a different way, namely, we first obtain the complete system of equations in variational derivatives (the equations of motion) for the Γ functionals, and then we construct the desired diagrammatic expansions as the iteration solutions of the system of equations of motion. This procedure considerably simplifies the combinatorial analysis of the diagrams of Γ; it is also important that the complete system of equations of motion for Γ determines this functional independently of its diagrammatic expansion, just as the Schwinger equations determine the Green functions. We note that such a procedure cannot be followed for the transformations (6.6), because it is impossible to write the complete system of equations in terms of time-independent potentials.

Let us conclude this section with a brief historical sketch. The technique of functional Legendre transforms was developed almost completely within the framework of statistical physics with the goal of obtaining a technique suitable for describing phase transitions. The first example of the use of this technique was the derivation in the 1930s of the well known virial expansion in the statistics of a classical nonideal gas [7,8] (see Sec. 6.3). The technique of working with anomalous averages like α_1 (the first Legendre transform) was developed by Belyaev [53] for describing a quantum Bose gas when a condensate is present. Objects like the second transform in various variational formulations of quantum statistics were studied by Yang and Lee [54], Luttinger and Ward [55], and de Dominicis [56]. Finally, the fundamental studies of de Dominicis and Martin [52] were the first in which the general scheme of the variational principle for quantum statistics was clearly

* We shall use the term "skeleton" to refer to any graph in which at least one of the elements is dressed. The term "m-skeleton" would be more precise but is not often used.

formulated and the diagrammatic representations of the first four Legendre transforms were explicitly constructed. Jona-Lasinio immediately noted [57] that the variational methods of [52] could also be used in quantum field theory and were very convenient for describing spontaneous symmetry breakdown. Dahmen and Jona-Lasinio [58] were the first to obtain the equation of motion for one of the Legendre transforms, and they attempted to solve this equation noniteratively.

We should note here that the authors of [52,58] started from the operator formulation in defining the various functionals. In this approach it is not completely obvious that the corresponding functional constructions of quantum statistics and field theory are identical, which led the authors of [58] to reformulate the basic equations of [52] in the language of quantum field theory. If, as in the present text, one starts from functional integrals, it is immediately seen that there is no significant difference between statistics and quantum field theory regarding the variation principle.

The method of analyzing the graphs of Γ by means of the equations of motion was first used in [59] for the first transform. The general equations of motion for Legendre transforms of any order were obtained in Refs. [60,61]. In Refs. [62–64] these equations were used to reproduce and generalize the results of [52] for the first four Legendre transforms; in [65] and [66] the diagrammatic representations of transforms of arbitrarily high order were constructed in the same way.

We note that the method developed in [59,62–66] of analyzing Legendre transforms using the equations of motion has also been successfully applied to objects of a different type: Legendre transforms of the logarithm of the S-matrix generating functional, which are discussed in the following section.

6.2.2 *The equations of motion in connected variables [60]*

The starting point is the equations of motion for the functional $G(A)$, which are derived directly from the definition (6.8) and have already been given in Sec. 1.7.3. These are the connection equations (1.194) with $n > 1$ [we have not introduced the trivial term A_0 into the action (6.7)] and the Schwinger equation (1.195), which is linearized in an obvious manner using the connection equations. In terms of the functional $W(A) = \ln G(A)$ the equations of motion take the form

$$A_1 + \sum_{n=2}^{\infty} A_n \frac{\delta W}{\delta A_{n-1}} = 0; \quad \frac{\delta W}{\delta A_n} = \frac{1}{n!} H_n, \quad n > 1 . \tag{6.14}$$

To shorten the notation we shall omit the arguments x_i, but it should be remembered that there is one free argument x in the Schwinger equation, and in the connection equation with given n there are n arguments $x_1 \ldots x_n$. The right-hand sides of the connection equations involve the quantities H_n, which have the meaning of the full Green functions without vacuum loops represented as polynomial forms of the connected Green functions $\beta_k \equiv \delta^k W/\delta A_1{}^k$, $k > 0$. The explicit expressions for the polynomials H_n are given by (1.201).

Our problem is to rewrite the equations (6.14) in terms of the Legendre transform Γ and its variables. We shall assume that $m > 1$, since the first transform was studied in Chap. 1.

For important reasons which we shall immediately explain, it is desirable to treat Γ not as a functional of the variables α corresponding to unconnected Green functions, but as a functional of the equivalent set of first connected Green functions $\beta \equiv \{\beta_1 \ldots \beta_m\}$. The reason is that we want to use the equations of motion to construct diagrammatic representations of Γ, and the vertices and lines will, of course, be determined by the functional arguments of Γ. If we want to have graphs with connected vertices and lines, we must work not with the variables α, even though they are directly conjugate to the potentials A, but with their connected components β.

Therefore, having first defined the functional $\Gamma(\alpha; A'')$ by equations (6.9) and (6.10), we now want to substitute the independent variables $\alpha \to \beta$. It is clear from the definitions of α and β and equations (6.14) and (1.201) that the variables α are expressed in terms of β as

$$\alpha_n = \frac{1}{n!} H_n = \frac{1}{n!} \left(\beta_1 + \frac{\delta}{\delta A_1}\right)^n \cdot 1, \quad \delta\beta_n/\delta A_1 = \beta_{n+1}, \tag{6.15}$$

i.e., $\alpha_1 = \beta_1$, $\alpha_2 = (\beta_2 + \beta_1\beta_1)/2$, and so on. We recall that the first connected function $\beta_1 = \alpha_1$ is the average value of the field, the second function β_2 is the full propagator, and β_k is the connected "k-leg".

Let us derive several expressions which will be useful in making the substitution $\alpha \to \beta$. We introduce the auxiliary field variable $\varphi = \varphi(x)$ and the functionals

$$\alpha(\varphi) = \sum_{n=0}^{\infty} \alpha_n \varphi^n, \quad \beta(\varphi) = \sum_{n=1}^{\infty} \frac{1}{n!} \beta_n \varphi^n. \tag{6.16}$$

The variables α_n are assumed to be defined by (6.15) for any n; in particular, $\alpha_0 = 1$. The entire system of equations (6.15) is obviously equivalent to the following functional equation:

$$\alpha(\varphi) = \exp \beta(\varphi). \tag{6.17}$$

We differentiate this equation with respect to β_k, assuming that the functions β are independent variables and that α are functionals of them:

$$\delta\alpha(\varphi)/\delta\beta_k = \alpha(\varphi)\,\delta\beta(\varphi)/\delta\beta_k = \alpha(\varphi)\,\varphi^k/k!. \tag{6.18}$$

The field φ is an arbitrary parameter, and we can equate the coefficients of each power of φ on the two sides of the equation. This gives

$$\delta\alpha_n/\delta\beta_k = \alpha_{n-k}/k!. \tag{6.19}$$

Here and below it is understood that $\alpha_s = 0$ for $s < 0$.

To avoid misunderstandings, it should be noted that both sides of (6.19) contain $n + k$ arguments x. On the left are n arguments of α_n and k arguments of β_k. The right-hand side should be understood as follows: k arguments of β_k and $k \le n$ arguments of α_n (which ones are not important owing to the symmetry of the α_n) enter into the right-hand side of (6.19) in the form of a product of k δ functions representing the kernel of the unit operator on the space of symmetric functions of k variables. The remaining $n - k$ arguments of α_n are the arguments of the function α_{n-k} on the right-hand side of (6.19).

We shall not explicitly write out the arguments x and the kernels of unit operators unless absolutely necessary; otherwise, the expressions become so awkward that it is difficult to understand anything. We hope that the interested reader will not have any serious difficulty in inserting the arguments, guided by the simple "conservation law": the total number of arguments x must be the same on both sides of the equation.

Differentiating (6.18) with respect to β_s and again equating the coefficients of the φ^n, we obtain

$$\delta^2\alpha_n/\delta\beta_k\delta\beta_s = \alpha_{n-k-s}/k!\,s!. \tag{6.20}$$

The derivatives $\delta\beta/\delta\alpha$ do not have a simple form; their calculation requires inversion of the matrix (6.19). The only inverse derivatives that we shall need in what follows are $\delta\beta_n/\delta\alpha_m$ for $n \le m$. These derivatives are easily calculated owing to the triangularity of the matrix (6.19):

$$\delta\beta_n/\delta\alpha_m = m!\,\delta_{nm} \quad \text{for} \quad n \le m. \tag{6.21}$$

We shall give another useful expression which follows from (6.19):

$$\frac{\delta}{\delta\beta_k} = \sum_{n=1}^{\infty} \frac{\delta\alpha_n}{\delta\beta_k} \cdot \frac{\delta}{\delta\alpha_n} = \frac{1}{k!} \sum_{n=k}^{\infty} \alpha_{n-k} \frac{\delta}{\delta\alpha_n}. \tag{6.22}$$

If the functional on which this derivative acts depends only on α_n with $n \le m$ (as will be true for us), it is necessary to sum up to $n = m$.

Let us now return to the equations of motion (6.14) and try to rewrite them in terms of Γ. For the linear Schwinger equation this can be done directly using Eqs. (6.9), (6.11), and (6.13):

$$-\frac{\delta\Gamma}{\delta\alpha_1} - \sum_{n=2}^{m} \alpha_{n-1} \frac{\delta\Gamma}{\delta\alpha_n} + A_{m+1}\alpha_m + \sum_{n=m+2}^{\infty} A_n \frac{\delta\Gamma}{\delta A_{n-1}} = 0. \tag{6.23}$$

In this equation Γ is assumed to be a functional of the first functions $\alpha = \{\alpha_1 \ldots \alpha_m\}$ and the higher potentials A''. Comparing this with (6.22), we see that the sum of all the derivatives with respect to α is collected to form the derivative with respect to β_1, so that for Γ in the variables $\beta = \{\beta_1 \ldots \beta_m\}$ and A'', Eq. (6.23) takes the form

$$\frac{\delta\Gamma}{\delta\beta_1} = A_{m+1}\alpha_m + \sum_{n=m+2}^{\infty} A_n \frac{\delta\Gamma}{\delta A_{n-1}}. \tag{6.24}$$

The unconnected variable α_m on the right-hand side can be expressed in terms of the independent variables β by (6.15) with $n = m$.

Let us now turn to the connection equations (6.14) and split them into two groups: equations with $n \le m$ and equations with $n > m$. First we consider the equations of the first group. Their right-hand sides H_n are known polynomials of the connected functions $\beta_1 \ldots \beta_m$, which are our independent variables. The left-hand sides of these equations reduce to α according to (6.9); expressing α in terms of β using (6.15), we obtain simple identities instead of equations with content.

At first glance this appears strange, but the point is that we have essentially used the connection equations to obtain (6.15), so naturally the combined use of the former and (6.15) leads to identities. From this it is clear that equations (6.15) with $n \le m$ are actually equivalent to the connection equations of the first group, and the problem amounts to representing the information contained in (6.15) in the form of equations for the functional Γ. We begin by writing the identities

$$\delta_{1k} = \frac{\delta A_k}{\delta A_1} = \sum_{n=1}^{m} \frac{\delta \beta_n}{\delta A_1} \cdot \frac{\delta A_k}{\delta \beta_n}, \quad k = 1, 2, \ldots, m. \tag{6.25}$$

We stress the fact that these equations do not contain any information about the form of the functional Γ; these are identities which only express the fact that the variables $A' \equiv \{A_1 \ldots A_m\}$ are replaced by $\beta \equiv \{\beta_1 \ldots \beta_m\}$. However, if we manage to combine the identities (6.25) with equations (6.15), we thereby introduce the desired information into the system and can obtain nontrivial equations for Γ.

This is how we shall proceed, using in (6.25) the recursion relation (6.15): $\delta\beta_n/\delta A_1 = \beta_{n+1}$. Since $n \le m$ in (6.25), only the single derivative $\delta\beta_m/\delta A_1$ is not an independent variable. We can eliminate it by solving (6.25) with $k = m$ for β_{m+1} and substituting the results into the other equations with $k < m$:

$$\beta_{m+1} = \delta\beta_m/\delta A_1 = -\sum_{n=1}^{m-1} \frac{\delta \beta_n}{\delta A_1} \cdot \frac{\delta A_m}{\delta \beta_n}\left(\frac{\delta A_m}{\delta \beta_m}\right)^{-1}, \tag{6.26}$$

$$\delta_{1k} = \sum_{n=1}^{m-1} \beta_{n+1}\left[\frac{\delta A_k}{\delta \beta_n} - \frac{\delta A_m}{\delta \beta_n}\left(\frac{\delta A_m}{\delta \beta_m}\right)^{-1}\frac{\delta A_k}{\delta \beta_m}\right]. \tag{6.27}$$

Now we express the variables A_k in terms of Γ using (6.11):

$$A_k = -\frac{\delta \Gamma}{\delta \alpha_k} = -\sum_{s=1}^{m} \frac{\delta \beta_s}{\delta \alpha_k}\frac{\delta \Gamma}{\delta \beta_s}. \tag{6.28}$$

In the special case $k = m$, taking into account (6.21) we obtain

$$A_m = -m! \cdot \delta\Gamma/\delta\beta_m. \tag{6.29}$$

Differentiating (6.28) and (6.29), we find

$$\frac{\delta A_k}{\delta \beta_n} = -\sum_{s=1}^{m}\left[\Gamma_{ns}\frac{\delta \beta_s}{\delta \alpha_k} + \Gamma_s \frac{\delta}{\delta \beta_n}\left(\frac{\delta \beta_s}{\delta \alpha_k}\right)\right], \tag{6.30}$$

$$\frac{\delta A_m}{\delta \beta_n}\left(\frac{\delta A_m}{\delta \beta_m}\right)^{-1}\frac{\delta A_k}{\delta \beta_m} = -\sum_{s=1}^{m}\Gamma_{nm}\Gamma^{-1}_{mm}\left[\Gamma_{ms}\frac{\delta \beta_s}{\delta \alpha_k} + \Gamma_s \frac{\delta}{\delta \beta_m}\left(\frac{\delta \beta_s}{\delta \alpha_k}\right)\right], \tag{6.31}$$

where $\Gamma_n \equiv \delta\Gamma/\delta\beta_n$ and $\Gamma_{nk} \equiv \delta^2\Gamma/\delta\beta_n\delta\beta_k$.

It is clear from the triangularity of the matrix of derivatives $\delta\beta/\delta\alpha$ that β_s is expressed in terms of α_r with $r \le s$, so that the derivative $\delta\beta_s/\delta\alpha_k$ is expressed in terms of α_r with $r \le s - k$, and, in the variables β, in terms of β_r with $r \le s - k$. Using the fact that $s - k < m$ in (6.31), we conclude that the derivative with respect to β_m on the right-hand side of (6.31) is zero, so that the term containing Γ_s can be dropped. Then, substituting (6.30) and (6.31) into (6.27), we obtain

$$-\delta_{1k} = \sum_{n=1}^{m-1} \beta_{n+1} \sum_{s=1}^{m} \left[Q_{ns}\frac{\delta\beta_s}{\delta\alpha_k} + \Gamma_s \frac{\delta}{\delta\beta_n}\left(\frac{\delta\beta_s}{\delta\alpha_k}\right)\right], \tag{6.32}$$

where $Q_{ns} \equiv \Gamma_{ns} - \Gamma_{nm}\Gamma^{-1}_{mm}\Gamma_{ms}$. The system (6.32) contains $m - 1$ independent equations, i.e., exactly the same as the number of connection equations in the first group. For $k = m$, (6.32) becomes an identity.

To simplify Eqs. (6.32) we contract them with the corresponding derivative $\delta\alpha_k/\delta\beta_r$ and then sum over k from 1 to m. On the left-hand side, taking into account (6.19), we obtain $-\delta_{1r}$, and the right-hand side involves the sums

$$\sum_{k=1}^{m} \frac{\delta\beta_s}{\delta\alpha_k}\cdot\frac{\delta\alpha_k}{\delta\beta_r} = \delta_{sr} \quad \text{and} \quad \sum_{k=1}^{m} \frac{\delta\alpha_k}{\delta\beta_r}\cdot\frac{\delta}{\delta\beta_n}\left(\frac{\delta\beta_s}{\delta\alpha_k}\right).$$

To calculate the second of these we move the derivative $\delta/\delta\beta_n$ onto the first factor. We obtain

$$\sum_{k=1}^{m} \frac{\delta}{\delta\beta_n}\left[\frac{\delta\alpha_k}{\delta\beta_r}\cdot\frac{\delta\beta_s}{\delta\alpha_k}\right] - \sum_{k=1}^{m} \frac{\delta^2\alpha_k}{\delta\beta_n\delta\beta_r}\cdot\frac{\delta\beta_s}{\delta\alpha_k}.$$

The first term does not contribute because the expression in the square brackets reduces to δ_{sr}, and the second term is calculated using (6.20) and (6.22):

$$\sum_{k=1}^{m} \frac{\delta^2\alpha_k}{\delta\beta_n\delta\beta_r}\cdot\frac{\delta\beta_s}{\delta\alpha_k} = \frac{(n+r)!}{n!\,r!}\cdot\frac{\delta\beta_s}{\delta\beta_{n+r}} = \frac{(n+r)!}{n!\,r!}\delta_{s,n+r}.$$

Finally, we reduce (6.32) to the form

$$\delta_{1r} = \sum_{n=1}^{m-r} \frac{(n+r)!}{n!r!} \beta_{n+1}\Gamma_{n+r} - \sum_{n=1}^{m-1} \beta_{n+1}Q_{nr}. \qquad (6.33)$$

These are the connection equations of the first group for the functional $\Gamma(\beta;A'')$.

Let us now turn to the equations of the second group, i.e., to the connection equations (6.14) with $n > m$. According to the second of Eqs. (6.13), the left-hand sides of the equations can be written as $\delta\Gamma/\delta A_n$, $n > m$. The right-hand side of the equation with a given n is a polynomial in the connected functions β_k with $k \leq n$. The first of these functions with $k \leq m$ are our independent variables, and the higher functions β_k with $k > m$ must be expressed in terms of the functional Γ or its derivatives, which leads to the desired equations. Owing to the recursion relation (6.15) between the connected functions β_k, the problem reduces to putting the *raising operator* $\mathcal{D} \equiv \delta/\delta A_1$ (the partial derivative in the variables A) into the needed form. This is easily done, writing

$$\mathcal{D} \equiv \delta/\delta A_1 = \sum_{k=1}^{m} \frac{\delta\beta_k}{\delta A_1} \cdot \frac{\delta}{\delta\beta_k} = \sum_{k=1}^{m} \beta_{k+1}\frac{\delta}{\delta\beta_k}$$

and substituting into this the expression for the derivative β_{m+1} obtained using (6.26) and (6.29):

$$\mathcal{D} = \sum_{k=1}^{m-1} \beta_{k+1}\left[\frac{\delta}{\delta\beta_k} - \Gamma_{km}\Gamma^{-1}_{mm}\frac{\delta}{\delta\beta_m}\right]. \qquad (6.34)$$

By acting successively on the independent variable β_m with the raising operator $\mathcal{D}$, we can express any of the connected functions β_n with $n > m$ in terms of derivatives of the functional Γ. We note that for $n < m$ from (6.34) we obtain $\mathcal{D}\beta_n = \beta_{n+1}$, as we should.

The connection equations (6.14) of the second group ($n > m$) can be written as

$$\frac{\delta\Gamma}{\delta A_n} = \frac{1}{n!}\exp\,[-W]\,\mathcal{D}^n\exp\,[W] = \frac{1}{n!}(\beta_1 + \mathcal{D})^n \cdot 1, \qquad (6.35)$$

with the understanding that $\mathcal{D}$ is the operator (6.34). We note that for the first

Legendre transform the raising operator (1.213) in the notation of the present section has the form $\mathscr{D} = -\Gamma_{11}^{-1}\delta/\delta\beta_1$.

Let us depict the equations we have obtained graphically, using the following notation for the connected Green functions:

$$\beta_1 \equiv \text{\small ⌁⌁}, \quad \beta_2 \equiv \text{——}, \quad \beta_k \equiv \text{✕} \;(k > 2) \tag{6.36}$$

and the derivatives of the functional Γ:

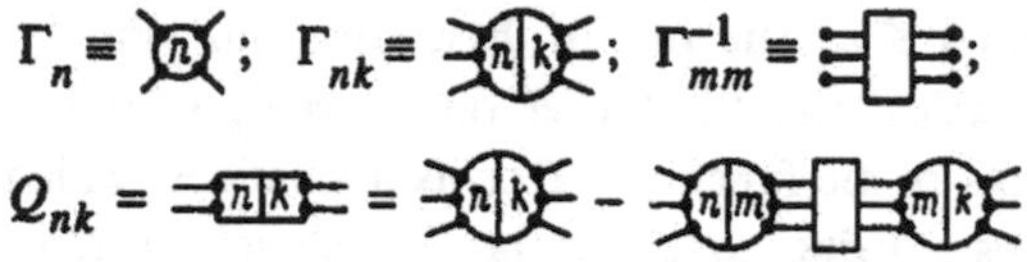

$$\Gamma_n \equiv \text{⊅⃝} ; \quad \Gamma_{nk} \equiv \text{⊅⟨n⟩⟨k⟩⟨} ; \quad \Gamma_{mm}^{-1} \equiv \text{≡⃞≣}; \tag{6.37}$$

$$\mathcal{Q}_{nk} = \text{═⟨n⟩⟨k⟩═} = \text{⊅⟨n⟩⟨k⟩⟨} - \text{⊅⟨n⟩⟨m⟩⃞⟨m⟩⟨k⟩⟨} .$$

Objects with external lines, like connected Green functions, will be distinguished from objects without external lines, like amputated functions, by the arrangement of the points in the blocks representing derivatives of Γ. The arrangement used in (6.37) can be justified only after we analyze the diagrammatic representations of Γ, but it is not difficult to guess it beforehand using a simple rule: the derivative of Γ with respect to an object with external lines, for example, with respect to β_n, will be an object without external lines, and vice versa; a quantity which is the inverse of an object with external lines will be an object without external lines, and vice versa. Later we shall verify that these simple rules are always true.

In the notation of (6.36) and (6.37) equations (6.33) take the form

$$\delta_{1r} = \sum_{n=1}^{m-r} \frac{(n+r)!}{n!\,r!} \; \text{⟨n·r⟩} - \sum_{n=1}^{m-1} \text{⟨n⟩⟨r⟩}. \tag{6.38}$$

For $n = 1$ the symbol $\text{⟨}\, n$ should be understood as $\text{—⟩} 1$.

The raising operator (6.34) is represented graphically as

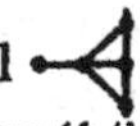

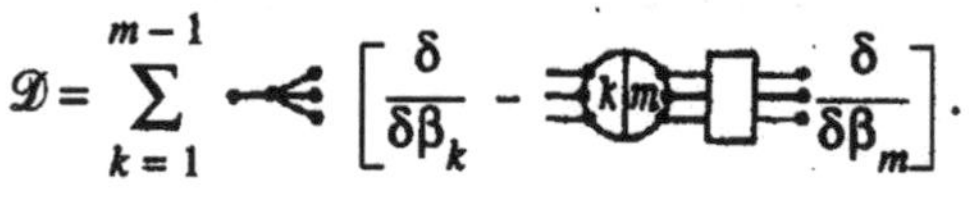

$$\mathscr{D} = \sum_{k=1}^{m-1} \text{⟨} \left[\frac{\delta}{\delta\beta_k} - \text{⟨k⟩⟨m⟩⃞≣} \frac{\delta}{\delta\beta_m} \right]. \tag{6.39}$$

6.2.3 The equations of motion in 1-irreducible variables [61]

In this book we shall study the first four Legendre transforms in detail. It turns out that when analyzing the third transform it is desirable, and for the fourth it is necessary, to make yet another substitution, going from the connected variables $\beta \equiv \beta_1 \ldots \beta_m$ to 1-irreducible variables, which we shall denote as $\gamma \equiv \gamma_1 \ldots \gamma_m$.

We define $\gamma_1 = \beta_1$, and as the other variables γ_k with $k > 1$ we use the 1-irreducible Green functions defined by Eq. (1.218) in Sec. 1.8.1. We recall that according to the definition, $\gamma_2 = -\beta_2^{-1}$ is the inverse full propagator with minus sign, γ_3 is the amputated function β_3, γ_4 is the amputated function β_4 minus all graphs containing one-particle singularities [see (1.217)], and so on. Connected functions are expressed in terms of 1-irreducible ones by Eqs. (1.214), the inverses of (1.218); to aid the reader, we shall rewrite these equations in the notation of the present section. We define the functional

$$\bar{\beta}(\varphi) = \sum_{n=2}^{\infty} \frac{1}{n!} \beta_n \varphi^n = \beta(\varphi) - \beta_1 \varphi \tag{6.40}$$

and construct its Legendre transform with respect to the variable $\varphi(x)$:

$$\psi(x) = \delta\bar{\beta}(\varphi)/\delta\varphi(x), \quad \gamma(\psi) = \bar{\beta}(\varphi) - \varphi\psi. \tag{6.41}$$

From this in the usual manner we obtain (omitting the arguments x)

$$\delta\gamma(\psi)/\delta\psi = -\varphi; \quad \frac{\delta^2\gamma}{\delta\psi^2} \cdot \frac{\delta^2\bar{\beta}}{\delta\varphi^2} = -1. \tag{6.42}$$

Because the expansion (6.40) begins with φ^2, the correspondence between the variables φ and ψ is $0 \leftrightarrow 0$, and the expansion of $\gamma(\psi)$ also begins with ψ^2. The functions γ_n of interest to us are the coefficients in the expansion of the functional $\gamma(\psi)$:

$$\gamma(\psi) = \sum_{n=2}^{\infty} \frac{1}{n!} \gamma_n \psi^n. \tag{6.43}$$

These expressions are equivalent to the definitions in Sec. 1.8.1; Eqs. (1.218) and (1.214) in the new notation take the form

$$\gamma_n = \left(\frac{\delta}{\delta\psi}\right)^n \gamma(\psi)\Big|_{\psi=0} = -\left[\left(\frac{\delta^2\bar{\beta}}{\delta\varphi^2}\right)^{-1} \frac{\delta}{\delta\varphi}\right]^{n-1} \varphi\Big|_{\varphi=0}, \tag{6.44}$$

$$\beta_k = \left(\frac{\delta}{\delta\varphi}\right)^k \bar{\beta}(\varphi)\Big|_{\varphi=0} = \left[-\left(\frac{\delta^2\gamma}{\delta\psi^2}\right)^{-1}\frac{\delta}{\delta\psi}\right]^{k-1}\psi\Big|_{\psi=0}. \tag{6.45}$$

These equations allow γ to be expressed explicitly in terms of β and vice versa. They can also be used to calculate various derivatives of the type $\delta\gamma/\delta\beta$ needed in making the variable substitution $\beta \to \gamma$ in the equations of motion. We shall omit the complicated algebra, and refer the interested reader to [61]. We give only the final result, namely, the equations of motion for the functional $\Gamma(\gamma; A'')$. We introduce the following graphical notation for the variables γ:

$\gamma_1 = \beta_1 \equiv$ , $\beta_2 = -\gamma_2^{-1} \equiv$ ; $\gamma_k \equiv$  $(k > 2)$ (6.46)

and the derivatives of the functional Γ ($\Gamma_n \equiv \delta\Gamma/\delta\gamma_n$, $\Gamma_{nk} \equiv \delta^2\Gamma/\delta\gamma_n\delta\gamma_k$):

$\Gamma_n \equiv$ ; $\Gamma_{nk} \equiv$ ; $\Gamma^{-1}_{mm} \equiv$ . (6.47)

The arrangement of the points in the blocks of (6.47), which indicates the presence or absence of external lines for a given object, is determined by the same considerations as in the preceding section. The variable $\gamma_1 = \beta_1$, which is a connected Green function, differs from the other variables γ_k, which are like amputated functions. Therefore, the representations (6.47) need to be supplemented by the following convention: if any of the indices of the derivatives Γ_n, Γ_{nk} is equal to unity, in the representation (6.47) the point should be moved from the end of the line to the block itself. For example, $\Gamma_{11} =$ .

The equations of motion involve the quantities

$Q_{ns} = \Gamma_{ns} - \Gamma_{nm}\Gamma^{-1}_{mm}\Gamma_{ms} =$  $-$ , (6.48)

$\bar{Q}_{ns} \equiv$  $= Q_{ns} - m^2\delta_{n2}\delta_{s2}$ 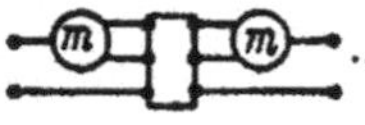 $+$

$+\, m\delta_{n2}$ 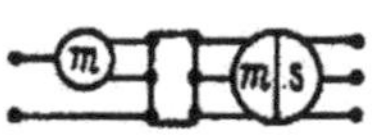 $+\, m\delta_{s2}$ 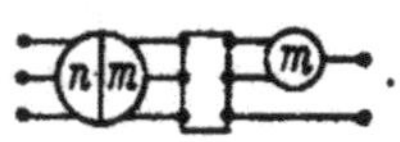. (6.49)

where δ_{n2} and δ_{s2} are the Kronecker deltas. The quantities $\bar{Q}_{ns}$ and Q_{ns} differ only when at least one of the indices n, s is equal to two.

Now we are in a position to write down the equations of motion for $\Gamma(\gamma; A'')$. The form of the linear Schwinger equation (6.24) does not change, because the variable $\gamma_1 = \beta_1$ is not mixed with the functions β_n, $n > 1$, by the substitution $\beta \rightarrow \gamma$; the unconnected function α_m on the right-hand side of (6.24) should now be expressed in terms of the independent variables γ, rather than β, as in (6.24).

For transforms of order higher than second, the first of the connection equations (6.33), (6.38), i.e., the equation with $r = 1$, takes the form [61]:

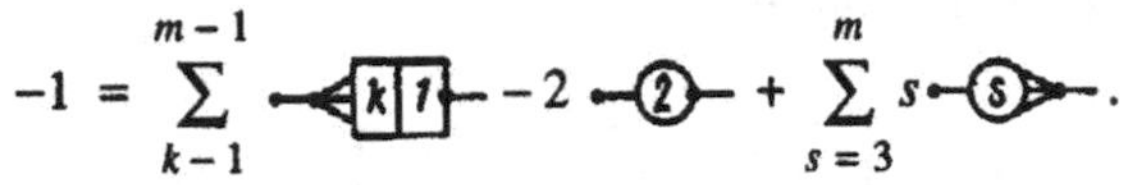

$$-1 = \sum_{k-1}^{m-1} \underset{k|1}{\blacktriangleleft} - 2 \underset{2}{\longrightarrow} + \sum_{s=3}^{m} s \underset{s}{\longrightarrow}. \tag{6.50}$$

As in (6.38), the symbol $\underset{k}{\blacktriangleleft}$ for $k = 1$ is understood as $\multimap\negmedspace:1$. The block $\multimap\negmedspace\underset{2}{\bigcirc}\negmedspace\multimap$ in (6.50) stands for $\Gamma_2\beta_2^{-1}$, i.e., the derivative $\Gamma_2 = \multimap\negmedspace\underset{2}{\bigcirc}\negmedspace\multimap$ with amputated right-hand external line.

The remaining equations (6.33) with $2 \leq r \leq m - 1$ take the form

$$0 = \sum_{k=1}^{m-1} \underset{k|r}{\blacktriangleleft} \Xi - (r+1) \underset{r+1}{\longrightarrow}\Xi - \sum_{sn_1\dots n_k} A^{sr}_{n_1\dots n_k} \underset{}{\bigcirc}\Xi. \tag{6.51}$$

The summation over s runs from 2 to m, and the numbers n_i, each of which can take values from 3 to m, indicate the degree of the vertices lying on the solid lower line in the last term in (6.51). The summation runs over all possible arrangements of these vertices with the condition that the total number of lines going from them to the block Γ_s ($n_i - 2$ from each) is equal to the number of free arguments of this block, i.e., $s - r + 1$ [Γ_s has s arguments altogether, but $r - 1$ of them were moved to the right in (6.51)]. The quantities $A^{sr}_{n_1\dots n_k}$ in (6.51) are numerical symmetry coefficients:

$$A^{sr}_{n_1\dots n_k} = s! \left[(r-1)! \prod_{i=1}^{k} (n_i - 2)! \right]^{-1}. \tag{6.52}$$

For the third transform there is only a single equation (6.51) with $r = 2$, and the sum over s, $n_1\dots n_k$ has the form

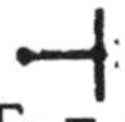

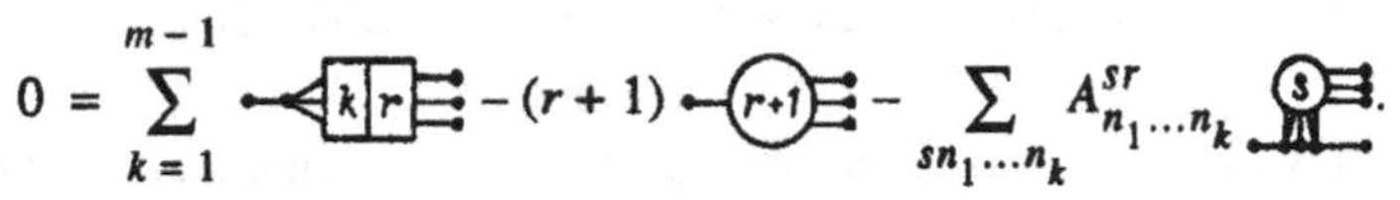

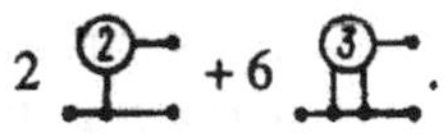

$$2 \underset{2}{\overset{}{\bigtriangleup}} + 6 \underset{3}{\overset{}{\bigtriangleup}}. \tag{6.53}$$

For the fourth transform for $r = 2$ the sum has the form

$$2 \underset{2}{\overset{}{\bigtriangleup}} + 6 \underset{3}{\overset{}{\bigtriangleup}} + 3 \underset{3}{\overset{}{\bigtriangleup}} + 24 \underset{4}{\overset{}{\bigtriangleup}} + 12 \underset{4}{\overset{}{\bigtriangleup}} + 12 \underset{4}{\overset{}{\bigtriangleup}} \tag{6.54}$$

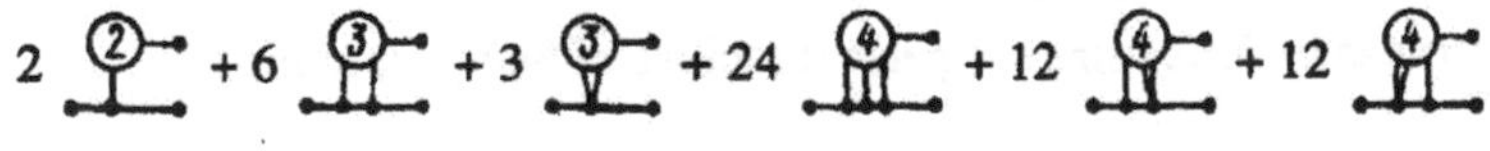

and for the second possible value $r = 3$ we obtain

(6.55)

We still need to give the connection equations of the second group (6.35), which is equivalent to writing the raising operator $\mathscr{D}$ in the variables γ. The desired representation has the form

$$\mathscr{D} = \beta_2\left[\frac{\delta}{\delta\gamma_1} + \sum_{k=2}^{m} \gamma_{k+1}\frac{\delta}{\delta\gamma_k}\right], \tag{6.56}$$

where $\beta_2 = -\gamma_2^{-1}$, and γ_{m+1} should be understood as

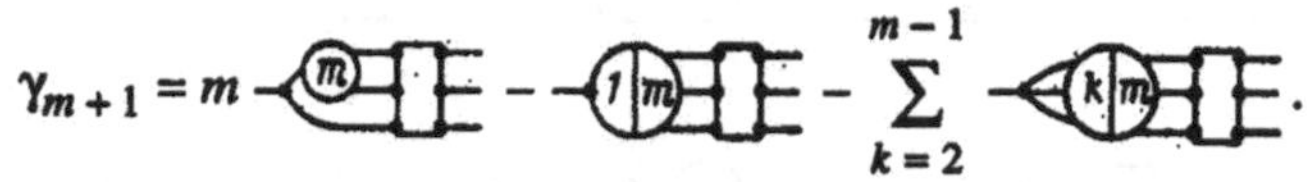

(6.57)

According to (6.15), $\mathscr{D}$ is the raising operator for the connected functions β: $\mathscr{D}\beta_k = \beta_{k+1}$. Acting with it successively on $\beta_m = \beta_m(\gamma)$, we find all the higher functions β_n as functionals of γ and A''.

The raising operator for 1-irreducible functions is defined by (1.219). In the notation of the present section

$$\gamma_{n+1} = \mathscr{D}_\gamma\gamma_n = \beta_2^{-1}\mathscr{D}\gamma_n, \quad n > 1. \tag{6.58}$$

For $m \geq 3$ the operator $\mathscr{D}_\gamma$ coincides with the expression in the square brackets in (6.56) (in the formulas of this section it is assumed that $m \geq 3$), and $\mathscr{D}$ should be taken in the form (6.34) in constructing $\mathscr{D}_\gamma$ for the second transform.

6.2.4 *Linear equations and their general solutions*

The Schwinger equation (6.24) is a linear inhomogeneous equation in partial variational derivatives, and its general solution is the sum of the particular solution of the inhomogeneous equation and the general solution of the homogeneous equation, an arbitrary functional of its first integrals.

Before analyzing the Schwinger equation in this manner, let us derive another linear equation for Γ. Of course, if such an equation exists, it is a consequence of the complete system of equations of motion, but in our case it is simpler and more useful to derive it anew.

This is easily done using the general recipes of Sec. 1.7, writing

$$0 = \int dx \int D\varphi \frac{\delta}{\delta\varphi(x)} [\varphi(x) \exp A(\varphi)] =$$

$$= \left[\operatorname{tr} 1 + \sum_{n=1}^{\infty} nA_n \frac{\delta}{\delta A_n} \right] G(A) . \qquad (6.59)$$

The number $\operatorname{tr} 1 = \int dx \delta(x - x)$ is the trace of the unit operator on the space of fields $\varphi(x)$ (the "dimension" of this space). In obtaining (6.59) we took the factor $\varphi(x)\delta A(\varphi)/\delta\varphi(x)$ outside the functional integral in the form of a differential operator.

Changing to $W = \ln G$, we find

$$\operatorname{tr} 1 + \sum_{n=1}^{\infty} nA_n \frac{\delta W}{\delta A_n} = 0, \qquad (6.60)$$

from which, using (6.9), (6.11), and (6.13), we obtain the equation for $\Gamma(\alpha; A'')$:

$$\operatorname{tr} 1 - \sum_{n=1}^{m} n\alpha_n \frac{\delta \Gamma}{\delta \alpha_n} + \sum_{n=m+1}^{\infty} nA_n \frac{\delta \Gamma}{\delta A_n} = 0. \qquad (6.61)$$

Now it is necessary to make the substitution $\alpha \to \beta \to \gamma$. We shall perform the necessary calculations here, since they serve as a good illustration of the technique used to derive the equations of motion for $\Gamma(\gamma; A'')$. We have

$$\sum_{n=1}^{m} n\alpha_n \frac{\delta \Gamma}{\delta \alpha_n} = \sum_{n=1}^{\infty} n\alpha_n \sum_{k=1}^{m} \frac{\delta \gamma_k}{\delta \alpha_n} \frac{\delta \Gamma}{\delta \gamma_k}. \qquad (6.62)$$

We have extended the sum over n to infinity, because in (6.62) $k \le m$ and $\delta\gamma_k/\delta\alpha_n = 0$ for $n > k$. For $k = 1$ only the term with $n = 1$ contributes to (6.62). Since $\gamma_1 = \beta_1 = \alpha_1$, it is clear that the derivative of Γ with respect to γ_1 enters into the sum (6.62) as $\gamma_1 \delta\Gamma/\delta\gamma_1$.

The final task is to calculate the coefficients of the derivatives $\delta\Gamma/\delta\gamma_k$ with $k > 1$. Using the definitions (6.41) and (6.43), we can write

$$\sum_{n=1}^{\infty} n\alpha_n \frac{\delta \gamma_k}{\delta \alpha_n} = \left(\frac{\delta}{\delta\psi}\right)^k \sum_{n=1}^{\infty} n\alpha_n \frac{\delta \gamma(\psi)}{\delta \alpha_n} \bigg|_{\psi=0} =$$

$$= \left(\frac{\delta}{\delta\psi}\right)^k \sum_{n=1}^{\infty} n\alpha_n \frac{\delta}{\delta\alpha_n} [\bar{\beta}(\varphi) - \varphi\psi] |_{\psi=0}. \qquad (6.63)$$

In this expression ψ is assumed to be an independent variable, and $\varphi = -\delta\gamma/\delta\psi$ is a functional of ψ depending parametrically on γ_n and thereby on α_n. It follows from (6.40) and (6.17) that $\bar{\beta}(\varphi) - \varphi\psi = \ln\alpha(\varphi) - \beta_1\varphi - \varphi\psi$. Differentating this quantity with respect to α_n, we obtain the following expression:

$$\frac{1}{\alpha(\varphi)}\varphi^n - \frac{1}{\alpha(\varphi)}\frac{\delta\alpha(\varphi)}{\delta\varphi}\frac{\delta\varphi}{\delta\alpha_n} - \delta_{n1}\varphi - \beta_1\frac{\delta\varphi}{\delta\alpha_n} - \psi\frac{\delta\varphi}{\delta\alpha_n}. \qquad (6.64)$$

In differentiating $\alpha(\varphi)$ we have used the fact that this functional depends on α_n both explicitly [hence the first term in (6.64)] and implicitly through φ. The independent variable ψ is not differentiated, and $\delta\beta_1/\delta\alpha_n = \delta_{n1}$. Furthermore,

$$\frac{\delta\alpha(\varphi)}{\delta\varphi} = \alpha(\varphi)\frac{\delta\beta(\varphi)}{\delta\varphi} = \alpha(\varphi)\left[\frac{\delta\bar{\beta}(\varphi)}{\delta\varphi} + \beta_1\right] = \alpha(\varphi)\,[\psi + \beta_1]\,. \qquad (6.65)$$

We have used Eqs. (6.17) and (6.40) and have taken into account the fact that in the substitution $\varphi = -\delta\gamma(\psi)/\delta\psi$ the quantity $\delta\bar{\beta}(\varphi)/\delta\varphi$ becomes ψ according to the definition of the Legendre transform (6.41), (6.42). Substituting (6.65) into (6.64), we see that the terms involving the derivatives $\delta\varphi/\delta\alpha_n$ cancel out, and Eq. (6.64) becomes $\varphi^n/\alpha(\varphi) - \delta_{n1}\varphi$. Substitution of this result into (6.63) gives a series in n, which can be summed:

$$\sum_{n=1}^{\infty} n\alpha_n\left[\frac{\varphi^n}{\alpha(\varphi)} - \delta_{n1}\varphi\right] = \frac{1}{\alpha(\varphi)}\varphi\frac{\delta\alpha(\varphi)}{\delta\varphi} - \beta_1\varphi, \qquad (6.66)$$

because the multiplication of the coefficients of the power series in φ by n is equivalent to performing the operation $\varphi\delta/\delta\varphi$. Then using Eqs. (6.17), (6.40)–(6.42) again, we reduce (6.66) to the form $\varphi\delta\bar{\beta}(\varphi)/\delta\varphi = -\psi\delta\gamma(\psi)/\delta\psi$. In the end we obtain the following simple result (we recall that $k > 1$):

$$\sum_{n=1}^{\infty} n\alpha_n\frac{\delta\gamma_k}{\delta\alpha_n} = \left(\frac{\delta}{\delta\psi}\right)^k\left[-\psi\frac{\delta\gamma(\psi)}{\delta\psi}\right]\Bigg|_{\psi=0} = -k\gamma_k \qquad (6.67)$$

and the desired linear equation is

$$\operatorname{tr} 1 - \gamma_1 \frac{\delta\Gamma}{\delta\gamma_1} + \sum_{k=2}^{m} k\gamma_k \frac{\delta\Gamma}{\delta\gamma_k} + \sum_{n=m+1}^{\infty} nA_n \frac{\delta\Gamma}{\delta A_n} = 0. \tag{6.68}$$

Let us now describe the solution of Eqs. (6.24) and (6.68). It is easily verified by direct calculation that as the particular solution of the inhomogeneous equation (6.68) we can take $\operatorname{tr}\ln\beta_2$ with coefficient 1/2, and the general solution of the homogeneous equation will be an arbitrary functional of the first integrals:

$$\bar{\gamma}_1 \equiv \gamma_1\lambda^{-1};\ \ \bar{\gamma}_k \equiv \lambda^k\gamma_k,\ \ 3 \le k \le m;\ \ \bar{A}_n \equiv \lambda^n A_n,\ \ n > m, \tag{6.69}$$

where $\lambda \equiv \beta_2{}^{1/2}$ is the "square root of the propagator," given by the equation* $\lambda^T\lambda = \beta_2$, and products $\lambda^k\gamma_k$ and so on are understood as contractions of the type (1.131) [see the remark following Eq. (1.131)]. The first integrals in (6.69) will henceforth be referred to as *invariant vertices.*

Let us consider an arbitrary graph consisting of vertices, lines, and legs γ_1, which are connected directly to vertices without an intermediate line. We shall assume that a line is associated with the full propagator $\beta_2 = -\gamma_2{}^{-1}$, and a vertex to which n lines and k legs γ_1 are connected is associated with γ_{n+k} if $n+k \le m$ and A_{n+k} if $n+k > m$. It is clear that the resulting expression will be a functional of the invariant vertices (6.69). This becomes obvious if each line is split in half and each half is associated with a vertex, and if the factor $\lambda^{-1}\lambda = 1$ is inserted at the points where legs γ_1 are connected to vertices, with the first factor associated with γ_1 and the second with the vertex. Therefore, any sum of graphs of this type automatically satisfies the homogeneous equation (6.68). Of course, the reverse does not follow: the ability to represent the solution by graphs is equivalent to the additional assumption that the functional can be expanded in a power series in invariant vertices.

Let us now turn to the Schwinger equation (6.24). The inhomogeneous term in this equation contains the function α_m, which is assumed to be expressed in terms of the independent variables $\gamma_1 \ldots \gamma_m$. In order to construct the particular solution of the inhomogenous equation, we define the functional

---

*This definition presupposes the equality $\beta_2 = \beta_2^T$, which is valid for a bosonic field. We also note that the equation $\lambda^T\lambda = \beta_2$ determines λ only up to a transformation $\lambda \rightarrow u\lambda$, where u is an arbitrary orthogonal operator: $u^T u = 1$.

$$\alpha^{(m)}(\varphi) = \exp \sum_{n=1}^{m} \frac{1}{n!} \beta_n \varphi^n = \sum_{n=0}^{\infty} \alpha_n^{(m)} \varphi^n. \tag{6.70}$$

The coefficient functions $\alpha_n^{(m)}$ differ from the ordinary functions α_n in that all contributions containing at least one of the higher functions β_k with $k > m$ have been discarded in them. The quantities $\alpha_n^{(m)}$ are known functionals of the independent variables γ, and $\alpha_n^{(m)} = \alpha_n$ for $n \leq m$.

Recalling the derivation of (6.19), it is easy to understand that it will also be valid for the functions $\alpha_n^{(m)}$, in particular, $\delta\alpha_n^{(m)}/\delta\beta_1 = \alpha_{n-1}^{(m)}$. Using this expression, it can be shown that the functional $\mathcal{F} \equiv \Sigma_{n>m} A_n \alpha_n^{(m)}$ is a particular solution of the inhomogeneous equation (6.24). In fact,

$$\frac{\delta \mathcal{F}}{\delta \beta_1} = \sum_{n=m+1}^{\infty} A_n \frac{\delta \alpha_n^{(m)}}{\delta \beta_1} = A_{m+1} \alpha_m^{(m)} + \sum_{n=m+2}^{\infty} A_n \alpha_{n-1}^{(m)},$$

as required, since $\alpha_m^{(m)} = \alpha_m$ and $\alpha_{n-1}^{(m)} = \delta \mathcal{F} / \delta A_{n-1}$. It is just as easily verified that the first integrals of the homogeneous equation (6.24) will be the vertices $\omega_n(x_1 \ldots x_n)$ similar to the higher potentials with $n > m$:

$$\omega_n = \sum_{k=0}^{\infty} \frac{1}{k!} A_{n+k} \gamma_1^k =$$

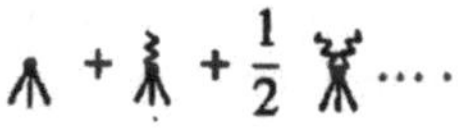 (6.71)

Equations (6.24) and (6.68) essentially eliminate γ_1 and $\gamma_2 = -\beta_2^{-1}$ as independent variables: it follows from (6.24) that γ_1 can enter only in certain combinations (6.71) with the higher potentials, and it follows from (6.68) that the line β_2 enters only in a certain combination with the vertices. If we consider a theory with a finite number of nonzero potentials A_n and perform a complete Legendre transform with respect to all the potentials, the right-hand side of (6.24) will vanish, and in this case the functional Γ will no longer depend at all on the variable γ_1. If the order of the Legendre transform is one unit lower than that of the full transform, only the first term will remain on the right-hand side of (6.24) and the entire dependence on γ_1 will be contained in the particular solution $A_{m+1}\alpha_{m+1}^{(m)}$, where A_{m+1} is the highest potential.

Later we shall construct the iteration solution of the equations of motion, which is represented as the sum of $1/2 \cdot \mathrm{tr}\, \ln \beta_2$ and the graphs discussed above. Here the graphs will be classified according to the number of lines, and the starting point of the iteration procedure will be finding the graphs containing zero lines. Such graphs exist, and they are all contained in the particular solution $\mathscr{F}$: expressing the functions $\alpha_n^{(m)}$ in terms of the independent variables γ, we obtain, among other things, the term $\gamma_1^n/n!$, whereas all the other contributions to $\alpha_n^{(m)}$ necessarily contain at least one line β_2. From this it is clear that as the zeroth-order approximation in the number of lines it is necessary to use

$$\Gamma^{(0)}(\gamma;A'') = \frac{1}{2}\,\mathrm{tr}\ln\beta_2 + \sum_{n=m+1}^{\infty}\frac{1}{n!}A_n\gamma_1^n. \tag{6.72}$$

The general term of the sum represents the bare vertex A_n with $n > m$, all the arguments of which are contracted with legs γ_1. Any other graph necessarily contains at least one line.

6.2.5 *Iteration solution of the equations*

Starting from the zeroth-order approximation (6.72), we can construct the iteration solution of the equations of motion, representing the functional Γ as a sum of graphs with dressed vertices $\gamma_3 \ldots \gamma_m$ and bare vertices A_n, $n > m$. The total number of equations of motion coincides with the number of independent variables of Γ, but in constructing the iteration solution there is no need to consider the entire system of equations of motion. The point is that under ordinary conditions, when iterating, for example, an equation of the form $\delta\Gamma/\delta\gamma_k = \ldots$, at each step we can add to the solution an arbitrary functional independent of γ_k but having any dependence on the other variables. The complete definition of this functional requires additional equations containing derivatives with respect to the other variables. However, if we know *a priori* that the desired functional can be written as the sum of the zeroth-order approximation (6.72) and graphs containing at least one line β_2, for determining these graphs a single equation of the form $\delta\Gamma/\delta\gamma_2 = \ldots$ is sufficient. The arbitrary functional which we could add to the solution of this equation must be independent of $\beta_2 = -\gamma_2^{-1}$, so it cannot be represented by graphs.

To extract the desired equation from the system (6.50), (6.51), we multiply (6.50) on the right by $\beta_2 =$ •——• and each of the equations (6.51) by $\gamma_{r+1}\beta_2 \equiv$ >——•. In other words, we reduce all the r external lines on the right in the blocks (6.51) to a point and connect the line β_2 to this point. If we then sum all the resulting equations, the terms with first derivatives Γ_k, $k > 2$, cancel out and we arrive at an equation of the needed type:

$$2\Gamma_2 = -\gamma_2^{-1} + \sum_{k,r=1}^{m-1}$$  $$- \sum_{s,n_1\dots n_k} A^s_{n_1\dots n_k}$$  $n_1 \dots n_k$. (6.73)

In depicting the last term on the right-hand side we have included the additional right-hand vertex γ_{r+1} in the full set of vertices located on the lower solid line, so that now there are at least two of these vertices. The factor $(r-1)!$ in (6.52) exactly corresponds to the other factors $(n_i - 2)!$, because for the last vertex $n_k = r + 1$. Therefore, the symmetry coefficient $A^s_{n_1 \cdots n_k}$ in (6.73) is given by $s!\left[\prod_i (n_i - 2)!\right]^{-1}$. We recall that the summation over s in (6.73) runs from 2 to m, each of the numbers n_i can take values from 3 to m, and all arrangements of at least two vertices with the condition $\Sigma(n_i - 2) = s$ are summed.

Let us now discuss the procedure for the iteration solution of (6.73). First we note that the first term of the zeroth-order approximation (6.72) is generated by the bare term $-\gamma_2^{-1}$ on the right-hand side of (6.73), but the other terms in (6.72) cannot be obtained from (6.73) and must be inserted by hand as the zeroth-order approximation known from other considerations.

The left-hand side of (6.73) involves the derivative $\Gamma_2 \equiv \delta\Gamma/\delta\gamma_2$. For the zeroth-order approximation $2\Gamma_2^{(0)} = -\gamma_2^{-1}$, and the other graphs of Γ depend on γ_2 via the line $\beta_2 = -\gamma_2^{-1}$, for which $\delta\beta_2/\delta\gamma_2 = \beta_2^2$. In more detail, $\delta\beta_2(x, x')/\delta\gamma_2(y, y') = \mathrm{Sym}\, \beta_2(x, y)\beta_2(x', y')$, where Sym denotes symmetrization with respect to the replacement $x \leftrightarrow x'$ or, equivalently, $y \leftrightarrow y'$. Graphically, the differentiation of a line with respect to γ_2 is equivalent to cutting this line in half, and the derivative of a graph with respect to γ_2 is the sum of the contributions obtained when each of its lines is cut. We note that each graph of the derivative has two external lines, the halves of the line which is cut in the differentiation. We also note that differentiation with respect to β_2 rather than γ_2 would correspond to simple removal rather than cutting of the line, and the derivative would not have external lines. It is also useful to note that not every sum of graphs of the form of Γ_2 is actually the derivative with respect to γ_2 of some graph or sum of graphs: the latter presupposes definite relations between the types of graph and the coefficients in the original set of graphs. The meaning of these additional "integrability conditions" is simple: when we differentiate a graph, we must obtain the sum of the contributions from cutting each of the lines. If some of the contributions are discarded in this sum, the rest can no longer be the derivative of any graph, as long as it is not proportional to the entire original sum (as will occur, for example, when all the lines of the differentiated graph are equivalent and the cutting of any of them leads to the same graph).

By connecting the ends of a cut line in any of the graphs of the derivative Γ_2 we obviously return to the original graph of Γ which was differentiated. Therefore, the contraction $\Gamma_2 \beta_2^{-1}$ (the extra external line is amputated) differs from Γ by only the coefficients of the graphs (an additional factor of n for a graph with n lines). This means that using the known graphs of the derivative Γ_2 together with their coefficients, we can easily, without calculation, find the corresponding graphs of Γ together with their coefficients.

Let us assume that we know all the graphs of Γ through order N in the number of lines and that we want to find the graphs of the next order $N + 1$ [the starting point is knowledge of the zeroth-order graphs of (6.72)]. Using the known graphs of Γ, let us calculate all the blocks entering into the right-hand side of (6.73). These blocks are built from derivatives of Γ with respect to the variables γ in a known way [see (6.46)–(6.49)]. We have already discussed differentiation with respect to γ_2. Differentiation with respect to γ_1 corresponds graphically to removing a leg γ_1, and differentiation with respect to a vertex γ_k, $k > 2$, corresponds to removing this vertex from the graph in all possible ways. The kernel Γ_{mm}^{-1} entering into (6.73) can always be found in the form of a series, the bare term of which is generated by the graph $\gamma_m \beta_2^m \gamma_m = \Leftrightarrow$ (with m lines) contained in Γ. The contribution of this graph to the derivative Γ_{mm} is equal to $2c\beta_2^m = 2c \,\rightleftharpoons$, where c is the coefficient of the original graph of Γ; therefore,

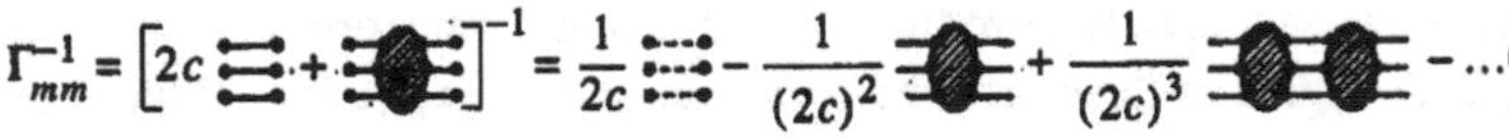

$$\Gamma_{mm}^{-1} = \left[2c \,\rightleftharpoons + \,\blacksquare \,\right]^{-1} = \frac{1}{2c} \,\cdots - \frac{1}{(2c)^2} \,\blacksquare + \frac{1}{(2c)^3} \,\blacksquare\blacksquare - \ldots (6.74)$$

A solid line denotes β_2, a dashed line denotes β_2^{-1}, and a shaded block together with the external lines denotes the contribution to Γ_{mm} from all graphs except the one singled out.

Therefore, from the graphs of Γ known through order N we can calculate all the blocks on the right-hand side of (6.73), thereby determining the "right-hand side in the Nth-order approximation." Owing to the nonlinearity of the right-hand side of (6.73) in Γ, this approximation does not correspond to any definite order in the number of lines, but by carefully studying Eq. (6.73) and evaluating the number of lines in the various blocks, it is easy to verify that the right-hand side in the Nth-order approximation must contain all the graphs obtained by differentiating the graphs of Γ through order $N + 1$ with respect to γ_2. This means that from the known right-hand side in the Nth-order approximation we can find all the graphs of Γ of order $N + 1$ in the number of lines. On the right-hand side of (6.73) these correspond to graphs having $N+2$ lines, two of which are external. We are not interested in the contributions of lower-order graphs, because they are assumed to be known, and the contributions of the higher-order graphs to the right-hand side of (6.73) must simply be discarded. It is useless to try to go farther than one order in the number of lines in one step of the iteration: although the right-hand side in the Nth-order approximation actually contains the contributions of graphs of order higher than $N+1$, not all of these contributions are present. The presence of all the needed graphs is guaranteed only through order $N+1$, but this, of course, is quite sufficient for iterating (6.73).

6.2.6 *The second Legendre transform*

We shall analyze the second transform using the equations of Sec. 6.2.2 (the equations of Sec. 6.2.3 are written assuming $m > 2$). For $m = 2$, the system (6.38) contains a single equation with $r = 1$:

$$2\Gamma_2 = \beta_2^{-1} + \text{[graph]} = \beta_2^{-1} + \text{[graphs]}, \qquad (6.75)$$

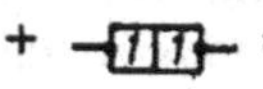

in which Γ_2 denotes the derivative of Γ with respect to β_2, and the blocks on the right-hand side depict the derivatives of Γ with respect to the variables β_1 and β_2 [see (6.37)]. Iterating this equation with the zeroth-order approximation (6.72) according to the general scheme of the preceding section, we represent Γ by graphs with dressed lines β_2, bare vertices A_n, $n > 2$, and legs $\beta_1 = \gamma_1$ connected directly to vertices. For the second transform the sum over n in (6.72) begins at three, and the bare term in the expansion of the type (6.74) of the inverse operator Γ_{22}^{-1} is generated by the term $1/2 \cdot \operatorname{tr} \ln \beta_2$ of the zeroth-order approximation (6.72):

$$\Gamma_2^{(0)} = \frac{1}{2}\beta_2^{-1}, \quad \Gamma_{22}^{(0)} = -\frac{1}{2}\beta_2^{-2} = -\frac{1}{2}\,\text{[graph]}\,.$$

We invite the reader to independently work out the first few steps of the iteration, and here give only the result for the simple and most important special case of a theory in which only the vertices A_3 and A_4 are nonzero. Through fifth order in the number of lines we have

$$\Gamma = \frac{1}{2}\operatorname{tr}\ln\beta_2 + \frac{1}{6}\,\text{[graph]} + \frac{1}{24}\,\text{[graph]} + \frac{1}{2}\,\text{[graph]} + \frac{1}{4}\,\text{[graph]} + \frac{1}{8}\,\text{[graph]} + \frac{1}{12}\,\text{[graph]} +$$

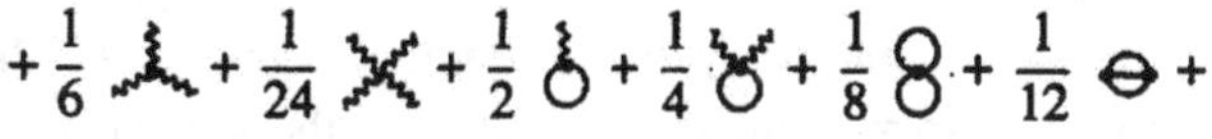

$$+ \frac{1}{6}\,\text{[graph]} + \frac{1}{12}\,\text{[graph]} + \frac{1}{48}\,\text{[graph]} + \frac{1}{8}\,\text{[graph]} + \frac{1}{4}\,\text{[graph]} + \frac{1}{8}\,\text{[graph]} + \ldots \qquad (6.76)$$

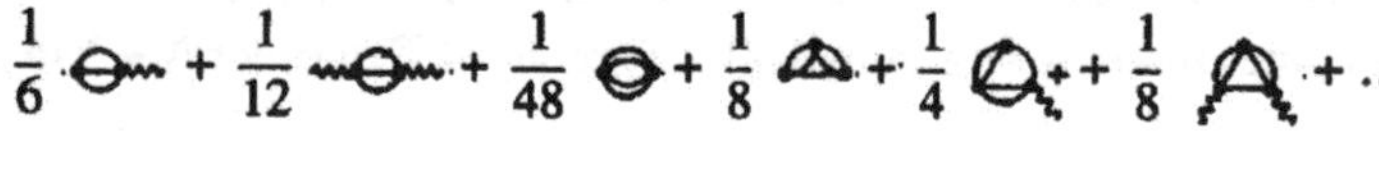

All the graphs given here are *2-irreducible*, i.e., they remain connected when any pair of lines is cut. This is not accidental: studying the right-hand side of (6.75) carefully, one easily sees that this equation possesses the property of *conservation of 2-irreducibility*, just as the equations (1.204) for the functional W possess the property of conservation of connectedness, and Eq. (1.223) for the first Legendre transform possesses the property of conservation of 1-irreducibility.

The conservation property amounts to the following: if all the graphs of Γ up to some order are 2-irreducible, the graphs of the next order obtained by iterating Eqs. (6.75) will also be 2-irreducible. Taking into account the known irreducibility of the first graphs of Γ, this observation proves by induction the 2-irreducibility of all the graphs of the second Legendre transform. Of course, this statement remains true for a theory with any number of nonzero potentials A_n.

Before discussing the question of the coefficients of the graphs, let us study the stationarity equations (6.11), which here take the form

$$-A_k = \delta\Gamma/\delta\alpha_k = \sum_{i=1}^{2} \Gamma_i \delta\beta_i/\delta\alpha_k, \tag{6.77}$$

where $i, k = 1, 2$, and Γ_i are the derivatives of Γ with respect to β_i. From (6.15) we have $\alpha_1 = \beta_1$, $\alpha_2 = (\beta_2 + \beta_1\beta_1)/2$. Solving these equations for β, we can calculate the derivatives $\delta\beta_i/\delta\alpha_k$ and write (6.77) in the form

$$-A_1 = \Gamma_1 - 2\Gamma_2\beta_1, \quad -A_2 = 2\Gamma_2. \tag{6.78}$$

In the variational approach these equations serve to determine the unknown dressed variables β_1 and β_2 in terms of the known bare potentials $A_{1,2}$.

Writing $2\Gamma = \mathrm{tr}\ln\beta_2 + 2\overline{\Gamma}$, $2\Gamma_2 = \beta_2^{-1} + 2\overline{\Gamma}_2$ and recalling that the potential A_2 is related to the bare propagator Δ as $\Delta = -A_2^{-1}$, we see that the second of the stationarity equations (6.78) is the usual Dyson equation $\beta_2^{-1} = \Delta^{-1} - 2\overline{\Gamma}_2$, in which the self-energy $\Sigma = 2\overline{\Gamma}_2$ is written not as a sum of bare graphs, but as a sum of graphs with bare vertices and dressed lines and legs β_1. Such equations are easily derived by summing the bare graphs of Σ, but it is much more difficult to show that the sum of the graphs of Σ obtained in this manner is actually the variational derivative with respect to β_2 of some functional, in other words, that the Dyson equation is actually a stationarity equation.

Returning to the analysis of the graphs of Γ, let us discuss the question of the coefficients of the graphs. First we rewrite the definition (6.10) for $m = 2$, expressing the variables α in terms of β:

$$\Gamma = W - A_1\beta_1 - A_2(\beta_2 + \beta_1^2)/2 \cdot \qquad (6.79)$$

We know (see Sec. 1.7.4) that the functional W is the sum of $1/2 \cdot \mathrm{tr} \ln \Delta$ and all connected graphs with line Δ and vertices A_n, $n \neq 2$. We write $W = 1/2 \cdot \mathrm{tr} \ln \Delta + 1/2 \cdot A_1 \Delta A_1 + \overline{W}$, isolating the graph $\frac{1}{2}$ •—• $= \frac{1}{2} A_1 \Delta A_1$ from W, and we substitute W in this form into the right-hand side of (6.79):

$$\overline{\Gamma} = \frac{1}{2} A_1 \Delta A_1 - A_1\beta_1 - \frac{1}{2} A_2(\beta_2 + \beta_1^2) + \frac{1}{2} \mathrm{tr} \ln \Delta - \frac{1}{2} \mathrm{tr} \ln \beta_2 + \overline{W}. \qquad (6.80)$$

We express the bare quantities A_1, A_2, and Δ in terms of β using (6.78):

$$A_1 = \beta_2^{-1}\beta_1 + 2\overline{\Gamma}_2\beta_1 - \overline{\Gamma}_1, \quad A_2 = -\beta_2^{-1} - 2\overline{\Gamma}_2,$$

$$\Delta = -A_2^{-1} = [\beta_2^{-1} + 2\overline{\Gamma}_2]^{-1} = \beta_2 - \beta_2 2\overline{\Gamma}_2\beta_2 + \dots. \qquad (6.81)$$

Using these expressions, it is easily verified that the sum of the first three terms on the right-hand side of (6.80) is

$$1/2 \cdot \mathrm{tr}\, 1 + \beta_2\overline{\Gamma}_2 + 1\text{–reducible graphs}. \qquad (6.82)$$

We are not interested in the explicit form of the 1-reducible graphs, because we plan to select the 2-irreducible part of (6.80), in other words, to equate the sums of the 2-irreducible graphs on each side of this equation. Here all 2-reducible and, of course, 1-reducible graphs will simply be discarded.

Collecting the logarithms in (6.80) and taking into account (6.81), we obtain

$$-1/2 \cdot \mathrm{tr} \ln (\beta_2 \Delta^{-1}) = -1/2 \cdot \mathrm{tr} \ln (1 + 2\beta_2\overline{\Gamma}_2). \qquad (6.83)$$

The first term in the expansion of the logarithm is 2-irreducible and exactly cancels the term $\beta_2\overline{\Gamma}_2$ in (6.82) (in our notation, $\beta_2\overline{\Gamma}_2 \doteq \mathrm{tr}\, \beta_2\overline{\Gamma}_2$), and all the other terms in the expansion of the logarithm (6.83) are 2-reducible, because they are represented graphically as rings with self-energy insertions.

These arguments together with the 2-irreducibility of all the graphs of $\overline{\Gamma}$ demonstrated earlier prove the equation $\overline{\Gamma} = 1/2 \cdot \mathrm{tr}\, 1$ + 2-irreducible part of $\overline{W}$, on the right-hand side of which the bare quantities A_1 and Δ are assumed to be expressed in terms of β by (6.81). Clearly, in selecting the 2-irreducible part the line Δ can simply be replaced by β_2, since the inclusion of additional terms in (6.81) for Δ leads to graphs with self-energy insertions, which are certainly 2-reducible. For the same reasons, the bare potential A_1, which enters into the graphs of $\overline{W}$ in the form of a vertex connected by a single line Δ to the rest of the graph, can simply be replaced by $\beta_2{}^{-1}\beta_1$. The final result can be written as

$$\Gamma = 1/2 \cdot \mathrm{tr}\, 1 + 1/2 \cdot \mathrm{tr} \ln \beta_2 +$$

$$+ \text{ 2–irreducible part of } \overline{W}(A_1 = \beta_2^{-1}\beta_1,\ \Delta = \beta_2)\,. \tag{6.84}$$

This equation, which is the analog of (1.229) for the first transform, establishes a one-to-one correspondence between the graphs of Γ and the 2-irreducible graphs of W. We can use it to construct the graphs of Γ without iterating Eq. (6.75), by simply selecting the needed graphs in the known diagrammatic expansion of W. We note that the replacement $\Delta \to \beta_2$, $A_1 \to \beta_2^{-1}\beta_1$ amputates the line connecting the vertex A_1 with the rest of the graph in the graphs of W, i.e., after this replacement the leg β_1 will be connected directly to the vertices A_n, $n > 2$, without intermediate lines.

We essentially need Eq. (6.75) only to the extent that it can be used for a simple proof of the 2-irreducibility of all the graphs of Γ. Without this it would be difficult to obtain (6.84), but after that equation is obtained there is no longer any sense in constructing the graphs of Γ by iterations of the equations of motion. We recall that the equations of motion played the same role in both the analysis of W (the proof of connectedness) and the analysis of the first Legendre transform (the proof of 1-irreducibility).

6.2.7 *The self-consistent field approximation*

As already mentioned in Sec. 1.8.3, for the first Legendre transform the simplest is the loopless approximation (1.230), which is equivalent to first-order perturbation theory in the vertices A_n, $n > 2$, for the functional Γ. This is what is used in specific models in seeking anomalous solutions for β_1. The anomaly is usually that $\beta_1 \neq 0$, while a symmetry of the theory requires $\beta_1 = 0$. All relativistic field models with spontaneous symmetry breaking of the type $\beta_1 = (0|\hat{\varphi}|0) \neq 0$ are versions of the well known Goldstone model [20]. Other examples are the Higgs [67] and Kibble [68] models. In these studies it was shown that the massless Yang–Mills field (in [67] the group is Abelian, and in [68] it is non-Abelian) interacting with a multiplet of scalar fields φ_a acquires a mass upon the appearance of anomalous averages $(0|\hat{\varphi}_a|0) \neq 0$.

From the practical point of view, the most important of the nonrelativistic models of this type is the quantum Bose gas (see Sec. 2.2.1), where the appearance of nonzero anomalous averages for the complex field ψ, ψ^+ leads to superfluidity. We have already noted that the technique for working with anomalous averages like β_1 was first developed in [53] in connection with the problem of the superfluidity of helium. The analogous relativistic models appeared somewhat later.

The anomaly of the next level of complexity after β_1 is that of the propagator β_2, and its study requires the use of the second rather than the first Legendre transform. As a rule, anomalies of this type are sought in "even" theories, in particular, in fermionic theories, for which all the functional variables A, α, β, and γ with odd labels are equal to zero. In such theories it would be simpler and more natural to formulate the problem *ab initio* in the language of only even variables, which is quite possible starting from the complete set of equations of motion in even variables given in Sec. 1.7.3. If the usual formulation of Sec. 6.2.1 is used, where odd potentials A_n with $n \leq m$ are assumed to be arbitrary, then the changeover to the even theory is done only at the time the stationarity equations (6.11) are solved: the odd potentials in these equations are set equal to zero, and a solution is sought with zero values of the odd Green functions α, β, and γ. In the special case of the second transform, the first of the stationarity equations (6.78) for $A_1 = 0$ has a trivial solution, $\beta_1 = 0$, while the second coincides with the usual Dyson equation: $\beta_2^{-1} = \Delta^{-1} - \Sigma(\beta_2)$, in which the self-energy Σ is written as a sum of graphs with dressed lines β_2. Of course, this is true for both bosonic and fermionic theories. In generalizing the variational formulation of Sec. 6.2.1 to the fermionic theory, all functional variables with odd indices should, of course, be treated as anticommuting quantities of the fermionic type, and even variables should be considered bosonic.

The analog of the loopless approximation for the first Legendre transform will in the case of the second transform be the Hartree–Fock self-consistent field approximation, in which only graphs of first order in the vertices (i.e., in the potentials A_n, $n \geq 3$) are kept in the functional Γ:

$$\Gamma|_{\rm scf} = \text{const} + 1/2 \cdot \text{tr} \ln \beta_2 + \sum_{n=3}^{\infty} A_n \alpha_n^{(2)} . \tag{6.85}$$

The sum appearing here is the particular solution of the Schwinger equation (6.24) for $m = 2$, and the functions $\alpha_n^{(2)}$ are defined in Sec. 6.2.4. In the approximation (6.85), Eq. (6.24) is satisfied exactly, while the connection equations (6.33), (6.35) are satisfied only in lowest order in the vertices. For the connection equations of the second group (6.35) the lowest order is the zeroth, and in this approximation the functions α_n on the right-hand sides of (6.35) are replaced by $\alpha_n^{(2)}$, which is equivalent to discarding the contributions of all the higher connected functions β_k with $k \geq 3$.

Keeping a finite number of graphs in Γ, we can construct the raising operators $\mathcal{D}$ and $\mathcal{D}_\gamma$ for connected and 1-irreducible functions, respectively (see the remark at the end of Sec. 6.2.3). Owing to the nonlinearities in Eqs. (6.34) and (6.56), these operators involve infinite series of graphs even when the number of graphs in Γ is finite. However, it would not be true to say that the higher functions β_n, γ_n obtained using Eqs. (6.15) and (6.58) with the approximate operator $\mathcal{D}$ will directly be the connected and the 1-irreducible functions, respectively, in this approximation. This would imply an impossible increase of the accuracy: as if in going to the second Legendre transform we gave up the use of perturbation theory for β_1 and β_2, with no effect on the higher functions β_n, γ_n with $n > 2$. It is nevertheless assumed that the latter can be represented in the form of the diagrammatic series of perturbation theory, but with dressed lines β_2 and legs β_1. Constructing these functions by means of the raising operator, we obtain reliable answers only up to the order to which Γ is calculated, and in higher orders we obtain only some and not all of the graphs. This absence of some of the graphs causes the approximate functions β_n, γ_n not to be fully symmetric even in their arguments $x_1 \ldots x_n$.

Let us now return to the self-consistent field approximation (6.85). In the stationarity equation for the propagator it corresponds to first-order perturbation theory in the self-energy Σ. For an even theory with a single nonzero vertex A_4 this equation has the form

$$\beta_2^{-1} = \Delta^{-1} - \frac{1}{2}$$  . (6.86)

In the case of a complex field $\varphi = (\psi, \psi^+)$ (for example, for electrons), the propagators Δ and β_2 should be considered as 2×2 matrices, where for the normal solution only their off-diagonal elements, i.e., the Green functions $\psi\psi^+$ and $\psi^+\psi$, are nonzero. For comparison with the usual form of the Hartree–Fock approximation for the Green functions (see [19], for example), we note that in the case of the two-body interaction (2.29) the symmetrized potential A_4 in (6.86) includes both direct and exchange contributions. With the theory of the atom in mind, we also note that the solution of (6.86) is always nonunique owing to the arbitrariness in the choice of signs $\pm i0$ in the denominators of the propagator. Different variants of the level filling correspond to the different choices. The atomic energy is determined by the value of the varied functional at a given stationarity point. In the zeroth-order approximation this quantity is proportional to $\mathrm{tr}\ln\Delta = \ln\det\Delta$ and, naturally, takes different values for different variants of the level filling (see the calculation at the end of Sec. 5.1.11).

In the theory of superconductivity one seeks an anomalous solution for β_2 with nonzero Green functions $\psi\psi$ and $\psi^+\psi^+$, and (6.86) in this case coincides with the Gor'kov equations [21]. The criterion for the stability of a solution (stationarity point) is correct fixed sign of the second variation of Γ at the given point (see Sec. 6.2.13 for more details). This requires a certain fixed sign of the second derivative Γ_{22}, which is equivalent to fixed sign of Γ_{22}^{-1}. In the self-consistent field approximation the inverse kernel Γ_{22}^{-1} is an infinite series:

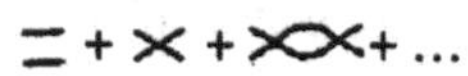

(omitting the coefficients), which is transformed into a progression of the form  when each pair of arguments is contracted with a function of the type $\delta(x-x')f(x)$. The instability of a solution is judged from the violation of the property of correct fixed sign in this progression [21]. We stress the fact that the infinite series of graphs arose only owing to the use of the inverse kernel.

Let us conclude by mentioning the phenomenological theory of the self-consistent Ginzburg–Landau field [69]. In this theory it is postulated that the desired anomalous average (the "order parameter") can be found as the stationarity point of some functional (which is self-evident in the language of the variational principle), and the form of this functional, the simplest form in a known sense, is postulated. In a rigorous approach the varied functional must be found from the microscopic theory; this is the only difference.

6.2.8 *The third Legendre transform*

In analyzing the graphs of the third transform we shall start from Eq. (6.73), which for $m = 3$ takes the form

$$2\Gamma_2 = -\gamma_2^{-1} + \sum_{k,r=1}^{2}$$  $$-2$$  $$-6$$  (6.87)

The various blocks on the right-hand side are given by Eqs. (6.46)–(6.49).

The iteration solution of this equation is constructed according to the general scheme of Sec. 6.2.5 without any fundamental difficulties. The graphs of the first eight orders in the number of lines in the theory with vertices A_3 and A_4 are

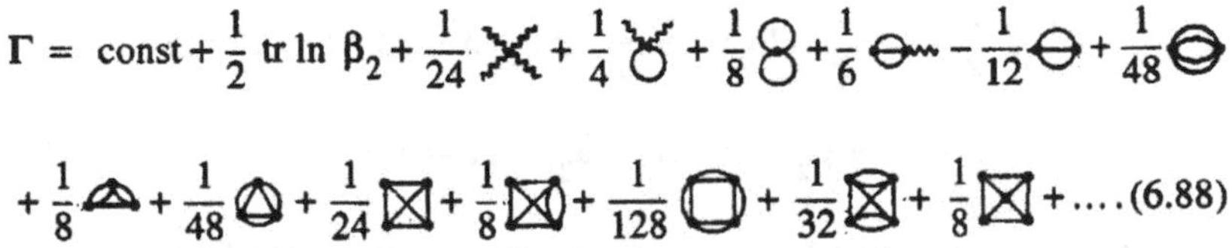
(6.88)

According to the general rules, the lines, legs, and vertex with $n = 3$ in these graphs are dressed, and the vertex with $n = 4$ is bare.

We shall use the term *k-section* of a graph to denote any set of k of its lines, the simultaneous cutting of which causes the graph to be split into two disconnected parts. Connectedness is the absence of 0-sections, 1-irreducibility is the absence of 0- and 1-sections, and 2-irreducibility is the absence of 0-, 1-, and 2-sections. Beginning at $k = 3$, it no longer makes sense to define irreducibility as the absence of all $s \leq k$-sections, since, for example, any graph with vertices γ_3 always has 3-sections: sets of the three lines which connect the vertex γ_3 with the rest of the graph. Therefore, beginning at $k = 3$ one introduces the concept of *nontrivial k-section* [52]: by definition, the section is nontrivial if the cutting of the corresponding lines causes the graph to be split into two nontrivial parts. For $k = 3$ the part which is the simple vertex $\gamma_3 \equiv$  is considered trivial, and for $k = 4$, first, the simple vertex $\gamma_4 \equiv$  and, second, the "Born graph" $\gamma_3\beta_2\gamma_3 \equiv$   are considered trivial. For $k > 4$ the trivial part already includes an infinite number of graphs [66], but in this book we shall deal only with 3- and 4-irreducibility.

Therefore, *k-irreducibility is the absence of nontrivial s-sections for all $s \leq k$.*

Studying the graphs (6.88), we easily see that they are all 3-irreducible and the first few are obtained from (6.76) by the following recipe: (1) all 3-reducible graphs are discarded; (2) the two *singular* 3-irreducible graphs   and   are discarded [these are special because their derivative with respect to the vertex is unconnected]; (3) the sign of the graph  is changed; and (4) the other graphs in (6.76) and their coefficients remain unchanged, except that the vertex with $n = 3$ should now be considered dressed, i.e., $A_3 \rightarrow \gamma_3$.

These rules are actually valid for all orders of perturbation theory. To formulate a precise statement, we isolate from W the prototypes of the graphs which undergo significant changes:

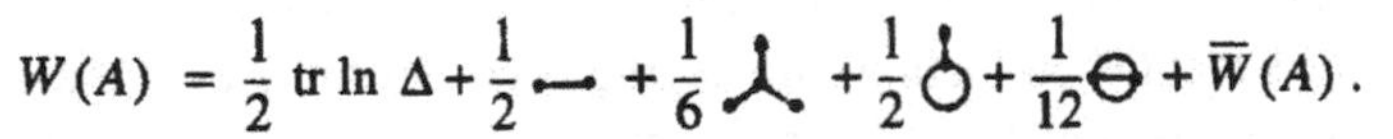
(6.89)

Then

$$\Gamma = \frac{1}{2}\,\mathrm{tr}\,1 + \frac{1}{2}\,\mathrm{tr}\ln\beta_2 - \frac{1}{12}$$  $$+ \text{3-irreducible part of}$$

$$\overline{W}(A_1 \to \beta_2^{-1}\beta_1;\ \ \Delta \to \beta_2;\ A_3 \to \gamma_3)\,. \tag{6.90}$$

The proof is constructed following the same scheme: in the first stage the equations of motion are used to prove the 3-irreducibility of all the graphs of Γ, and in the second stage the 3-irreducible graphs of Γ and W are compared by selecting the 3-irreducible part of (6.10) for $m = 3$.

This is not the only possible method. A complete analysis of the graphs of Γ can be carried out, as first done by de Dominicis and Martin [52], directly from the definitions (6.9) and (6.10). For this it is necessary to substitute into (6.9) the functional W in the form of a diagrammatic series. Then the resulting equations for the bare variables A_n are solved by iteration, so that they are thereby represented as diagrammatic series in dressed variables. Finally, the series for A_n are substituted into the right-hand side of (6.10) and it is verified that all the reducible graphs cancel out. The exact analog of this proof for 0-irreducibility (i.e., connectedness) is the direct verification of the cancellation of all the unconnected graphs in the expansion of $\ln G(A)$, which is equivalent to the proof of connectedness of the logarithm of $R(\varphi)$ given in Sec. 1.4.8.

The drawback of this method is the need to know many nontrivial properties of the symmetry coefficients, on which the proof of cancellation of the reducible graphs on the right-hand side of (6.10) is based. We recall that the proof of the connectedness of the logarithm in Sec. 1.4.8, which is similar in spirit, required knowledge of the relation between the symmetry coefficient of an unconnected graph and the coefficients of its connected components. The proof of the cancellation of the reducible graphs becomes more and more complicated with increasing order of irreducibility, and the results of de Dominicis and Martin [52] are based on numerous combinatorial lemmas taken from other studies. A fairly complete idea of this combinatorics can be obtained from the review by Bloch [70]; there one can also find a great deal of other useful information about the variational principle in statistical physics.

We think that the use of the equations of motion greatly simplifies the proof of irreducibility, making it unnecessary to deal with problems related to the symmetry coefficients.

Returning to our problem, let us prove the 3-irreducibility of all the graphs of Γ. The presence of the last two graphs on the right-hand side of (6.87) clearly shows that the problem we are facing is considerably more difficult than the earlier proofs of 1- and 2-irreducibility. There, in the iterations each of the terms on the right-hand side of the iterated equation generated only graphs with the needed irreducibility property, and the entire proof reduced to verifying this fact. Now this is certainly not so, because in iterations the last terms on the right-hand side of (6.87) generate graphs of the form

(6.91)

with nontrivial 3- and even 2-sections (shown by the dashed line). From this it is clear that if all the graphs of Γ still turn out to be 3-irreducible, it is only because the reducible graphs (6.91) generated by the last terms on the right-hand side of (6.87) cancel with graphs of the same type generated by the other terms on the right-hand side of (6.87).

The direct verification of the cancellation of the reducible graphs is a very complicated matter, but a simple trick was proposed in [63] to prove these cancellations. It amounts to the following. We assume that Γ contains a 3-reducible graph, i.e., a graph of the form  with two nontrivial blocks. Then the derivative Γ_2 necessarily also contains graphs which are generated by cutting one of the lines of the 3-section, i.e., graphs of the form  with nontrivial blocks. We shall refer to these as *graphs with nontrivial 2-section in the direct channel*, or simply *dangerous graphs*. If we can show that the right-hand side of (6.87) calculated in some approximation contains no dangerous graphs, it will then follow that the graphs of the next highest order arising in the iterations will be 3-irreducible. The idea is simple: if there were no cancellation of graphs of the type (6.91), then the right-hand side of (6.87) would necessarily contain graphs with nontrivial 2-section in the direct channel, and if we verify that the latter are not present, we thereby prove indirectly the cancellation of graphs of the type (6.91).

As always, we shall prove by induction. We assume that all the graphs of Γ up to some order are 3-irreducible, and we show that the right-hand side of (6.87) constructed using these graphs cannot contain dangerous graphs with nontrivial 2-section in the direct channel, so that the graphs of the next highest order will also be 3-irreducible.

The proof reduces to the systematic checking of the various possible cases. For example, we assume that the last term on the right-hand side of (6.87) contains a dangerous graph of the form . Clearly, this is possible only when the block $\Gamma_3 \equiv \delta\Gamma/\delta\gamma_3$ contains a graph of the type . However, this contradicts the assumption of the 3-irreducibility of the graphs used to calculate the derivative Γ_3. In fact, differentiation of a graph with respect to γ_3 is equivalent to removing the vertex γ_3 in all possible ways, from which we see that by connecting the external lines of the derivative Γ_3 with the vertex γ_3, we return to the original graphs up to coefficients. If the derivative Γ_3 contains graphs of the form , among the graphs from differentiation of Γ there must be graphs of the form , the 2-reducibility of which contradicts the original assumption.

We have thereby proved that the last term on the right-hand side of (6.87) cannot contain dangerous graphs. Let us now consider the next-to-last term. It has 2-section separating the Γ_2 block from the lower line, but this section is not in the direct channel and so we are not interested in it. A nontrivial 2-section can arise in the direct channel only from graphs of the type  in the derivative 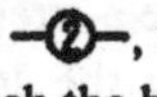, but this again contradicts the induction assumption, according to which the block  must consist only of 3-irreducible graphs.

Let us now look at the terms containing the blocks , which are given by (6.46)–(6.49) and have a rather complicated structure. As an example, we consider the contribution to (6.87) from the block with $k = 1$ and $r = 2$:

(6.92)

The quantity Γ_{33}^{-1} entering here is given in the form of a series in (6.74).

The checking of the various cases is quite tedious, but involves no fundamental difficulties, and in the end one finds that the presence of nontrivial 2-sections in the direct channel in the graphs (6.92) always contradicts the induction assumption that the graphs from which the blocks in (6.92) are constructed are 3-irreducible. The sections shown by the dashed line in (6.92) are not dangerous, because the right-hand cut block, the vertex γ_3, is trivial. A dangerous 2-section in the block  can arise only from graphs of the form  in the derivative $\Gamma_{12} \equiv$ , but the presence of such graphs contradicts the induction assumption guaranteeing the 3-irreducibility of all the graphs of the block . The other terms of (6.92) can be analyzed similarly, and it is important not to overlook the possibility that there exist 2-sections in which one of the cut lines belongs to one block, and the other to the other block. However, the total number of possible types of 2-section is finite, and the result will be the same: the presence of dangerous 2-sections on the right-hand side of (6.87) contradicts the assumption of the 3-irreducibility of the graphs from which the blocks entering into (6.87) are constructed. We take this as conclusive proof of the 3-irreducibility, and refer the interested reader to [63] for more details.

The second part of the proof of (6.90), comparison of the coefficients of the 3-irreducible graphs of Γ and W, is also completely standard. It is necessary to select the 3-irreducible part of Eq. (6.10) for $m = 3$ after isolating from W the singular graphs (6.89) and from Γ the term $1/2 \cdot \text{tr} \ln \beta_2$, and expressing the potentials A_1, A_2, and A_3 on the right-hand side of (6.10) in terms of Γ using the stationarity equations (6.11).

Let us present without derivation a generalization of Eqs. (1.229), (6.84), and (6.90). In changing from universal to ordinary notation, terms of the type $\text{tr} \ln \Delta$, $\text{tr} \ln \beta_2$ enter from each of the fields (for fermionic components they have the opposite sign). If a field is complex and the propagator is taken to be the Green function $\psi\psi^+$, the coefficient of the $\text{tr} \ln$ is doubled. For example, for the Yukawa interaction studied in Sec. 1.4.6 (including quantum electrodynamics), instead of $1/2 \cdot \text{tr} \ln \beta_2$ we must write $1/2 \cdot \text{tr} \ln \beta_2 - \text{tr} \ln \beta'_2$, where β_2 is the propagator of the bosonic field φ and β'_2 is the fermionic propagator $\psi\psi^+$.

The rules for constructing the graphs of Γ generalize directly. The Legendre transform contains all the irreducible skeleton graphs with the usual (as in W) coefficients and signs with the following exceptions, concerning only completely dressed (i.e., not containing bare variables) graphs. First, all irreducible *singular graphs* with unconnected derivative with respect to any of the dressed variables are discarded; second, the sign of graphs quadratic in the dressed vertices is changed.

6.2.9 The fourth transform

Equation (6.73) for the fourth transform has the form [64]

$$2\Gamma_2 = -\gamma_2^{-1} + \sum_{k,\, r = 1}^{3} \; \blacktriangleleft\!\boxed{k|r}\!\blacktriangleright \; - x, \tag{6.93}$$

where

$$x = 2 \; \text{\char} + 6 \; \text{\char} + 3 \; \text{\char} + 3 \; \text{\char} + 12 \; \text{\char} + \\ + 12 \; \text{\char} + 12 \; \text{\char} + 6 \; \text{\char} + 24 \; \text{\char} . \tag{6.94}$$

The graphs obtained by iterating (6.93) through eighth order in the number of lines for the theory with vertices A_3 and A_4 have the form

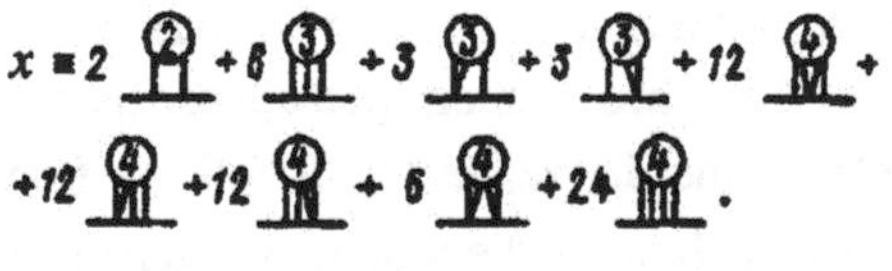

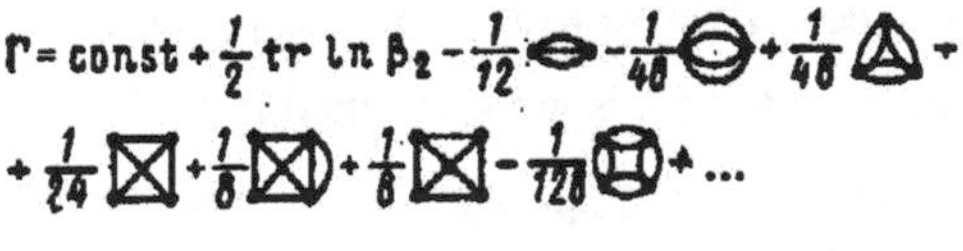

$$\Gamma = \text{const} + \tfrac{1}{2} \text{tr} \ln \beta_2 - \tfrac{1}{12} \, \ominus - \tfrac{1}{48} \, \ominus + \tfrac{1}{48} \, \triangle + \\ + \tfrac{1}{24} \, \boxtimes + \tfrac{1}{8} \, \boxtimes + \tfrac{1}{8} \, \boxtimes - \tfrac{1}{128} \, \ominus + \dots . \tag{6.95}$$

In such a theory the fourth transform is complete, so Γ does not depend on $\gamma_1 = \beta_1$, and all the vertices and lines in (6.95) will be dressed.

All the graphs in (6.95), except for the last one, are 4-irreducible, i.e., they do not have nontrivial 4-subgraphs. We recall that for $n = 4$ the simple vertex and the graph are considered trivial.

The irreducibility properties of the graphs of Γ are analyzed just as in the preceding sections. By assuming that the first graphs of Γ are 3-irreducible and verifying that the right-hand side of (6.93) calculated from these graphs cannot contain dangerous 2-sections of the form with two nontrivial (i.e., different from the simple vertex γ_3) blocks, we thereby prove the 3-irreducibility of all the graphs of Γ.

Systematic analysis of all the cases shows that for the induction assumption that the first graphs are 3-irreducible, the blocks and the first five terms in (6.94) cannot contain either 2- or 3- sections of the form with two nontrivial blocks. As an example, let us consider the fifth term in (6.94) and assume that it contains the 3-section shown by the dashed line in the graph (6.96*a*) and in more detail in the graph (6.96*b*):

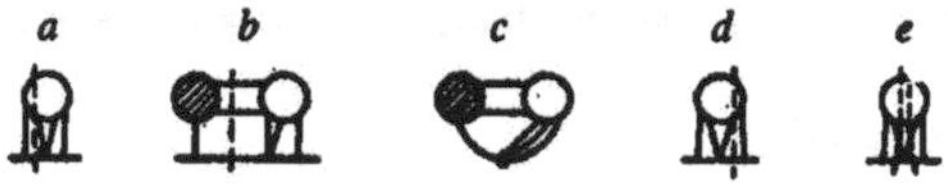

(6.96)

According to the induction assumption, the graph (6.96*c*), obtained by reducing all the external lines of the derivative Γ_4 to a point, must be 3-irreducible, i.e., all its 3-sections must be trivial. This implies that the shaded block in the graphs (6.96*b*) and (6.96*c*) must be a simple vertex γ_3. However, then the 3-section of (6.96*b*) is not dangerous, since the cut block on the left is the trivial graph .

Other variants of the 3-sections of this graph are shown in (6.96*d*) and (6.96*e*). The section of (6.96*d*) is like that of (6.96*a*), and we only need to consider the graphs (6.96*e*). These sections presuppose the presence of graphs of the form in the derivative Γ_4. It follows from the induction assumption of the 3-irreducibility of the first graphs that the two blocks of the graph must be trivial, because the block must be 3-irreducible. Accordingly, the only dangerous graph in the derivative Γ_4 is the graph . It is the derivative with respect to γ_4 of the fifth-order graph , which, as seen from (6.95), is not contained in Γ. Therefore, there are no dangerous graphs of the form at all in the derivative Γ_4, which proves the absence of dangerous 3-sections in the graphs considered. The absence of dangerous 3-sections can be proved in the same way for the other terms on the right-hand side of (6.93), except for the last four graphs of (6.94), to which we now turn.

Let us first discuss the possible presence in these graphs of dangerous 2-sections. Clearly, they can arise only from graphs of the form ▬●▬●▬ in the derivative Γ_4, but, as we have just proved, there are none. Therefore, under the conditions of the induction assumption that the first graphs of Γ are 3-irreducible, there can be no graphs with dangerous 2-sections on the right-hand side of (6.93). This proves the 3-irreducibility of all the graphs of Γ.

Turning now to 4-irreducibility, let us consider the possible presence in the last four terms of (6.94) of graphs with dangerous 3-sections. Clearly they can arise only from graphs of the form ▬●▬●▬ in the derivative Γ_4. There actually are such graphs; in particular, the sixth-order graph in (6.95) gives a contribution ✕○✕ to Γ_4.

If we had made an induction assumption about 4-irreducibility, which is in fact valid for the first graphs of (6.95), the only thing that would follow would be that the blocks entering into the dangerous graphs ▬●▬●▬ of the derivative Γ_4 must be trivial. When we were dealing with 3-irreducibility, the dangerous graph with trivial blocks was absent, but now this is not so; we have the following three 4-irreducible graphs:

$$\tfrac{1}{48}\,\triangle\! + \tfrac{1}{8}\,\boxtimes + \tfrac{1}{8}\,\boxtimes , \qquad\qquad (6.97)$$

which generate in Γ_4 graphs of the form ▬●▬●▬. When substituted into the last terms of (6.94) these lead to dangerous 3-sections, so that in the next order they generate 4-reducible graphs. The latter in turn also generate contributions of the form ▬●▬●▬ to Γ_4, which leads to the appearance of new 4-reducible graphs, and so on.

Therefore, Γ contains an infinite number of 4-reducible graphs, and the problem can be viewed as how to completely characterize them. To do this, we select from both sides of (6.93) graphs with dangerous 3-section in the direct channel, which, we know, can be contained only in the last four terms of (6.94):

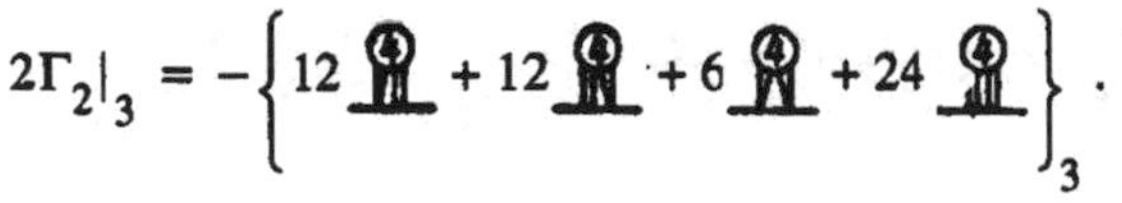

$$2\Gamma_2|_3 = -\left\{ 12\,\substack{\oplus\\\hspace{0.5pt}} + 12\,\substack{\oplus\\\hspace{0.5pt}} + 6\,\substack{\oplus\\\hspace{0.5pt}} + 24\,\substack{\oplus\\\hspace{0.5pt}} \right\}_3 . \qquad (6.98)$$

The symbol $|_3$ indicates that only graphs of the needed type are selected.

Substituting the contributions of the graphs (6.97) into the right-hand side of (6.98), we find the first 4-reducible graphs, and all the others are obtained by iterating (6.98).

Let us introduce the asymmetric vertex without external lines:

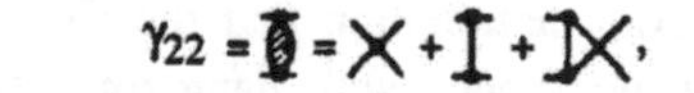
(6.99)

which differs from the amputated function β_4 only by the absence of the graph  in the direct channel. Taking into account the full symmetry of Γ_4, Eq. (6.98) can then be rewritten as

(6.100)

Let us consider the first steps of the iteration procedure. Taking Γ to be the sum of the three graphs in (6.97), we find the derivative $\Gamma_4 \equiv \delta\Gamma/\delta\gamma_4$:

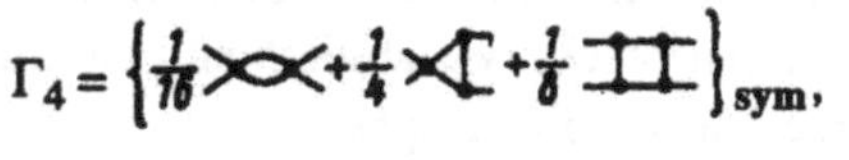
(6.101)

where sym denotes complete symmetrization in the arguments 1–4. In substituting (6.101) into (6.100) it is necessary to keep only those contributions to Γ_4 which contain a dangerous 2-section of the form . For example, after symmetrization

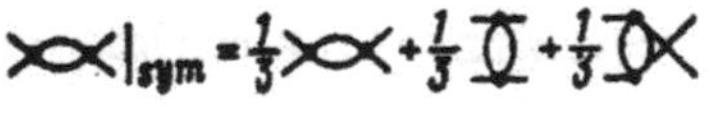
(6.102)

it is necessary to keep only the first term on the right-hand side. For the second of the graphs in (6.101) the dangerous 2-section occurs, as in (6.102), in only one of the three channels, and for the last graph in (6.101) it occurs in two of the three channels. Therefore, the "dangerous part" of (6.101) is written as

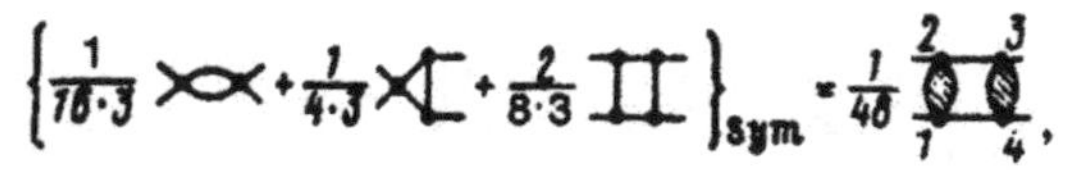
(6.103)

where Sym denotes not complete symmetrization in the arguments 1–4, but the partial symmetrization characteristic of the kernel (6.99): $1 \leftrightarrow 2$, $3 \leftrightarrow 4$,

(12) $\leftrightarrow$ (34). Then, substituting (6.103) into (6.100), we obtain the equation

$$\Gamma_2|_3 = -\frac{1}{16}$$  , (6.104)

which must determine the first 4-reducible graphs of Γ.

Clearly, the right-hand side of (6.104) can be otained by differentiating the eight ring lines of the following graph with respect to γ_2:

$$-\frac{1}{16\cdot 8}$$  (6.105)

(we recall that differentiation with respect to γ_2 is equivalent to cutting a line). However, it would not be true to say that the graph (6.105) is the 4-reducible part of Γ in this approximation. The point is that in the derivative of (6.105) with respect to γ_2, the dangerous graphs with 3-sections come not only from differentiation with respect to ring lines, but also from differentiation of the "crossbars" γ_{22}, which gives in Γ_2 the additional term

$$-\frac{4\cdot 2}{16\cdot 8}$$  (6.106)

not appearing on the right-hand side of (6.104). The factor of 4 in (6.106) is the number of crossbars, and the factor of 2 is inserted because each crossbar contains [see (6.99)] two graphs . We note immediately that this problem arose only because there are four crossbars. If there were more, differentiation with respect to them would not produce dangerous graphs.

Therefore, we must add to the graph in (6.105) a graph or graphs to compensate for the additional dangerous graphs (6.106) in the derivative Γ_2, which are not present on the right-hand side of (6.104).

The dangerous part of (6.106) has the form

$$-\frac{4\cdot 2\cdot 4}{16\cdot 8}$$  (6.107)

since only the graphs  of the vertex (6.99) contribute to it. They generate $2^3 = 8$ terms, of which only the four graphs of the form (6.107) have 3-section in the direct channel. Clearly, the graph (6.107) is obtained by differentiating the graph in the shape of a cube with respect to γ_2, so to cancel the graph (6.107) in the derivative Γ_2 it is necessary to add to (6.105) a cube with coefficient $4\cdot 2\cdot 4/16\cdot 8\cdot 12 = 1/48$ (all twelve lines of the cube are equivalent).

Thus, in this approximation the 4-reducible part of Γ is the sum of the graph (6.105) and the cube with coefficient 1/48.

The following steps of the iteration procedure are easier than this, because now the differentiation with respect to crossbars no longer leads to dangerous graphs. For example, finding Γ_4 from (6.105) (the anomalous cube has no vertex γ_4, so is not involved in the subsequent iterations), we obtain

$$\Gamma_4 = -\frac{4}{16 \cdot 8}$$  $\Big|_{\text{sym}}$. (6.108)

In substituting this expression into (6.100) it should be remembered that only one of the three channels in (6.108) leads to dangerous graphs, so in this approximation

$$\Gamma_2|_b = \frac{1}{32}$$  ,

from which we see that here the 4-reducible part of Γ has the form of a ring with five crossbars γ_{22} with coefficient 1/32·10. In the next step of the iterations we obtain a ring with six crossbars, and so on. The final result is

$$\text{4-reducible part of } \Gamma = \frac{1}{48}$$  $$+ \sum_{n=4}^{\infty} \frac{(-1)^{n+1}}{n2^{n+1}}$$  . (6.109)

The coefficients of the rings correspond to the series for the logarithm, so the sum of rings can be written compactly as $1/2 \cdot \mathrm{tr}\, \ln_3(1+ \gamma_{22}\beta_2{}^2/2)$, where $\ln_3$ denotes the logarithm minus the first three terms of its series expansion.

The 4-irreducible graphs of Γ are analyzed, as usual, by selecting the 4-irreducible part of (6.10). Here we present only the result. Γ contains all the 4-irreducible graphs of the functional $W(A_1 \to \beta_2{}^{-1}\gamma_1,\ \Delta \to \beta_2,\ A_3 \to \gamma_3,\ A_4 \to \gamma_4)$ with their coefficients, except for *singular graphs*, which have unconnected or 1-reducible derivatives with respect to any of the variables γ_i. These are, first, the graph  $=\gamma_1\beta_2{}^{-1}\gamma_1$, obtained from the graph  $= A_1\Delta A_1$ by the above substitution of the arguments of W, and, second, the singular 4-irreducible graphs

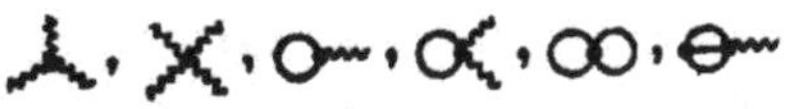

having unconnected derivative with respect to one of the vertices γ_3 or γ_4, and the graph 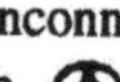, having 1-reducible derivative with respect to γ_4. Regarding the coefficients, the exceptions are the graphs  and  quadratic in the vertices, which enter Γ with a minus sign, while all the graphs enter W with a plus sign. We note that exactly the same "rules for exceptions" applied for all the lower-order Legendre transforms.

We also note that these rules, like the statement (6.109), are valid for all theories with any number of nonzero potentials A_n.*

6.2.10 *Stationarity equations, renormalization, and parquet graphs*

According to the general scheme of Sec. 6.2.1, for known Γ the specific values of the Green functions (α, β, γ) corresponding to given potentials A are found by solving the stationarity equations (6.11). For the complete transform the potentials enter only into the right-hand sides of these equations, and the functional Γ itself is independent of A.

In Sec. 1.9 it was shown that the renormalization transformation of the field is equivalent to the transformation (1.238) of the action functional and the normalization coefficient of the functional integral (6.8). The coefficient changes only the additive normalization constant in the functionals W and Γ, which is unimportant in the stationarity equations (6.11), as they contain only derivatives of Γ. (The renormalization transformation of Γ taking into account additive constants is discussed in more detail in the following sections.) Therefore, in the stationarity equations the renormalization transformation reduces only to a change of their right-hand sides: if the original bare potentials are substituted as the A_n, the solution will be the Green functions corresponding to these potentials, and if the transformed (renormalized) potentials are substituted as the A_n, the solution will be the transformed (renormalized) Green functions. The graphs of Γ and its derivatives with respect to dressed variables on the left-hand sides of (6.11) themselves do not change.

If Γ is written as a diagrammatic series and (6.11) is solved by iteration, constructing the desired Green functions as series in bare vertices, we of course arrive at the usual diagrammatic expansions of perturbation theory. In theories with local potentials it is very simple to get rid of all the bare and

---

* It is useful to keep this in mind even when studying a theory with zero A_n for $n > 4$. Higher ($n > 4$) potentials can be introduced at an intermediate stage, and later set equal to zero. Then one can easily obtain skeleton-graph representations for the higher functions β_n by selecting the connected part of the derivative $\delta\Gamma/\delta A_n = \alpha_n$. In these representations all skeleton graphs will be 4-irreducible (i.e., they will have no nontrivial 1-, 2-, 3-, and 4-subgraphs), because the reducible part of Γ (6.109) does not contain higher potentials A_n with $n > 4$ and therefore does not contribute to derivatives with respect to these potentials. When constructing the higher functions β_n using the raising operator (6.56), the irreducibility of the resulting graphs is not at all obvious *a priori*.

renormalization constants in (6.11) by making the number of subtractions needed for the right-hand sides of these equations to vanish. As a result, we obtain closed equations satisfied by both the unrenormalized and the renormalized Green functions. Iterating these equations in the appropriate way, we arrive at the diagrammatic series of renormalized perturbation theory; the renormalized parameters appear in the solution as subtraction constants.

Let us consider in more detail the theory with four potentials $A_1 \ldots A_4$, for simplicity assuming that it is even: $A_1 = A_3 = 0$. For this theory the fourth transform is complete, and the stationarity equations (6.11) always have an even solution with $\gamma_1 = \gamma_3 = 0$, since Γ is even in these variables (the complete transform is completely independent of γ_1). It is this normal even solution that we shall now consider, leaving aside models of the type in [20] with spontaneous symmetry breaking $\varphi \to -\varphi$.

Owing to the invariance of the stationarity condition under any change of variable, Eqs. (6.11) can be written as $\delta\Phi/\delta\gamma_k = 0$, where Φ is the functional (6.12). In the case of interest to us, only two stationarity equations will have content, namely,

$$\Gamma_4 + A_4 \frac{\delta\alpha_4}{\delta\gamma_4} = 0, \quad \Gamma_2 + A_2 \frac{\delta\alpha_2}{\delta\gamma_2} + A_4 \frac{\delta\alpha_4}{\delta\gamma_2} = 0, \tag{6.110}$$

where, as usual, $\Gamma_k \equiv \delta\Gamma/\delta\gamma_k$. From the definitions of α, β, and γ we easily obtain $\delta\alpha_4/\delta\gamma_4 = \beta_2{}^4/24$, so that

$$-A_4 = 24\,\Gamma_4\beta_2^{-4} = 24$$  . (6.111)

The block on the right-hand side denotes the derivative Γ_4 with amputated external lines. The second of the stationarity equations (6.110) determines the self-energy $\Sigma = \Delta^{-1} - \beta_2{}^{-1}$ as a functional of the dressed variables β_2 and γ_4. It is equivalent to the ordinary Schwinger equation for Σ:

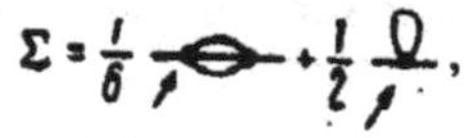 (6.112)

in which the bare vertices A_4, indicated by the arrows, are assumed to be expressed in the variables β_2, γ_4 via (6.111).

Equation (6.111) represents A_4 by skeleton graphs

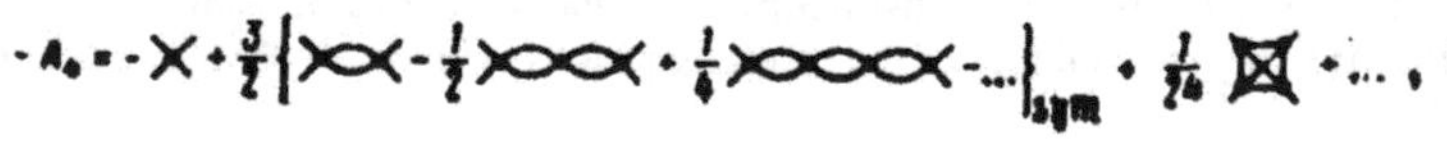 (6.113)

where sym denotes complete symmetrization in the arguments 1–4. The first term on the right-hand side, the vertex γ_4, is generated by the graph , the first term in the curly brackets is generated by the first of the 4-irreducible graphs (6.97), and the other terms are generated by the 4-reducible graphs (6.109). We note that in an even theory the kernel (6.99) coincides with γ_4. The terms following the curly brackets in (6.113) represent the contribution of the 4-irreducible graphs of Γ above sixth order in the number of lines. In an even theory the first such contribution is the "envelope" in (6.113), which in Γ corresponds to a complete graph with five vertices. We recall that a *complete graph is one in which any pair of vertices is connected by a single line*. The symmetry group of a complete graph contains all permutations of its n vertices, and the symmetry number is $n!$.

Therefore, using functional Legendre transforms it is possible to automatically obtain representations of the type: *a dressed vertex is a bare vertex plus the sum of skeleton graphs*. Here the explicit definition of the reducible part of Γ contains the solution of the combinatorial problem of overlap in the vertices, because it is the reducible part of Γ and only this part which generates graphs with nontrivial vertex subgraphs on the right-hand side of (6.113) and equations analogous to it. Overlap in the vertex appears beginning at $n = 4$, so in the third Legendre transform there was no 3-reducible part.

Solving (6.113) iteratively, we represent the dressed vertex γ_4 as a sum of graphs with bare vertices. If in (6.113) we discard the contributions of all nontrivial 4-irreducible graphs, we obtain the equation of the parquet approximation [after symmetrization, the progression in the curly brackets in (6.113) enters into each of the three channels with coefficient 1/2]. The iterations of this equation correctly reproduce all the parquet graphs of γ_4 together with their coefficients. A parquet graph is a graph whose minimal 4-subgraphs are only  and ; a minimal 4-subgraph is a graph which does not contain any other 4-subgraphs. We note that in the literature (see [71], for example) the term parquet is often used to refer to other equations which reproduce not all the parquet graphs, but only the part of them which gives the leading contribution in some approximation.

In relativistic field theory with the interaction $(\lambda/24)\int dx\varphi^4(x)$, the potential A_4 has the form $i\lambda\delta\,(x_1 - x_2)\delta(x_2 - x_3)\delta(x_3 - x_4)$. By transforming to momentum space in (6.113) and making a single subtraction, we make the left-hand side vanish; in the equation for the self-energy (6.112) the bare potential $A_2 = -\Delta^{-1}$ is removed by two subtractions. The iteration solution of the resulting equations leads to the series of renormalized perturbation theory. The single subtraction in (6.113) does not remove the overlapping divergences present in graphs with nontrivial vertex subgraphs, but in the iteration process the overlapping divergences of different graphs will cancel in each order of perturbation theory.

6.2.11 *Symmetry properties of the complete Legendre transform and the "spontaneous interaction"*

Let us consider the group of renormalization transformations of the field φ defined in Sec. 1.9, i.e., the set of all linear inhomogeneous transformations of the form $\varphi = Z\varphi' + c$ with nonzero Jacobian $D\varphi/D\varphi' = \det Z$.

Let $A(\varphi)$ be the series (6.7) plus the zeroth potential A_0. The equation $A(\varphi) = A'(\varphi')$ determines the induced transformation of the potentials $A \to A'$. For a polynomial theory with a finite number of nonzero potentials the transformation $A \to A'$ does not change the degree of the polynomial. Changing the integration variable in (6.8), we obtain $G(A) = G(A')\det Z$, from which $W(A) = W(A') + \mathrm{tr}\ln Z$.

Let $\Gamma(\alpha)$ be the Legendre transform of $W(A)$ with respect to all the potentials A. The constant A_0 enters into $W(A)$ additively, so that the conjugate variable $\alpha_0 = \partial W/\partial A_0$ is identically equal to unity. The constant A_0 cancels on the right-hand side of (6.10).

The potential transformation $A \to A'$ induces a transformation of the conjugate variables α: $\alpha(A) \to \alpha(A') \equiv \alpha'(A)$. Using (6.8) and (6.9), it is easily verified that $\alpha \to \alpha'$ is the ordinary renormalization transformation of the full Green functions without vacuum loops (see Sec. 1.9), and the induced transformations $\beta \to \beta'$, $\gamma \to \gamma'$ are the renormalization transformations of the connected and 1-irreducible Green functions, respectively: $\beta'_1 = Z^{-1}(\beta_1 - c)$ and $\beta'_n = (Z^{-1})^n\beta_n$, $\gamma'_n = (Z^{\mathrm{T}})^n\gamma_n$ for $n > 1$. The vertices $\bar{\gamma}_n$ defined by (6.69) are invariant under renormalization transformations if the renormalization transformation of the quantity $\beta_2^{1/2} \equiv \lambda$ is given by $\lambda' = \lambda Z^{-1\mathrm{T}}$, consistent with the definition $\beta_2 = \lambda^{\mathrm{T}}\lambda$ and the transformation law $\beta'_2 = (Z^{-1})^2\beta_2 = Z^{-1}\beta_2 Z^{-1\mathrm{T}}$. Then $\bar{\gamma}'_n = (\lambda Z^{-1\mathrm{T}})^n(Z^{\mathrm{T}})^n\gamma_n = \bar{\gamma}_n$, as stated.

Let us now consider the transform $\Gamma(\alpha)$. The quantity

$$\sum_n A_n\alpha_n = \sum_n A_n\delta W/\delta A_n = \int D\varphi A(\varphi)\exp A(\varphi) \Big/ \int D\varphi \exp A(\varphi) \tag{6.114}$$

is explicitly invariant under transformations $A \to A'$. Taking α to be the independent variables in the definition (6.10) and using (6.114) and the rule of transformation of W obtained earlier, it is easy to show that $\Gamma(\alpha') = \Gamma(\alpha) - \mathrm{tr}\ln Z$. The same conclusion can be arrived at by a different route: in Sec. 6.2.4 it was shown that for the complete Legendre transform the functional Γ is independent of $\gamma_1 = \beta_1$ and can be written as $1/2 \cdot \mathrm{tr} \ln \beta_2 + F(\gamma)$, where F is a functional of the invariant vertices (6.69). The vertices $\bar{\gamma}$ do not change in a renormalization transformation, and $\mathrm{tr} \ln \beta_2$ is replaced by $\mathrm{tr} \ln \beta'_2 = \ln \det [Z^{-1}\beta_2 Z^{-1\mathrm{T}}] = \mathrm{tr} \ln \beta_2 - 2 \mathrm{tr} \ln Z$.

It follows from the transformation formulas obtained above that the functionals W and Γ are invariant under renormalization transformations which have unit Jacobian $D\varphi/D\varphi' = \det Z$ (in the terminology of Sec. 1.7.3, transformations from the group of motions of the field φ). Translations, reflections, and rotations of the field φ are among such transformations, as are gauge transformations.

The desired Green functions are defined as the stationarity points of the functional $\Phi(\alpha; A) = \Gamma(\alpha) + \Sigma_n A_n \alpha_n$ with respect to variations of α for fixed A. The functional Φ is always invariant under simultaneous transformations $A \to A'$, $\alpha \to \alpha'$ with $\det Z = 1$, but invariance under transformations $\alpha \to \alpha'$ for fixed A will occur only when the potentials A are themselves invariant, i.e., when $A'_n = A_n$ for all n. Invariance of the potentials is equivalent to invariance of the action functional under the transformation of the field φ in question. In this case $\Phi(\alpha; \mathrm{A}) = \Phi(\alpha'; A') = \Phi(\alpha'; A)$, and if spontaneous symmetry breakdown does not occur, the stationarity point $\alpha \equiv \alpha(A)$ will also be invariant, i.e., $\alpha' = \alpha$.

Let us now briefly discuss the "theory without a seed." We assume that all the potentials A_n in (6.11) are set equal to zero, in other words, the variational method is used to seek the Green functions for a theory with zero action functional. The desired Green functions are then defined as the stationarity point of the functional $\Gamma = 1/2 \cdot \mathrm{tr} \ln \beta_2 + F(\bar{\gamma})$. The stationarity condition is invariant under the choice of independent variables; taking them to be β_1, β_2, and $\bar{\gamma}$, we obtain

$$\delta\Gamma/\delta\beta_1 = 0; \ \delta\Gamma/\delta\beta_2 = 1/2 \cdot \beta_2^{-1} = 0; \ \delta F/\delta\bar{\gamma}_n = 0. \qquad (6.115)$$

The first of these equations is an identity because Γ is independent of β_1, and the second has the formal solution $\beta_2 = \infty$, from which we see that the functional Γ for finite β has no stationarity points at all, and a nontrivial theory with zero action does not exist.

Now let us consider a theory with nonzero potentials A_1 and A_2, but without an interaction: $A_n = 0$ for all $n \geq 3$. In this case

$$\Phi(\alpha;A) = \mathrm{const} + \frac{1}{2}\mathrm{tr} \ln \beta_2 + F(\bar{\gamma}) + A_1\beta_1 + \frac{1}{2}A_2(\beta_2 + \beta_1^2), \qquad (6.116)$$

and the stationarity conditions in the variables β_1, β_2, and $\bar{\gamma}$ take the form

$$A_1 + A_2\beta_1 = 0;\ \beta_2^{-1} + A_2 = 0;\ \delta F(\bar{\gamma})/\delta\bar{\gamma}_n = 0. \qquad (6.117)$$

The first two equations are easily solved: $\beta_2 = -A_2^{-1} = \Delta$, $\beta_1 = \Delta A_1$. This solution coincides with the Green functions of the free theory, so it can be stated that even if the last equations (6.117) have nontrivial solutions with $\bar{\gamma} \neq 0$, this "spontaneous interaction" which arises does not at all affect the first two connected Green functions, which remain exactly the same as in the free theory. This property of the spontaneous interaction was pointed out in [16].

The equations $\delta F/\delta\bar{\gamma}_n = 0$ determining the spontaneous-interaction vertices are typical "equations without a seed," which are often discussed in connection with the problem of scaling (see, for example, [72]–[74]) and in other contexts ("the complete bootstrap"). Here it is worth noting that although these equations possess a very high degree of symmetry, it would be dangerous to conclude that their solutions, if they exist at all, are also symmetric.* The reason is that when there is a very high degree of symmetry, spontaneous breakdown of this symmetry is the rule rather than the exception. There exist symmetry groups which will always be spontaneously broken simply because there are no stationarity, i.e., invariant points at all for these groups in the space of the variables α. A simple example is the transformations induced by translation $\varphi \to \varphi + c$ of the field φ. More interesting examples of such groups can also be given.

We recall that the functional $F(\bar{\gamma})$ is represented as a sum of graphs of the vacuum-loop type with dressed lines and vertices. Each such graph is a simple "polynomial" in the invariant vertices. For example,

$$= \iiint dx_1 dx_2 dx_3 \bar{\gamma}_3(x_1x_2x_3)\,\bar{\gamma}_3(x_1x_2x_3)\,. \qquad (6.118)$$

Let us consider an arbitrary transformation $x' = f(x)$ with nonzero Jacobian $J_f(x) = dx'/dx$, which does not change the region of integration in (6.118). The set of such transformations forms the group of "all coordinate transformations"; for four-dimensional x it contains the Poincaré group as a subgroup.

* If all the functional arguments are replaced by numbers and the functional integral (6.8) is replaced by an ordinary integral, we arrive at the so-called *zero-dimensional theory*, which correctly reproduces the number of graphs, but does not distinguish their values. In this theory functional transformations become numerical transformations, and there are certainly no anomalous solutions, because the functions in question are strictly convex.

Making the substitution $x_i \to x_i'$ for all the arguments in (6.118), we see that the graph (6.118) is invariant under the transformation $\bar{\gamma}_3 \to \bar{\gamma}_3^f$, where by definition

$$\bar{\gamma}_n^f(x_1 \ldots x_n) = \prod_{i=1}^{n} [J_f(x_i)]^{1/2} \bar{\gamma}_n(f(x_1) \ldots f(x_n)) . \qquad (6.119)$$

Obviously, any graph is invariant under such transformations, so that the functional $F(\bar{\gamma})$ itself is also invariant. Using the terminology of relativity theory, it can be said that the equations without a seed (6.117) are generally covariant. However, it is clear that nonzero solutions of these equations, if they exist, cannot be invariant, because functions $\bar{\gamma}_n \neq 0$ which are invariant under all the transformations (6.119) do not exist, in general.* This means that a symmetry expressed as general covariance of the functional $F(\bar{\gamma})$ will always be spontaneously broken.

The group of all coordinate transformations can, of course, be shrunk to some sufficiently small subgroup such that there are invariant points in the space of the variables $\bar{\gamma}$; in this case, the assumption about the symmetry of the solution will no longer be formally inconsistent. This is essentialy what was done in [72]–[74], devoted to justifying the similarity hypothesis in statistical physics, which in the language of the vertices $\bar{\gamma}$ is formulated as their invariance under the subgroup of transformations (6.119) induced by rescalings $x' = \lambda x$ (see also [47]). It is assumed that the bare potentials in the equations can be neglected at the phase-transition point, and then any symmetry of the equations (for example, scale or conformal) will automatically be transferred to their solutions.

However, if we take into account the superhigh symmetry of the "equations without a seed,"** then any conclusion about the compatibility of some symmetry group with the equations becomes meaningless: there are arbitrarily many such groups, but it is known *a priori* that most of them will be spontaneously broken. Each assumption about invariance of the solution (if one exists at all) must be based on arguments which explain exactly why this symmetry will not be spontaneously broken.

---

* If we exclude from consideration symmetrized δ-function products of the form $\delta(x_1 - x_2)\delta(x_3 - x_4)\ldots$ with an even number of arguments x, which are clearly unsuitable owing to their structure. The existence of such "generally covariant" functions was pointed out to the author by A. A. Andrianov.

** We note that the functional $F(\bar{\gamma})$ is invariant also under transformations $\bar{\gamma}_n \to u^n \bar{\gamma}_n$, where u is an arbitrary orthogonal ($u^T u = 1$) operator.

6.2.12 *The ground-state energy*

According to the definitions of Sec. 6.2.1, in the pseudo-Euclidean theory $A(\varphi) = iS(\varphi)$, where S is the action functional. In considering the Legendre transform, we shall first assume that the potentials A_n are arbitrary symmetric functions. However, in practice we are always interested in a theory with potentials of the Lagrangian type, i.e., potentials for which the action functional is written as the integral over time of some real t-local functional, the Lagrangian. In the absence of nonstationary external fields the Lagrangian does not depend explicitly on time, i.e., the potentials A_n are invariant under time translations. If this symmetry is not broken spontaneously, as we shall assume, then all the Green functions will also possess this property.

According to the general rules of Sec. 1.6.5, the integral (6.8) is proportional to the vacuum expectation value of the S matrix, and this quantity is related by (1.74) to the ground-state energy shift. Therefore,

$$W(A) - W_0 = -i\varepsilon(A)\int dt, \tag{6.120}$$

where W_0 is the value of $W(A)$ for the theory, which we assume to be free, and $\varepsilon(A)$ is the difference of the ground-state energies for the complete and the free theory. We note that owing to time-translational invariance, both quantities W and W_0 are proportional to $\int dt$, so that this infinite factor in (6.120) cancels.

We know from Sec. 1.7.4 that $W(A)$ is the sum of $A_0 + 1/2 \cdot \mathrm{tr}\ln\Delta$ and all connected graphs with line $\Delta = -A_2^{-1}$ and vertices A_n, $n \neq 0, 2$. If we take the free theory to be the theory with zero vertices A_n, $n \neq 0, 2$, then $W_0 = A_0 + 1/2 \cdot \mathrm{tr}\ln\Delta$, and these contributions cancel on the left-hand side of (6.120).

In the variational approach the quantity $W(A)$ can be found as the value of the varied functional (6.12) at the stationarity point $\alpha = \alpha(A)$:

$$W(A) = A_0 + \Gamma - \sum_{n=1}^{m} \alpha_n \delta\Gamma/\delta\alpha_n. \tag{6.121}$$

We have taken into account the addition of the constant A_0 to (6.7) and have used the stationarity equations (6.11) to express the bare variables A_n in terms of the dressed variables α.

Equation (6.121) contains the solution of the nontrivial combinatorial problem of resumming the bare vacuum loops [the graphs of $W(A)$] to form skeleton vacuum loops, where the "degree of dressing" is determined by the order of the Legendre transform used. If Γ is a complete transform, then all potentials A_n, $n \neq 0$, are eliminated on the right-hand side of (6.121), and the quantity $W(A)$ is expressed only in terms of α, i.e., in terms of the Green functions of the theory.

When using (6.121) in practice, it is desirable to transform from the unconnected variables α to at least the connected variables β. The sum over n on the right-hand side of (6.121) can be rewritten as $\Sigma_k a_k \Gamma_k$, where $\Gamma_k \equiv \delta\Gamma/\delta\beta_k$ and $a_k = \Sigma_n \alpha_n \delta\beta_k/\delta\alpha_n$. Using the expressions of Sec. 6.2.2, it is easily shown that $a_k = -\delta^k \alpha^{-1}(\varphi)/\delta\varphi^k|_{\varphi=0}$, where $\alpha(\varphi) = \exp\beta(\varphi)$ is the functional (6.16). From this it is clear that the coefficients a_k coincide with the complete functions without vacuum loops in which the sign of all terms containing an even number of connected factors β is changed: $a_1 = \beta_1$, $a_2 = \beta_2 - \beta_1^2$, $a_3 = \beta_3 - 3\beta_2\beta_1 + \beta_1^3$ (symmetrization in the arguments a_k is understood), and so on.

In the case of the complete transform, Γ is independent of the first connected function β_1, which enters only into the coefficients a_k. The entire dependence on β_1 can be isolated explicitly by a renormalization transformation $\varphi = \varphi' + \beta_1$. It induces the transformation $A(\varphi) = A'(\varphi')$ of the potentials A in the functional (6.7), and here the quantity W is not changed (see the preceding section): $W(A) = W(A')$. $W(A')$ can be calculated using (6.121), replacing all quantities in it by their transforms: $A_0 \to A'_0$, $\alpha \equiv \alpha(A) \to \alpha(A') \equiv \alpha'$, where $A'_0 = A(\beta_1)$ is the transformed zeroth potential and α' are the transformed Green functions. It follows from Sec. 1.9 that the transform we are considering does not change the connected functions β_n with $n \neq 1$, but the first connected function vanishes: $\beta'_1 = 0$, $\beta'_n = \beta_n$ for $n \neq 1$. Therefore, after the transformation the terms a_k containing β_1 vanish, and the entire dependence on β_1 turns out to be concentrated in the transformed zeroth potential $A'_0 = A(\beta_1) = iS(\beta_1)$ (we recall that S is the action functional):

$$W(A) = A(\beta_1) + \Gamma - \sum_{k=2}^{m} a_k \Gamma_k \Big|_{\beta_1 = 0}. \tag{6.122}$$

The sum over k begins at two, because $a_1 = \beta_1$ vanishes for $\beta_1 = 0$.

As an example, let us consider the third transform, assuming that it is complete. From the expressions given above for the a_k it follows that for $\beta_1 = 0$, $a_2 = \beta_2$ and $a_3 = \beta_3$, so that (6.122) takes the form $W(A) = A(\beta_1) + \Gamma - \beta_2\Gamma_2 - \beta_3\Gamma_3$. In our case $\Gamma = 1/2 \cdot \mathrm{tr}\, 1 + 1/2 \cdot \mathrm{tr} \ln \beta_2 + F$, where F is the known (see Sec. 6.2.8) sum of 3-irreducible skeleton vacuum loops. Therefore, $W(A) = A(\beta_1) + 1/2 \cdot \mathrm{tr} \ln \beta_2 + F - \beta_2 F_2 - \beta_3 F_3$.

The operators $\beta_2\delta/\delta\beta_2$ and $\beta_3\delta/\delta\beta_3$ acting on F change only the coefficients of the graphs: the first introduces an additional factor $-n_2$, and the second introduces n_3, where n_2 and n_3 are the numbers of lines and vertices, respectively (the minus sign with n_2 appears because in connected variables a line of a graph is associated with β_2^{-1} and not with β_2). For the vacuum loops of this theory $2n_2 = 3n_3$, so that the expression $F - \beta_2F_2 - \beta_3F_3$ is the sum of all the graphs of F with the additional combinatorial factor $1 + n_2 - n_3 = 1 + n_3/2$ for a graph with n_3 vertices. We note that the skeleton vacuum loops are invariant under renormalization dilatations, and tr ln transforms as $\mathrm{tr}\ \ln \beta_2 = \mathrm{tr} \ln Z\beta'_2Z^{\mathrm{T}} = \mathrm{tr} \ln \beta'_2 + 2\mathrm{tr}\ln Z$.

The expressions we have obtained can be used in the relativistic theory of the atom (quantum electrodynamics in the external Coulomb field of the nucleus), where the third transform is complete (see the remark at the end of Sec. 6.2.8). The graphs of W in the relativistic theory contain ultraviolet divergences, and the energy of the atom is the finite difference of the (infinite) energies of the n-electron and vacuum states. These two states correspond to choosing two solutions of the stationarity equations which differ in how the poles are circumvented in the fermionic propagator [see the remark following (6.86)]. The difference between the propagators also leads to a difference between the other Green functions, but both sets of Green functions are renormalized by identical dilatation constants Z, and the renormalization constants cancel when the difference of the values of W for the two solutions is taken.

In conclusion, we note that in statistical physics expressions like (6.120)–(6.122) determine the thermodynamic potential Ω.

6.2.13 *Stability and convexity properties of functional Legendre transforms*

Functional transforms can also be used to efficiently construct the other Legendre transforms mentioned earlier: transforms with respect to numerical parameters entering linearly into the action functional, or transforms with respect to time-independent potentials.

To be specific, let us consider the case of a set of real numerical parameters λ entering linearly into the functional (6.7): $A^{\lambda}(\varphi) = A^{(0)}(\varphi) + \Sigma_i\lambda_iA^{(i)}(\varphi)$. All the $A^{(i)}(\varphi)$ in this expansion are assumed to be fixed functionals with potentials of the Lagrangian type which are translationally invariant in time (see the definition in the preceding section).

The point λ in the space of numerical parameters will be referred to as singular if the point A^{λ} in potential space is singular. By definition (see Sec. 6.1.4), for nonsingular λ the functional variational problem with potentials A^{λ} has a unique solution $\alpha(A^{\lambda})$.

We shall assume that nearly all points λ are nonsingular, i.e., that the dimension of the set of singular points (if they exist) is smaller than the total dimension of the space of parameters λ. We also assume that the region of nonsingular points λ is connected, i.e., that any two nonsingular points can be connected by a trajectory wholly lying in the region of nonsingular points.

We define the numerical Legendre transform of the function $\overline{W}(\lambda) \equiv W(A^{\lambda})$ as

$$\mu_i = \partial\overline{W}(\lambda)/\partial\lambda_i; \ \overline{\Gamma}(\mu) = \overline{W}(\lambda) - \sum_i \lambda_i\mu_i. \tag{6.123}$$

In a time-translationally invariant theory each of the quantities μ_i, $\overline{W}$, and $\overline{\Gamma}$ is proportional to the total "time volume" $\int dt$, and if desired this infinite factor can be canceled out everywhere.

On the other hand, we have the functional Legendre transform

$$\alpha_n = \delta W(A)/\delta A_n; \ \Gamma(\alpha) = W(A) - \sum_n A_n\alpha_n. \tag{6.124}$$

If the point λ is nonsingular, the value of $\alpha(A^{\lambda})$ is determined uniquely. Using the fact that $\overline{W}$ depends on λ only through the potentials A^{λ}, we obtain

$$\mu_i = \frac{\partial\overline{W}}{\partial\lambda_i} = \sum_n \frac{\delta W(A^{\lambda})}{\delta A_n^{\lambda}}\frac{\partial A_n^{\lambda}}{\partial\lambda_i} = \sum_n A_n^{(i)}\alpha_n(A^{\lambda}). \tag{6.125}$$

From this it is clear that for nonsingular λ the value of $\mu(\lambda)$ is uniquely determined by the solution $\alpha(A^{\lambda})$ of the functional variational problem.

We know from Sec. 4.3.2 that $\overline{W}(\lambda)$ is proportional to the convex function $\varepsilon(A^{\lambda})$, and this shows that in the region of nonsingular λ Eqs. (6.125) are uniquely solvable for λ. Therefore, in this region there is a one-to-one correspondence between the variables λ and μ, which allows the solution of the functional variational problem to be treated as a single-valued function of μ: $\alpha(A^{\lambda}) \equiv \alpha^{\mu}$.

Substituting μ in the form (6.125) into the right-hand side of Eq. (6.123) determining $\overline{\Gamma}(\mu)$, we see that the function $\overline{\Gamma}(\mu)$ coincides with the value of the functional $\Gamma(\alpha) + \Sigma_n A_n^{(0)}\alpha_n$ for $\alpha = \alpha^{\mu}$, and for the varied function $\overline{\Phi}(\mu;\lambda) = \overline{\Gamma}(\mu) + \Sigma_i\lambda_i\mu_i$ of the numerical transform we obtain

$$\overline{\Phi}(\mu;\lambda) = \left[\Gamma(\alpha) + \sum_n A_n^\lambda \alpha_n\right]\Big|_{\alpha=\alpha^\mu} = \Phi(\alpha;A^\lambda)\Big|_{\alpha=\alpha^\mu}, \qquad (6.126)$$

where Φ is the varied functional of the functional transform. The right-hand side of (6.126) depends on μ via the Green functions α^μ, and the parameters λ in the potentials A^λ are assumed fixed in the variation with respect to μ. Variations of μ in the function $\overline{\Phi}(\mu;\lambda)$ correspond to variations of α^μ, and since the functional $\Phi(\alpha; A^\lambda)$ at the point α^μ is stationary with respect to any variations of α, it is also stationary with respect to any variations of a particular form which induce variations of the numerical parameters μ. This proves that the values of μ given by (6.125) are actually the solutions of the variational problem for the numerical Legendre transform. It is also clear that (6.125) can be extended to singular points λ by continuity.

Therefore, if we can calculate the functional Legendre transforms in some approximation and can solve the stationarity equations for them, we can effectively construct approximate expressions for the numerical Legendre transforms. It is not difficult to generalize this construction to the case of time-independent potentials (see Sec. 6.2.1).

In field theory with time-translationally invariant potentials of the Lagrangian type the value of Φ at the stationarity point determines the ground-state energy E up to the normalization: $W(A) = \Phi(\alpha(A); A) = -i(E - E_0)\int dt$, where E_0 is a normalization constant independent of the potentials. In quantum statistics the thermodynamic potential Ω plays the role of E, and the "time volume" becomes finite: $\int dt \rightarrow \beta \equiv 1/kT$. From Secs. 4.3.2 and 5.1.9 we know that E and Ω are convex-upwards functions of the parameters λ, so that their Legendre transforms will be convex-downwards functions of the conjugate variables μ. The correspondence established by (6.126) between the numerical and functional transforms allows the known properties of convexity upwards to be extended to the latter: the correctly normalized (in field theory, multiplied by i and divided by $\int dt$) functional $\Phi(\alpha; A)$ must have nonnegative second variation at the stationarity point for those directions of the variations $\delta\alpha$ which are induced by variations of the numerical parameters μ (or, equivalently, λ). If instead of a numerical transform we consider a more general transform with respect to time-independent potentials, which also possesses the universal convexity properties (see Secs. 4.3.2 and 5.1.9), we obtain an expression analogous to (6.126), and in exactly the same way we prove the convexity of the general functional transform with respect to variations $\delta\alpha$ induced by variations of the time-independent potentials. We recall that the convexity properties of which we are speaking follow from the standard spectral representations for the Green functions, so their violation would imply violation of the spectral representations.

The physical meaning of these convexity requirements is very simple: *in field theory the desired Green functions, which are the solution of the variational problem, reach the minimum ground-state energy within a certain class of allowed variations.* We note that from (6.126) and the general convexity properties of the numerical transforms (see Sec. 6.1.3) it also follows that when the solution of the functional variational problem is degenerate in the class of time-translationally invariant functions, one and the same ground-state energy is associated with different solutions α corresponding to a singular point A in potential space.

The last statement is true only for an exact theory. In practice, we always deal with approximate expressions obtained by selecting the first few graphs of the functional transforms, and these approximate expressions do not possess the correct convexity properties.

When degeneracy is present, the stationarity points of approximate functionals are classified as either stable or unstable, depending on whether or not the convexity requirements formulated above are satisfied at the given point. Usually, violation of convexity is judged from violations in certain spectral representations (a typical example is the proof of the instability of the normal solution for the superconductor in [21]).

In turn, stable points are classified as *absolutely stable* points, which correspond to the absolute minimum of the ground-state energy, and *metastable* points having higher energy. An exact functional possessing the correct convexity properties cannot have either unstable or metastable stationarity points, from which we see that when working with approximate functionals we should select only solutions which are absolutely stable. This, of course, does not apply to nonstationary or nonequilibrium systems, but we shall not consider such systems here.

6.3 Legendre Transforms of the Logarithm of the S-Matrix Generating Functional

6.3.1 *Definitions and general properties*

Up to now we have dealt with Legendre transforms of the generating functional of connected Green functions. In this section we shall consider objects of a different type: Legendre transforms of the connected part of the

S matrix, i.e., of the functional $W \equiv \ln R$, where R is the S-matrix generating functional (1.84). We write (1.84) in the form

$$R(\varphi) = \exp W(\varphi) = \exp\left[\frac{1}{2}\cdot\frac{\delta}{\delta\varphi}\Delta\frac{\delta}{\delta\varphi}\right]\exp \mathcal{M}(\varphi), \qquad (6.127)$$

and, as usual, refer to the functional $\mathcal{M}(\varphi)$ as the generating vertex.

Legendre transforms of the logarithm of R are interesting mainly for the statistics of spin systems and the nonideal classical gas. For such systems the partition function in an arbitrary external field takes the form of an S-matrix generating functional, which essentially distinguishes them from field-type systems, where the partition function is similar to the generating functional of full Green functions. According to the general rules of Sec. 1.4, the functional $W = \ln R$ can be represented as the sum of all connected graphs with lines Δ and generating vertex $\mathcal{M}(\varphi)$ or as the sum of all connected graphs without self-contracted lines and with the reduced generating vertex (1.99). We shall use the diagrammatic technique with the reduced vertex, which we denote as $\mathcal{M}'(\varphi)$, and shall classify the graphs according to the number of lines.

The functional W depends on φ and Δ. In this section we study its Legendre transform in the variable φ.

The contribution of zeroth order in the number of lines in $W(\varphi)$ is the vertex $\mathcal{M}'(\varphi)$, shown graphically as a separate point. We write $W(\varphi) = \mathcal{M}'(\varphi) + \overline{W}(\varphi)$, isolating the contribution of the zeroth-order approximation, and give the graphs of $\overline{W}$ through three orders (see Appendix 2):

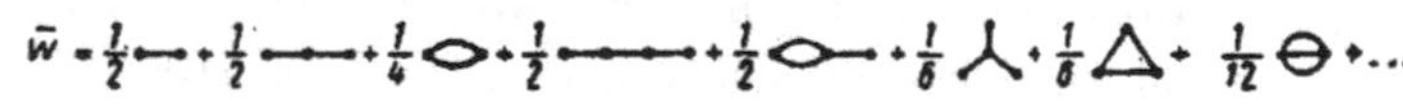
(6.128)

We recall that in these graphs, any vertex at which n lines converge is associated with a universal vertex factor

$$\mathcal{M}'_n(\varphi) = \delta^n \mathcal{M}'(\varphi)/\delta\varphi^n \qquad (6.129)$$

(we have omitted the arguments $x_1 \ldots x_n$).

We shall now define the Legendre transform of $W(\varphi)$ in the variable $\varphi(x)$, introducing the conjugate variable $\alpha(x) \equiv \delta W(\varphi)/\delta\varphi(x)$ and the functional

$$\Gamma(\alpha) = W(\varphi) - \alpha\varphi. \qquad (6.130)$$

The usual stationarity equation for Γ follows from this definition (see Sec. 6.2.1), as does the fact that the second derivatives of W and Γ with respect to φ and α are the inverses of each other up to a sign. We shall write these expressions in abbreviated form, dropping the arguments x and using the subscripts φ, α to label the partial derivatives with respect to the corresponding variables:

$$W_{\varphi} = \alpha, \quad \Gamma_{\alpha} = -\varphi, \quad W_{\varphi\varphi}\Gamma_{\alpha\alpha} = -1. \tag{6.131}$$

We use $F(\alpha)$ to denote the zeroth-order approximation for $\Gamma(\alpha)$. It is clear from (6.130) that F is the Legendre transform of the reduced generating vertex $\mathcal{M}'(\varphi)$, which plays the role of the zeroth-order approximation for $W(\varphi)$:

$$F(\alpha) = \mathcal{M}'(\varphi) - \alpha\varphi, \quad \alpha = \delta\mathcal{M}'(\varphi)/\delta\varphi. \tag{6.132}$$

We write $\Gamma(\alpha) = F(\alpha) + \bar{\Gamma}(\alpha)$ and rewrite the definition (6.130), separating the zeroth-order approximations from Γ and W:

$$F(\alpha) + \bar{\Gamma}(\alpha) = \mathcal{M}'(\varphi) + \bar{W}(\varphi) - \alpha\varphi. \tag{6.133}$$

We express the variable φ in terms of α using the stationarity equation $\varphi = -\Gamma_{\alpha} = \psi - \bar{\Gamma}_{\alpha}$, where $\psi(\alpha)$ represents the contribution to $\varphi(\alpha)$ from the zeroth-order approximation F. Substituting φ in this form into the right-hand side of (6.133) and expanding $\mathcal{M}'(\varphi)$ in a Taylor series, we obtain

$$F(\alpha) + \bar{\Gamma}(\alpha) = \sum_{n=0}^{\infty} \frac{1}{n!} u_n (-\bar{\Gamma}_{\alpha})^n + \bar{W}(\varphi) - \alpha(\psi - \bar{\Gamma}_{\alpha}), \tag{6.134}$$

where we have written

$$u_n(\alpha) \equiv \delta^n \mathcal{M}'(\psi)/\delta\psi^n = \mathcal{M}'_n(\varphi)\big|_{\varphi = \psi(\alpha)}. \tag{6.135}$$

It is clear from the definition of ψ that this quantity coincides with $\varphi = \varphi(\alpha)$ in (6.132). From this it follows that $u_0 = F + \psi\alpha$ and $u_1 = \alpha$, so that the contributions of the zeroth-order approximation and the terms linear in $\bar{\Gamma}_{\alpha}$ in (6.134) cancel and we obtain

$$\bar{\Gamma}(\alpha) = \sum_{n=2}^{\infty} \frac{1}{n!} u_n (-\bar{\Gamma}_{\alpha})^n + \bar{W}(\varphi)\Bigg|_{\varphi = \psi(\alpha) - \bar{\Gamma}_{\alpha}}. \tag{6.136}$$

For what follows we note that all the $u_n(\alpha)$ with $n \geq 1$ can be constructed from $u_1 = \alpha$ by means of the recursion relation $u_{n+1} = \delta u_n/\delta\psi$ following from (6.135), which in the variables α takes the form

$$u_1 = \alpha, \quad u_{n+1} = -F_{\alpha\alpha}^{-1}\delta u_n/\delta\alpha. \qquad (6.137)$$

Equation (6.136) obtained above is valid for all theories, and it can be used to recursively construct the graphs of $\bar{\Gamma}$ in terms of the known graphs of $\bar{W}$. In fact, the functionals $\bar{\Gamma}$ and $\bar{W}$ do not contain the zeroth-order approximation, so that their expansions begin at first order in Δ. Therefore, the contribution of lowest (first) order in Δ on the right-hand side of (6.136) is contained in the known graphs of $\bar{W}(\varphi)$, where the variable $\varphi = \psi - \bar{\Gamma}_\alpha$ in them must be taken in zeroth order in Δ, i.e., approximated as $\varphi = \psi$. Therefore, in first order the graphs of $\bar{\Gamma}$ and $\bar{W}$ coincide, but the vertex factors (6.129) of the graphs $\bar{W}$ become the factors (6.135) for the graphs of $\bar{\Gamma}$. Determining $\bar{\Gamma}$ in first order in this manner and knowing $\bar{W}$ to second order, we can use (6.136) to find the second-order graphs of $\bar{\Gamma}$, and so on. Thus, for the first three orders we obtain

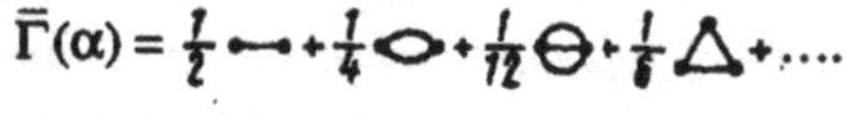

$$\bar{\Gamma}(\alpha) = \tfrac{1}{2}\,\rule[2pt]{14pt}{0.6pt}\; + \tfrac{1}{4}\,\bigcirc\!\!-\; + \tfrac{1}{12}\,\ominus\; + \tfrac{1}{8}\,\triangle\; + \dots. \qquad (6.138)$$

In these graphs a line is Δ and vertices are associated with the universal factors (6.135); in particular, in the first-order graph $u_1 = \alpha$.

The graphs given above for the first three orders are the same for all theories, i.e., for any vertex $\mathcal{M}'(\varphi)$. Differences in the graphs of Γ for different theories begin to appear only in fourth order.

Let us now introduce some new topological concepts. The graphs of the Legendre transforms of the generating functional of connected Green functions studied earlier possessed definite irreducibility properties, which characterized the degree of connectedness of a graph when its lines are cut. Now this role will be played by the properties of "vertex irreducibility" characterizing the degree of connectedness of a graph when its vertices are removed.

We shall say that a graph is *1-irreducible in the vertices* if it remains connected when any one of its vertices is removed. Such graphs are also called *stars* [8]. Of the graphs given in (6.128), only those which remained in (6.138) are 1-irreducible.

Generalizing this definition, it is natural to introduce the concept of the *vertex k-section*, by which we mean the set of k vertices of a graph whose simultaneous removal causes the graph to break up into two (or more) unconnected parts. There are trivial and nontrivial k-sections. For $k = 2$ the trivial k-section is that which splits the graph into two parts, one of which is a simple line. Such 2-sections are contained in any graph: the removal of a pair of vertices connected by a line causes the line to be cut. Therefore, 2-irreducibility must be defined as the absence of 1-sections and nontrivial 2-sections.

The concepts of 1- and 2-irreducibility defined above are sufficient for analyzing the Legendre transforms of $W(\varphi)$. The stronger properties of 3-, 4-, etc. irreducibility, which no one has yet studied, must apparently be properties of objects of the type of Legendre transforms with respect to the multiparticle potentials of the logarithm of the partition function for the classical gas with many-body forces.

In what follows we shall often refer to the vertices of a graph which are 1-sections of it as *anomalous vertices*; they are also called *articulation points* [8].

Let us now return to (6.136) and select its "star part," in other words, we equate the sums of the graphs 1-irreducible in the vertices on both sides of the equation. The α dependence in the graphs of $\bar{\Gamma}$ is concentrated in the vertex factors, so that the derivative $\bar{\Gamma}_\alpha$ corresponds to graphs with a single labeled vertex, the one differentiated with respect to α. We shall represent such graphs as  and call them *petals*.

The general term of the Taylor series in (6.136) is represented graphically as a "flower" with $n \geq 2$ petals connected at their labeled vertices. Clearly, the connection point is an anomalous vertex of that graph. Therefore, all such graphs are 1-reducible, and must be simply discarded when the star part is selected.

In the graphs of $\bar{W}(\varphi)$ the φ dependence is contained in the vertex factors (6.129). Expressing φ in terms of α via $\varphi = \psi - \bar{\Gamma}_\alpha$ and taking into account the definitions (6.129) and (6.135), we obtain

$$\mathcal{M}_n'(\varphi) = \sum_{k=0}^{\infty} \frac{1}{k!} u_{n+k} (-\bar{\Gamma}_\alpha)^k = \qquad (6.139)$$

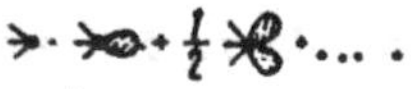

Clearly, in selecting the star part we should keep only the first term in the expansions (6.139), because all the others lead to graphs which are certainly 1-reducible. This proves that for any theory

$$\text{star part of } \bar{\Gamma}(\alpha) = \text{star part of } \bar{W}(\varphi)\big|_{\varphi = \psi(\alpha)}. \qquad (6.140)$$

This means that the star graphs of $\bar{\Gamma}$ differ from the star graphs of $\overline{W}$ only in that the universal vertex factors (6.129) are now expressed in terms of the variable α conjugate to φ and they become (6.135).

Therefore, the specific features of a theory are only manifested explicitly in the vertex factors (6.135) and in the structure of the nonstar graphs. Since the graphs (6.138), which are the same for all theories, are stars, nonstar graphs can arise only beginning at fourth order in the number of lines. Equation (6.136), in principle, determines all the graphs of $\bar{\Gamma}$, including the nonstar ones, and for any particular theory in any given order (6.136) can be used to find all the graphs of $\bar{\Gamma}$, even though, of course, everything gets more and more complicated as the order increases. At the present time, the complete description of the graphs of $\bar{\Gamma}$ in all orders is known for only two theories: for the classical gas [7,8] and the Ising model [6,75]. For the gas the answer is very simple: there are no nonstar graphs at all in $\bar{\Gamma}$. In the following section we shall prove this statement, using, as usual, the equations of motion for the object studied.

6.3.2 The classical nonideal gas and the virial expansion

For a gas with two-body forces the reduced generating vertex is given by (5.68): $\mathcal{M}'(\varphi) = \int d\mathbf{x} \exp \varphi(\mathbf{x})$. This interaction is special because all the vertex factors (6.129) are the same up to δ functions. The zeroth-order approximation (6.132) for Γ is easily found:

$$\alpha(\mathbf{x}) = \delta\mathcal{M}'(\varphi)/\delta\varphi(\mathbf{x}) = \exp\varphi(\mathbf{x}),$$

$$F(\alpha) = \int d\mathbf{x}\ \alpha(\mathbf{x})\ [1 - \ln\alpha(\mathbf{x})], \tag{6.141}$$

and all the vertex factors (6.135) are equal to α up to δ functions.

Equation (5.63) will play a main role in what follows. To change to the notation of the preceding section we need to make the substitution $Z \to R = \exp W$, $A_1 \to \varphi$, $A_2 \to \Delta$, after which it can be rewritten as $2R_\Delta = R_{\varphi\varphi} - \mathbf{1}R_\varphi$. As usual, the subscripts denote partial derivatives with respect to the corresponding variables, and $\mathbf{1}$ is the kernel of the unit operator, i.e., $\delta(\mathbf{x} - \mathbf{x}')$. For $W = \ln R$ we obtain

$$2W_\Delta = W_{\varphi\varphi} + W_\varphi^2 - \mathbf{1}\cdot W_\varphi. \tag{6.142}$$

This equation still needs to be rewritten in terms of Γ, which is easily done using (6.131) and the equation $W_\Delta = \Gamma_\Delta$ analogous to (6.4) and (6.13):

$$2\Gamma_{\Delta} = -\Gamma_{\alpha\alpha}^{-1} + \alpha^2 - 1 \cdot \alpha. \tag{6.143}$$

Isolating the zeroth-order approximation (6.141) from Γ, we can write $\Gamma_{\alpha\alpha}^{-1}$ as a series. We have $F_{\alpha} = -\ln\alpha$, $F_{\alpha\alpha} = -1\alpha^{-1}$, from which

$$-\Gamma_{\alpha\alpha}^{-1} = [1 \cdot \alpha^{-1} - \bar{\Gamma}_{\alpha\alpha}]^{-1} = 1 \cdot \alpha + \alpha\bar{\Gamma}_{\alpha\alpha}\alpha + \alpha\bar{\Gamma}_{\alpha\alpha}\alpha\bar{\Gamma}_{\alpha\alpha}\alpha + \ldots.$$

Substituting this series into (6.143), we obtain the equation

$$2\bar{\Gamma}_{\Delta} = \alpha^2 + \alpha\bar{\Gamma}_{\alpha\alpha}\alpha + \alpha\bar{\Gamma}_{\alpha\alpha}\alpha\bar{\Gamma}_{\alpha\alpha}\alpha + \ldots, \tag{6.144}$$

the iteration solution of which determines all the graphs of $\bar{\Gamma}$.

Denoting Δ by a line, α by a point, and the second derivative $\bar{\Gamma}_{\alpha\alpha}$ by the block  and contracting (6.144) with Δ, we obtain

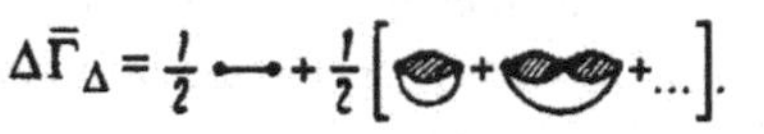 (6.145)

The first graph on the right-hand side determines $\bar{\Gamma}$ in first order in the number of lines. Calculating the blocks $\bar{\Gamma}_{\alpha\alpha}$ on the right-hand side from the known $\bar{\Gamma}$, we find $\bar{\Gamma}$ in second order, and so on. We note that determining $\Delta\bar{\Gamma}_{\Delta}$ is equivalent to determining $\bar{\Gamma}$, since these two quantities differ only in the coefficients of the graphs ($\Delta\bar{\Gamma}_{\Delta}$ has an additional factor of k for a graph with k lines). The general iteration scheme for (6.145) is the same as that described in Sec. 6.2.5.

Let us now turn to the proof of the main statement: all the graphs obtained by iterating (6.145) are stars. It is necessary to begin with an observation which is easily proved by induction: in all the graphs obtained by iteration, the same vertex factor α (up to δ functions) is associated with any vertex. Taking this into account, it is easily verified that (6.145) possesses the property of "star conservation," namely, if all the graphs of $\bar{\Gamma}$ up to some order are stars, the graphs of next order obtained by iteration will also be stars. Actually, if we assume that any of the graphs on the right-hand side of (6.145) has a vertex 1-section of the form , it will follow that this section is contained inside one of the blocks $\bar{\Gamma}_{\alpha\alpha}$, contradicting the induction assumption. Here it is important that the vertices in the graphs of $\bar{\Gamma}$ are associated with a simple factor of α, so that in the second derivative $\bar{\Gamma}_{\alpha\alpha}$ the two operations of differentiation with respect to α necessarily act on different vertices. Were this not so, among the graphs of $\overline{\Gamma}_{\alpha\alpha}$ there would necessarily be single-petal graphs of the form  with a single isolated vertex on which two derivatives $\delta/\delta\alpha$ act. Insertion of this graph into the chain (6.145) would lead to graphs of the form , which are obviously 1-reducible in spite of the fact that the blocks $\overline{\Gamma}_{\alpha\alpha}$ entering into them are calculated from 1-irreducible graphs.

It follows from the statement proved above and from (6.140) that for a gas

$$\overline{\Gamma}(\alpha) = \text{star part of } \overline{W}(\varphi \to \ln \alpha). \tag{6.146}$$

This means that $\overline{\Gamma}(\alpha)$ is the sum of all star graphs with line Δ and vertices $u_n(\alpha) = \alpha$. The fact that all the vertex factors u_n are the same makes it possible (see Sec. 1.4.5) to sum over the number of lines connecting given pairs of vertices and transform to Mayer lines $g = -1 + \exp\Delta$, leaving only Mayer graphs, for which any pair of vertices is connected by no more than one line.

The statement (6.146) is the second Mayer theorem, well known in the theory of a classical gas [7,8]. The method we have used to prove it was borrowed from [6]. The theorem can be directly generalized to the case of a gas with arbitrary many-body forces [46]. The concept of 1-irreducibility for the supergraphs with which it is necessary to deal in this theory (see Sec. 5.3.2) is defined exactly as for ordinary graphs.

For a translationally invariant (i.e., spatially uniform) system, φ and α are constants independent of $\mathbf{x}$, the line $\Delta(\mathbf{x}, \mathbf{x}')$ depends only on the difference $\mathbf{x} - \mathbf{x}'$, and W and Γ are proportional to the total volume of the system $V \equiv \int d\mathbf{x}$, so that it is necessary to deal with specific quantities (divide out the volume). For such systems the constant $a \equiv \exp\varphi$ is the activity, α is the particle number density (in the notation of Sec. 5.3.1, $\varphi = A_1$ and $\Delta = A_2$), and the specific value of W, as explained at the end of Sec. 6.1.3, is related to the pressure p as $V^{-1}W = \beta p$.

The usual diagrammatic expansion (Sec. 5.3.1) of the functional W represents the quantity βp as a power series in the activity $\exp\varphi$, which enters as a factor into each vertex of the graphs. If we wish to construct the expansion of βp in powers of the density α, it is necessary to express W in terms of Γ using (6.130) and (6.131):

$$\beta p V = W = \Gamma + \alpha\varphi = \Gamma - \alpha\Gamma_{\alpha}. \tag{6.147}$$

For a uniform system each of the graphs on the right-hand side is proportional to V, so that this infinite factor cancels in (6.147).

The contribution of the zeroth-order approximation (6.141) to the right-hand side of (6.147) is, as is easily checked, equal to αV, and if we restrict ourselves to it, (6.147) will be the ideal-gas equation of state. For a nonideal gas the contributions of the graphs of $\bar{\Gamma}$ are added; classifying them according to the number of vertices, we obtain the series

$$\bar{\Gamma}(\alpha) = V \sum_{n=2}^{\infty} \alpha^n S_n, \tag{6.148}$$

the general term of which represents the sum of the contributions of all star graphs with n vertices. Substituting (6.148) and the contribution of the zeroth-order approximation into (6.147), we obtain the desired expansion:

$$\beta p = \alpha + \sum_{n=2}^{\infty} \alpha^n (1-n) S_n. \tag{6.149}$$

This is called the *virial expansion*, and the quantities $(1-n)S_n$ are called the *virial coefficients* [7,8]. The numbers S_n depend only on the form of the line Δ, which is determined (see Sec. 5.3.1) by the two-body potential and the temperature.

6.3.3 *The Ising model [6]*

For the Ising model (see Sec. 5.2.1) the role of the argument x is played by the number of the lattice site i, and according to (5.48) the vertex $\mathcal{M}'(\varphi)$ has the form $\Sigma \ln 2\cosh \varphi_i$. We note that in this case the reduced and ordinary generating vertices are the same because the self-contracted line Δ is equal to zero (absence of an exchange self-interaction).

In zeroth order $\alpha_i = \delta\mathcal{M}'(\varphi)/\delta\varphi_i = \tanh \varphi_i$ and $F(\alpha) = \Sigma f(\alpha_i)$, where

$$f(\alpha) \equiv \ln 2 - \frac{1}{2}[(1-\alpha)\ln(1-\alpha) + (1+\alpha)\ln(1+\alpha)]. \tag{6.150}$$

For the universal vertex factors (6.135) we obtain

$$u_1(\alpha) = \alpha, \; u_{n+1}(\alpha) = (1-\alpha^2)\,\delta u_n(\alpha)/\delta\alpha, \tag{6.151}$$

i.e., $u_1 = \alpha$, $u_2 = 1 - \alpha^2$, $u_3 = -2\alpha(1-\alpha^2)$, and so on. In writing out the u_n we have omitted the δ symbols for the lattice sites. All the $u_n(\alpha)$ are polynomials of definite parity coinciding with the parity of the number n.

Equations (5.50) in the notation of Sec. 6.3.1 take the form

$$2\partial R/\partial\Delta_{ik} = \partial^2 R/\partial\varphi_i\partial\varphi_k,\ \ i \neq k;\ \ \partial^2 R/\partial\varphi_i\partial\varphi_i = R.$$

From this for $W = \ln R$ we have $2W_\Delta = W_{\varphi\varphi} + W_\varphi W_\varphi$, $i \neq k$; $W_{\varphi\varphi} + W_\varphi W_\varphi = 1$, $i = k$; and for the Legendre transform (6.130) we obtain

$$2\Gamma_\Delta = -\Gamma^{-1}_{\alpha\alpha} + \alpha\alpha,\ \ i \neq k;\ \ -\Gamma^{-1}_{\alpha\alpha} + \alpha\alpha = 1,\ \ i = k. \tag{6.152}$$

As usual, after isolating from Γ the zeroth-order approximation (6.150), for which $F_{\alpha\alpha} = -\mathbf{1}(1-\alpha^2)^{-1}$, we represent the inverse matrix $\Gamma^{-1}_{\alpha\alpha}$ by the series

$$-\Gamma^{-1}_{\alpha\alpha} = [\mathbf{1}(1-\alpha^2)^{-1} - \bar{\Gamma}_{\alpha\alpha}]^{-1} = \mathbf{1}(1-\alpha^2) + \ldots \tag{6.153}$$

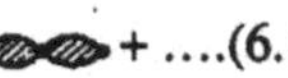

The point denotes the factor $1-\alpha^2$ and the shaded block is $\bar{\Gamma}_{\alpha\alpha}$.

When the series (6.153) is substituted into (6.152) with $i \neq k$, the term which is a multiple of the unit matrix, $\mathbf{1}\cdot(1-\alpha^2)$, does not contribute, and we obtain

$$2(\bar{\Gamma}_\Delta)_{ik} = \alpha_i\alpha_k + \ldots, \tag{6.154}$$

while (6.152) with $i = k$ takes the form

$$\ldots = 0. \tag{6.155}$$

(there is no summation over i). Equation (6.155) requires that the diagonal matrix elements of the progression of blocks be equal to zero.

By iterating these equations it is possible in principle to find all the graphs of $\bar{\Gamma}$, though it should be noted that in fact this is very difficult to do. We shall proceed differently: we shall use the equations of motion, as usual, only to establish certain topological properties of the graphs of $\bar{\Gamma}$. In the present case the properties of interest are those which allow a complete description of the nonstar graphs. In contrast to the gas, there are such graphs in the Ising model. They appear beginning at fourth order. In the first three orders the graphs of $\bar{\Gamma}$ for the Ising model, as for any other theory, have the form (6.138) and are star graphs. According to the general rules of Sec. 6.3.1, the universal factors (6.151) are associated with the vertices of these (and other star) graphs.

A complete description of the nonstar graphs will be given in the following section, and for now we briefly discuss the special case of a uniform system, which is most important from a practical point of view. In the notation of Sec. 5.2.1, $\varphi_i = \beta h_i$ and $\Delta_{ik} = \beta \mathcal{V}_{ik}$, where $\beta = 1/kT$, h is the external field, and $\mathcal{V}$ is the spin exchange-interaction matrix. In the classical Ising model one takes $\mathcal{V}_{ik} = J\lambda_{ik}$, where J is the "exchange integral" (a constant with the dimension of energy); λ is the nearest-neighbor interaction matrix with $\lambda_{ik} = 1$ if i and k are nearest neighbors on the lattice and $\lambda_{ik} = 0$ otherwise. It is also possible to consider more complicated interactions, for example, ones for which $\lambda_{ik} = 1$ not only for nearest, but also for next-to-nearest neighbors.

For a uniform system the field φ_i and the conjugate variable, the magnetization α_i, are independent of the site number, and each of the graphs of W and Γ is proportional to the total number of sites $V \equiv \Sigma\, 1$, so that it is necessary, as usual, to divide out the volume everywhere. Classifying the graphs of $\bar{\Gamma}$ according to the number of lines Δ, each of which contains the dimensionless factor βJ, we obtain the *high-temperature expansion* of the specific quantity

$$\gamma(\alpha, \beta) \equiv \Gamma/V = f(\alpha) + \sum_{n=1}^{\infty} (\beta J)^n P_n(\alpha), \tag{6.156}$$

in which $f(\alpha)$ is the contribution of the zeroth-order approximation (6.150), and the general term of the series is the sum of the contributions of all the graphs of $\bar{\Gamma}$ with n lines. The expansion (6.156) is referred to as a high-temperature expansion because $\beta \sim 1/T$.

The coefficient functions $P_n(\alpha)$ in (6.156) turn out to be polynomials in α^2, and the highest degree of the P_n is α^{2n}. This is obvious for the star graphs of $\bar{\Gamma}$ when the explicit form of the vertex factors (6.151) is taken into account, together with the fact that each graph contains an even number of odd vertices. In the following section we shall see that this is also true for nonstar graphs.

The coefficients of the polynomials P_n (they all become integers after multiplication by $n!$) have the meaning of structure constants determined by the lattice and the type of interaction. We have calculated the coefficients of all the polynomials P_n through $n = 8$ for two planar (square and triangular) and three three-dimensional cubic lattices (face-centered, volume-centered, and simple). Two versions of the latter were studied: one with only nearest-neighbor interactions, and the other with nearest- and next-to-nearest neighbor interactions. These results are used to analyze the critical behavior in [76,77].

6.3.4 *Analysis of nonstar graphs for the Ising model [75]*

In the ordinary graphs of W and Γ the vertices $1 \ldots n$ are assigned indices $i_1 \ldots i_n$ and are summed over independently, i.e., each i_α runs over the entire lattice. Here the sum over all $i_1 \ldots i_n$ will, of course, contain terms with coincident subgroups of indices.

It is sometimes more convenient to consider sums over noncoincident indices, i.e., sums with the additional condition $i_\alpha \neq i_\beta$ for any pair $\alpha \neq \beta$. The sum "over all" can be represented as a linear combination of sums "over noncoincident" vertices: the n indices are split in all possible ways into groups, inside which all the indices coincide, while the indices of different groups are different.

As already mentioned, for ordinary graphs the summation runs "over all." It is also possible to consider graphs for which by condition the summation runs only over noncoincident indices. For brevity we shall refer to these as ***graphs in $\mathcal{N}$-form***, and write the symbol $\mathcal{N}$ in front of the graph. Any graph D can be rewritten in $\mathcal{N}$ form:

$$D = \mathcal{N} \sum_s D_s. \tag{6.157}$$

The summation runs over all possible splittings s of the set of indices of graph D into groups of noncoincident indices (we shall say "over all contraction variants"), and D_s is a *contracted graph*, in which all the vertices belonging to the same group are contracted to a single point. Among the graphs D_s there is the original graph D, corresponding to the splitting where each group consists of a single element. Different contraction variants will be depicted by dashed "coincidence lines," which are associated with the unit matrix δ_{ik}, in contrast to a solid line associated with Δ_{ik}. All the vertices of a group contracted to a single point are connected by dashed lines, i.e., each such group corresponds to a complete graph of dashed lines.

In the Ising model it is possible to ignore contractions in which at least one pair of vertices is connected simultaneously by a dashed and a solid line; the contribution of such a graph is zero because the diagonal elements of Δ are zero. As an example, we give the $\mathcal{N}$-form for the square:

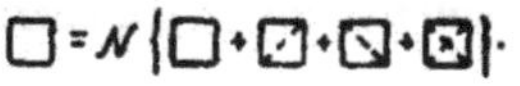

(6.158)

Let us now turn to the proof of the basic statement of this section: *in $\mathcal{N}$-form all the graphs of $\bar{\Gamma}$ are stars.*

To avoid misunderstandings, let us precisely define the meaning of the term *star* for the case of graphs in $\mathcal{N}$-form. It should be born in mind that the use of dashed lines is only for convenience in showing the contractions; the contracted vertices should actually be thought of as reduced to a single point. This point will be a normal or anomalous vertex (i.e., a 1-section) of the graph depending on whether or not the connectedness of the graph is preserved when this point is removed. When dashed lines are used the removal of this vertex implies the simultaneous removal of all the vertices connected together, i.e., connected by dashed lines.

Of the graphs of (6.158), only the squares with a single dashed diagonal are nonstar graphs.

We shall prove the above statement by induction using the equations of motion (6.154) and (6.155). We assume that all the graphs of $\bar{\Gamma}$ up to some order are stars in $\mathcal{N}$-form, and we shall show that the graphs of next highest order obtained by iterating (6.154) and (6.155) will also be stars in $\mathcal{N}$-form. We note that the induction assumption is satisfied for the graphs of the first three orders in (6.138).

We assume that $\bar{\Gamma}$ contains a nonstar graph with anomalous vertex of the type (6.159*a*):

$\bar{\Gamma} \ni$  $\rightarrow \partial\bar{\Gamma}/\partial\Delta_{ik} \ni$ . (6.159)

a *b*

The graphs of the derivative $\bar{\Gamma}_\Delta$ on the left-hand side of (6.154) are obtained from the graphs of $\bar{\Gamma}$ by removing a single solid line Δ in all possible ways. Therefore, in the derivative with respect to Δ_{ik} of the nonstar graph (6.159*a*) there necessarily will also be graphs in which the line connected to the anomalous vertex is removed, in other words, in which one of the isolated vertices i, k is anomalous. This argument is illustrated by the graphical equation (6.159). For definiteness, we have assumed that the vertex i is anomalous, but this does not lead to loss of generality owing to the symmetry of $\bar{\Gamma}_\Delta$ in the labels i, k.

Therefore, if we wish to prove that in some step of the iterations no nonstar (in $\mathcal{N}$-form) graphs appear, it is sufficient to show that after the right-hand side of (6.154) is reduced to $\mathcal{N}$-form it contains no dangerous graphs with petals at the vertex i, i.e., in the left-most vertex of the chains of blocks (6.154).

We recall that the chains in (6.154) involve diagonal matrices $\delta_{\underline{ik}}(1-\alpha_i^2) \equiv r_{ik}$, associated with the points singled out in (6.154), and matrices $\bar{\Gamma}_{\alpha\alpha}$ with elements $\partial^2 \overline{\Gamma}/\partial\alpha_i \partial\alpha_k$, shown as blocks  in (6.154). The indices of the extreme vertices of the chains (i on the left and k on the right) are fixed, and the indices of the other vertices are summed over, which corresponds to the ordinary matrix product. For example, the term with three blocks is the matrix element ik of the product $r\overline{\Gamma}_{\alpha\alpha}r\overline{\Gamma}_{\alpha\alpha}r\overline{\Gamma}_{\alpha\alpha}r$.

According to the induction assumption, the graphs of $\overline{\Gamma}$ whose differentiation produces the blocks $\overline{\Gamma}_{\alpha\alpha}$ are stars in $\mathcal{N}$-form, so that inside these blocks there are no anomalous vertices. We must show that after the right-hand side of (6.154) is reduced to $\mathcal{N}$-form, no dangerous graphs with a petal at the vertex i of the type (6.159*b*) can appear.

At first glance it appears that such graphs must arise. First, the extreme left-hand block of the chain $\partial^2 \overline{\Gamma}/\partial\alpha_i \partial\alpha_s$ (where s is the summation index) contains a contribution with $i = s$ obtained by double differentiation of the same vertex. There were no such graphs for the gas, because the vertices were associated with a simple factor of α, the second derivative of which is zero. This is not true here, more precisely, we have not verified that this is true, so we must take into account blocks with $i = s$. Such blocks are actually single-petal ones and generate dangerous graphs in chains: 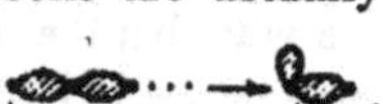.

There is yet another mechanism by which dangerous graphs can arise. The individual blocks $\overline{\Gamma}_{\alpha\alpha}$ are obtained by differentiating graphs in $\mathcal{N}$-form, so that the indices of the different vertices of a single block do not coincide. However, in a chain of such blocks, which is an ordinary matrix product, there are no restrictions on the coincidence of the indices of vertices of different blocks, in other words, such a chain has not yet been fully reduced to $\mathcal{N}$-form. If this reduction is performed, additional graphs appear with contractions of vertices from different blocks. Among these will be dangerous graphs like (6.159*b*). For us it is important that all dangerous contractions leading to the appearance of a petal at the vertex i must contain a dashed line connecting the vertex i with one of the connection points of the blocks $\overline{\Gamma}_{\alpha\alpha}$.

We shall now show that all the dangerous graphs on the right-hand side of (6.154) actually cancel. Representing the matrix with elements $1 - \delta_{ik}$ as a solid line with a slash through it (a noncoincidence line), we can represent (6.155) graphically as

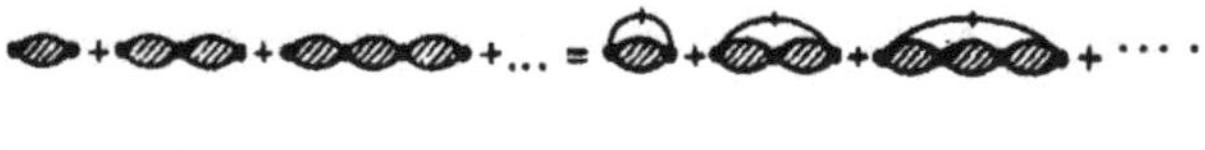 (6.160)

In (6.154) we have $i \neq k$ by condition, so that if desired we can connect the extreme vertices i and k by a noncoincidence line. Then, using the graphical

equation (6.160), we can draw an additional noncoincidence line from the vertex i to the connection point nearest the vertex k where the blocks $\bar{\Gamma}_{\alpha\alpha}$ are attached in (6.154) [however, this can only be done simultaneously in all the terms of the block progression (6.154)]. As a result, the sum of chains of blocks on the right-hand side of (6.154) takes the form

Then, using (6.160) again, we can add a noncoincidence line from the vertex i to the block connection point which is second from the right. Repeating this operation an infinite number of times, we connect the vertex i by noncoincidence lines with all the points where blocks $\bar{\Gamma}_{\alpha\alpha}$ are connected.

The final observation is that in transforming the right-hand side of (6.154) in this way, we have excluded both of the above-mentioned mechanisms by which dangerous graphs can appear. The extreme left-hand block $\bar{\Gamma}_{\alpha\alpha}$ is now accompanied by a noncoincidence line, which forbids the double differentiation of the same vertex. The second mechanism, that of dangerous contractions in the reduction to $\mathcal{N}$-form, is also eliminated, because all graphs with dangerous contractions vanish when noncoincidence lines are added: the vertex i is connected to some block connection point by a coincidence line and a noncoincidence line simultaneously.

Having proved the absence of dangerous graphs on the right-hand side of (6.154), we have thereby proved by induction the statement that all graphs of $\bar{\Gamma}$ in $\mathcal{N}$-form are stars.

If the graphs of $\bar{\Gamma}$ are represented in ordinary rather than $\mathcal{N}$-form, there will be nonstar graphs among them. The statement proved above makes it possible to characterize these graphs in a way which is complete and convenient for applications.

Let us assume that $\bar{\Gamma}$ is represented by ordinary graphs and that $\bar{\Gamma} = \Phi_1 + \Phi_2$, where Φ_1 is the star part and Φ_2 the nonstar part of $\bar{\Gamma}$. The term Φ_1 is known from (6.140), and Φ_2 must be found starting from the fact that after reduction to $\mathcal{N}$-form, the sum $\Phi_1 + \Phi_2$ contains only star graphs: $\Phi_1 + \Phi_2 = \mathcal{N}\psi$, where ψ are stars.

Let us first check that this condition determines Φ_2 uniquely. We assume that it does not, and that two different sets Φ_2 and Φ_2' of nonstar graphs satisfy the condition. Then $(\Phi_1 + \Phi_2) - (\Phi_1 + \Phi_2') = \mathcal{N}(\psi - \psi')$, i.e., after reduction to $\mathcal{N}$-form the set of nonstar graphs $\Phi_2 - \Phi_2'$ is represented by the star graphs $\psi - \psi'$. However, this is impossible, because after $\Phi_2 - \Phi_2'$ is reduced to $\mathcal{N}$-form we must obtain, among other graphs, all the original nonstar graphs of $\Phi_2 - \Phi_2'$ [see the remark following (6.157)].

We have thereby shown that our condition determines the nonstar part of $\bar{\Gamma}$ uniquely. This means that by finding, by any method, a set of nonstar graphs satisfying this condition, we thereby find the unique solution of the problem. This is how we shall proceed, proposing a "compensation prescription" for finding the nonstar graphs.

Let D be a star graph, σ be the set of all contraction variants s in this graph, and D_s be a contracted graph. We shall refer to a contraction s as anomalous if D_s is a nonstar graph; we use ν to denote the set of all anomalous contractions.

Let us order the set of all contractions, associating with each $s \in \sigma$ a number $p = p(s)$, equal by definition to the number of dashed lines of the given contraction. For $p = 0$ we have $D_s = D$, for $p = 1$ a single (any) pair of vertices is contracted, and for $p = 2$ two pairs are contracted. For $p = 3$ there are two possibilities: either three pairs of vertices are contracted, or one triplet of vertices is.

Denoting the set of all (anomalous) contractions with a given p as $\sigma_p(\nu_p)$, we rewrite (6.157) in more detail: $D = \mathcal{N}\{D + \Sigma_{s\in\sigma_1} D_s + \Sigma_{s\in\sigma_2} D_s + \ldots\}$. Clearly, after reduction to $\mathcal{N}$-form the difference $D^{(1)} \equiv D - \Sigma_{s\in\nu_1} D_s$ does not contain graphs with anomalous contractions of first order, because $D_s = \mathcal{N}\{D_s + \Sigma_{s'}[D_s]_{s'}\}$, and all graphs $[D_s]_{s'}$ with $p(s') \geq 1$ have more than one contraction (i.e., $p > 1$).

We shall refer to the change from D to $D^{(1)}$ as the first step in the compensation. By construction, $D^{(1)}$ in $\mathcal{N}$-form contains only anomalous contractions with $p > 1$.

Let us isolate from these the anomalous contractions of lowest ($p = 2$) order: $D^{(1)} = \mathcal{N}\{D^{(1)} + \Sigma_{s\in\nu_2} D_s^{(1)} + \ldots\}$, and form the difference (the second step in the compensation) $D^{(2)} = D^{(1)} - \Sigma_{s\in\nu_2} D_s^{(1)}$. By construction, $D^{(2)}$ in $\mathcal{N}$-form does not contain anomalous contractions with $p \leq 2$. In the third step of the compensation we subtract from $D^{(2)}$ graphs with anomalous contractions of third order, and so on. Repeating this operation the needed number of times, we arrive at the *fully compensated graph*, which does not contain any anomalous contractions after reduction to $\mathcal{N}$-form.

The compensation prescription can be briefly summarized as follows: from a given star graph of D we construct the expression

$$D_{\text{comp}} = D + \sum_s c_s D_s, \tag{6.161}$$

in which the summation runs over all topologically inequivalent variants of the anomalous contractions, and the numerical coefficients c_s are selected such that after reduction to $\mathcal{N}$-form, Eq. (6.161) does not contain nonstar graphs. The uniqueness theorem proved earlier allows us to state that for the Ising model the *functional* $\bar{\Gamma}$ *is the sum of all fully compensated star graphs.*

According to (6.140), the vertex factors of star graphs are universal and for the Ising model have the form (6.151). The factors associated with the contraction points in the compensated graphs D_s are products of the factors (6.151) of all the vertices contracted to a given point, so that they are certainly not universal, i.e., they are determined by the entire graph as a whole, and not simply by the number of lines converging at a vertex. It is also clear that all contraction points in the compensating graphs must be anomalous vertices, because from (6.136) and the subsequent discussion we see that the vertex factors can be nonuniversal only for anomalous vertices. The summation in (6.161) runs over all inequivalent contractions for which at least one contraction point is an anomalous vertex, but it follows from the above arguments that the coefficients c_s will actually be nonzero only for those graphs D_s in which any contraction point is an anomalous vertex.

The first of the star graphs having anomalous contractions is the square. From (6.158) we see that for compensation it is necessary to add a square with one dashed diagonal with coefficient –2. As another example, we give the expression for the fully compensated hexagon:

In the practical calculation of the contributions of various graphs in the case of a uniform field, the vertex factors (6.151) are independent of the site number and can be removed as an overall factor from the summation over vertex indices. It is important that this factor is the same for all the graphs on the right-hand side of (6.161), so the inclusion of all the compensating nonstar graphs reduces only to a change of the rule for calculating the contributions of star graphs with unit vertices.

For the classical Ising model a factor βJ is removed from each line Δ (see the discussion at the end of the preceding section). The main difficulty is to calculate the contribution of the graph with unit vertices and lines λ (we recall that λ_{ik} is equal to one or zero depending on whether or not the sites i and k interact). We note that for a unit line $\lambda_{ik}^n = \lambda_{ik}$, so that the contribution of the graph is not changed when any of its lines is doubled, tripled, etc. Therefore, it is necessary to calculate only the contributions of the star Mayer graphs, which are relatively few: only 5 in the first five orders, 4 in sixth order, 7 in seventh order, 16 in eighth order, 42 in ninth order, and 111 in tenth order (see Appendix 2).

Calculation shows that the compensating graphs are quantitatively important. For example, for a face-centered cubic lattice with nearest-neighbor interaction, the contribution of the heptagon without compensations is 403 200, and after compensations it is 52 416. For the octagon these numbers are respectively 4 038 300 and

330 588. For the octagon with nearest- and next-to-nearest neighbor interactions in the same lattice these numbers are respectively 68 383 350 and 15 420 582. For planar lattices the contribution of a graph after compensations sometimes is even negative.

6.3.5 *The second Legendre transform for the classical gas*

The Legendre transform considered in the preceding sections goes from bare to dressed variables in the vertices of the graphs of the partition function (or the S matrix in field theory). For the gas this problem was solved in the 1930s (see [7]), and for the Ising model it was solved only recently [6,75] (we also note the interesting study [78], which is close in spirit to [6]). The topological problem of describing the nonstar graphs has not yet been solved for any other system, including the quantum Heisenberg ferromagnet (the statement (6.140) completely characterizing the star graphs for any system was proved in [6]).

The natural next step should be to go from bare to dressed variables in the lines of graphs, which requires analysis of the Legendre transform of the functional $W(\varphi, \Delta)$ in the two variables φ and Δ.

At present, this problem has been solved only for the gas. This was first done in [79], then other versions of the proof appeared (see, for example, [56] and references therein). We shall give yet another version, based, like the other similar proofs in this chapter, on the equations of motion for the object in question.

The transform we shall deal with can be called the second one, because it is closely related to the property of vertex 2-irreducibility (see the definition in Sec. 6.3.1). In this classification, the transform (6.130) is the first.

Let us consider the Legendre transform of the logarithm of the partition function $W(\varphi, \Delta)$ with respect to the two variables φ, Δ, introducing the conjugate variables $\alpha = W_\varphi$, $s = W_\Delta$ and constructing the functional $\mathcal{F}(\alpha, s) = W(\varphi, \Delta) - \alpha\varphi - s\Delta$ (as before, the subscripts denote partial derivatives with respect to the corresponding variables). It follows from the definitions that $\mathcal{F}_\alpha = -\varphi$, $\mathcal{F}_s = -\Delta$, and the matrices of second derivatives of W and $\mathcal{F}$ with respect to the variables φ, Δ and α, s, respectively, are the inverses of each other up to a sign. The equation $W_\Delta = \Gamma_\Delta$, which is a special case of (6.13), allows $\mathcal{F}$ to be treated as the Legendre transform with respect to the variable Δ of the first transform (6.130):

$$s = \Gamma_\Delta, \quad \mathcal{F}(\alpha, s) = \Gamma(\alpha, \Delta) - s\Delta. \tag{6.162}$$

We note that this is a general property of all transforms of the type (6.10): they can be constructed recursively, i.e., each successive one is the transform of the preceding one with respect to the next variable.

We know from Sec. 6.3.2 that for a gas the functional $\Gamma(\alpha, \Delta)$ is the sum of the zeroth approximation (6.141) and all the star graphs with lines Δ and vertex factors α. The graphs of the first three orders have the form (6.138). Differentiating them with respect to Δ, we obtain $2s(x, x') = \alpha(x)\alpha(x') +$ 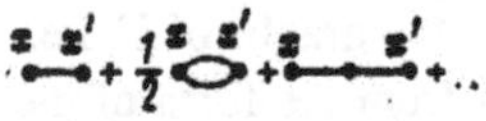, where vertices (points) are associated with a factor α and lines with Δ; the arguments $\mathbf{x}, \mathbf{x}'$ of the labeled vertices are fixed.

Instead of s, it is convenient to introduce the new variable $\omega \equiv \alpha^{-1}[2s - \alpha\alpha]\alpha^{-1}$, which in lowest order coincides with Δ:

$\omega(x,x')=$ 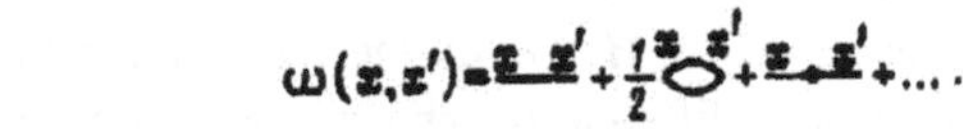 (6.163)

If we were considering not a gas, the ends of a line  would contain the vertex factors u_2 [see (6.135)], and the variable ω would be given by $2s = u_2\omega u_2 + \alpha\alpha$ if we wanted ω and Δ to coincide in lowest order.

For a gas the variable α is interpreted as the density, and ω is related to the correlation function of the density fluctuations $W_{\varphi\varphi}$ (see the definition in Sec. 5.3.1) by the following simple expression following from (6.142):

$$W_{\varphi\varphi} = 1 \cdot \alpha + \alpha\omega\alpha. \tag{6.164}$$

The addition $\alpha\omega\alpha$ to the δ-like term $1{\cdot}\alpha$ represents the contribution of the interaction to the correlation.

The stationarity equations $\mathcal{F}_\alpha = -\varphi$, $\mathcal{F}_s = -\Delta$ in the variables α, ω take the form $\mathcal{F}_\alpha(\mathbf{x}) - 2\alpha^{-1}(\mathbf{x})\int d\mathbf{x}'\mathcal{F}_\omega(\mathbf{x}, \mathbf{x}')[1 + \omega(\mathbf{x}, \mathbf{x}')] = -\varphi(\mathbf{x})$ and $2\mathcal{F}_\omega(\mathbf{x}, \mathbf{x}') = -\alpha(\mathbf{x})\,\Delta(\mathbf{x}, \mathbf{x}')\alpha(\mathbf{x}')$, respectively. We write them in abbreviated form as

$$\mathcal{F}_\alpha - 2\alpha^{-1}\mathcal{F}_\omega(1+\omega) = -\varphi,\ 2\mathcal{F}_\omega = -\alpha\Delta\alpha. \tag{6.165}$$

We see from (6.163) that the correspondence between the variables ω and Δ is $0 \leftrightarrow 0$; from this and from (6.162) it follows that the zeroth-order approximation for $\mathcal{F}$ is the same functional (6.141) as for Γ.

Solving the graphical equation (6.163) iteratively for Δ and using the second equation in (6.165), we explicitly find the graphs of the derivative $\mathcal{F}_\omega$ and from them the graphs of $\mathcal{F}$. In first orders from (6.163) we obtain $\Delta(\mathbf{x}, \mathbf{x}')$ $=$ , where the line is now ω instead of Δ. Substituting this expansion into the second equation in (6.165), we find $2\mathcal{F}_\omega(\mathbf{x}, \mathbf{x}') =$ 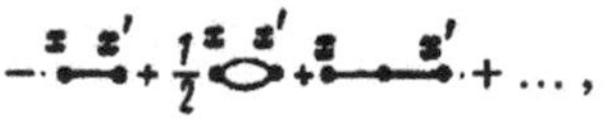, from which

$$\mathcal{F}(\alpha,\omega) = F(\alpha) - \frac{1}{4}$$ 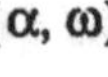 $$+\frac{1}{12}$$ 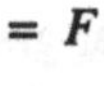 $$+\frac{1}{6}$$ 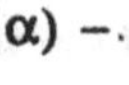 $$+\dots \qquad (6.166)$$

where $F(\alpha)$ is the zeroth-order approximation (6.141). In the graphs of (6.166) the line is ω and the vertices are the same as in the graphs of Γ, i.e., each is associated with a simple factor of α. It is clear from the form of the graphs of $\bar{\Gamma}$ for the gas and the procedure of constructing $\mathcal{F}$ that this will be true in all orders.

We note that in the first three orders the functional $\mathcal{F}$ will have the form (6.166) not only for a gas, but also for any other theory if the variable ω is defined such that in lowest order $\omega = \Delta$. However, the vertex factors u_2 and u_3 in the graphs (6.166) will then be different in accordance with the general rule (6.135). As for the first transform, the specific features of a particular theory are manifested in the graphs of $\mathcal{F}$ only beginning at fourth order.

Let us now obtain the analog of (6.136) for the second transform. For this we isolate the first graphs from Γ and $\mathcal{F}$,

$$\Gamma = F + \frac{1}{2}$$ $$+\frac{1}{4}$$  $$+\ \bar{\bar{\Gamma}}; \qquad \mathcal{F} = F - \frac{1}{4}$$  $$+\ \bar{\bar{\mathcal{F}}}\,, \qquad (6.167)$$

and substitute these expressions into (6.162). The zeroth-order approximation F then cancels, and the first-order graph $= \alpha\Delta\alpha$ is cancelled in Γ by one of the contributions of the form $s\Delta = [\alpha\omega\alpha + \alpha\alpha]\Delta/2$, so that we obtain

$$-\frac{1}{4}\alpha\omega^2\alpha + \bar{\bar{\mathcal{F}}}(\alpha,\omega) = \frac{1}{4}\alpha\Delta^2\alpha + \bar{\bar{\Gamma}}(\alpha,\Delta) - \frac{1}{2}\alpha\omega\Delta\alpha, \qquad (6.168)$$

where $\alpha\omega^2\alpha$ denotes the contribution of the graph with lines ω, and so on.

The variable Δ on the right-hand side of (6.168) must be expressed in terms of ω using the second equation (6.165), which can be rewritten as

$$\Delta = \omega - 2\alpha^{-1}\bar{\bar{\mathcal{F}}}_\omega\alpha^{-1} \equiv$$ $$+$$  $$. \qquad (6.169)$$

After substitution into (6.168) and cancellations, we obtain the desired expression:

$$\bar{\bar{\mathcal{F}}}(\alpha,\omega) = \alpha^{-1}(\bar{\bar{\mathcal{F}}}_\omega)^2\alpha^{-1} + \bar{\bar{\Gamma}}(\alpha,\Delta)\,, \qquad (6.170)$$

which can be used to recursively construct the graphs of $\bar{\bar{\mathcal{F}}}$ from the known graphs of Γ.

Let us now select the graphs which are 2-irreducible in the vertices on both sides of (6.170). There are no such graphs at all in the first term on the right-hand side, since it has the structure  and contains the nontrivial 2-section shown by the dashed line. If we express Δ in terms of ω in the second term, we must discard the contributions of $\bar{\bar{\mathscr{F}}}_\omega$ to Δ in selecting the 2-irreducible part, because the insertion of the nontrivial block  as a line into the graph of $\bar{\bar{\Gamma}}(\alpha, \Delta)$ necessarily leads to a graph which is 2-reducible in the vertices. Therefore,

$$\text{2-irreducible part of } \bar{\bar{\mathscr{F}}}(\alpha, \omega) = \text{2-irreducible part of } \bar{\bar{\Gamma}}(\alpha, \omega)\,, \quad (6.171)$$

which is the analog of (6.140) for the second transform. All these arguments and the result (6.171) are valid for any theory, not only for a gas. The only difference will be that the second vertex factor $u_2(\alpha)$ will enter into (6.168)–(6.170) instead of α.

Equation (6.171) completely characterizes the graphs which are 2-irreducible in the vertices for any theory, so the entire difficulty, as usual, is to describe the reducible graphs. Such graphs even exist for the gas, but they have a simple form and can be characterized completely: first, there are polygons, second, there are graphs of the form , which we shall call "melons." Below we shall show that for a gas

$$\text{2-reducible part of } \bar{\bar{\mathscr{F}}}(\alpha, \omega) = \sum_{n=4}^{\infty} \frac{(-1)^{n+1}}{2n}$$ 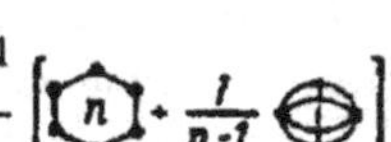 . (6.172)

The 2-irreducible graphs of third order given in (6.166) exactly correspond to the $n = 3$ contribution to the sum in (6.172).

As usual, we shall prove (6.172) using the equation of motion for $\mathscr{F}$. To obtain this equation it is necessary to express the second derivative $W_{\varphi\varphi}$ in (6.164) in terms of the derivatives of $\mathscr{F}$.

This is easily done knowing that the matrices of second derivatives of W and $\mathscr{F}$ in the variables φ, Δ and α, s are, respectively, the inverses of each other up to a sign. From this we obtain $W_{\varphi\varphi}^{-1} = -\mathscr{F}_{\alpha\alpha} + \mathscr{F}_{\alpha s}\mathscr{F}_{ss}^{-1}\mathscr{F}_{s\alpha}$. We substitute this into (6.164) after inverting the latter; then, expressing the derivatives of $\mathscr{F}$ in terms of the variables α, ω we obtain

$$2\mathscr{F}_\omega = \alpha[1\cdot\alpha + \alpha\omega\alpha]^{-1}\alpha + \alpha\mathscr{F}_{\alpha\alpha}\alpha - 2\omega\mathscr{F}_\omega - h_{\alpha\omega}\mathscr{F}_{\omega\omega}^{-1}h_{\omega\alpha}, \quad (6.173)$$

where

$$h_{\alpha\omega} \equiv \alpha\mathscr{F}_{\alpha\omega} - 2\mathscr{F}_{\omega} \cdot \mathbf{1}. \tag{6.174}$$

The arrangement of the arguments $\mathbf{x}, \mathbf{x}'$ in (6.173) will become clear later.

We warn the reader wishing to check these expressions that the algebra involved in the substitution $s \to \omega$ is quite tedious, although no serious difficulties arise. It is technically simpler to make the substitution in two steps, changing first from the variable $s(\mathbf{x}, \mathbf{x}')$ to the variable $\beta\,(\mathbf{x}, \mathbf{x}')$ given by the equation $2s = \beta + \alpha\alpha$ (here $\mathscr{F}_{\alpha\alpha} - \mathscr{F}_{\alpha s}\mathscr{F}_{ss}^{-1}\mathscr{F}_{s\alpha} \to \mathscr{F}_{\alpha\alpha} - \mathscr{F}_{\alpha\beta}\mathscr{F}_{\beta\beta}^{-1}\mathscr{F}_{\beta\alpha} - 2\mathscr{F}_{\beta}$), and then from β to $\omega = \alpha^{-1}\beta\alpha^{-1}$.

The left-hand side of (6.173) involves a derivative with respect to the line $\omega(\mathbf{x}, \mathbf{x}')$. The first term on the right-hand side is the progression $\alpha\,[\mathbf{1}\cdot\alpha + \alpha\omega\alpha\,]^{-1}\alpha =$  with fixed arguments $\mathbf{x}, \mathbf{x}'$ of the vertices at the ends of the chains. The zeroth-order contribution from the progression cancels with the contribution of the zeroth-order approximation in the term $\alpha\mathscr{F}_{\alpha\alpha}\alpha \equiv \alpha(\mathbf{x})\mathscr{F}_{\alpha\alpha}(\mathbf{x}, \mathbf{x}')\alpha(\mathbf{x}')$; the first-order contribution from the progression generates the graph  in $\mathscr{F}$, and the contributions of second and higher orders generate the triangle in (6.166) and the 2-reducible polygons of (6.172). The two-reducible melons of (6.172) are generated in iterations by the terms $\alpha\mathscr{F}_{\alpha\alpha}\alpha$ and $\omega\mathscr{F}_{\omega} \equiv \omega(\mathbf{x}, \mathbf{x}')\mathscr{F}_{\omega}(\mathbf{x}, \mathbf{x}')$ on the right-hand side of (6.173) from the 2-irreducible graph  [it will be shown below that the last term on the right-hand side of (6.173) does not generate either melons or polygons]. Equating the coefficients of the melon with n lines on both sides of (6.173), we obtain the recursion formula $(n + 1)c_{n+1} = -(n - 1)c_n$ for the coefficient c_n of the melon with n lines. Knowing that $c_2 = -1/4$, we thus obtain the coefficients of the melons given in (6.172). We note that these coefficients are anomalous, i.e., they differ from the inverse symmetry number, which for the melon with n lines is $2n!$ (the coincidence in second and third orders is accidental). We also note that up to a factor of $-1/2$ the coefficients of the melons coincide with the expansion coefficients of the function $(1 + x)\ln(1 + x)$, so that if desired the melon series in (6.172) can be summed explicitly. The polygons in (6.172) are easily summed to a logarithm in the spatially uniform theory after transforming the line $\omega(\,\mathbf{x} - \mathbf{x}')$ to momentum space.

In the last term on the right-hand side of (6.173) all the arguments of the subscripts ω are contracted, and $\mathbf{x}$ and $\mathbf{x}'$ are the arguments of the subscripts α in the factors $h_{\alpha\omega}$. The definition (6.174) written out in detail is

$$h_{\alpha\omega}(\mathbf{x};\mathbf{y},\mathbf{y}') = \alpha(\mathbf{x})\frac{\delta^2\mathscr{F}}{\delta\alpha(\mathbf{x})\,\delta\omega(\mathbf{y},\mathbf{y}')} -$$

$$-\frac{\delta\mathcal{F}}{\delta\omega(\mathbf{y},\mathbf{y}')}\left[\delta(\mathbf{x}-\mathbf{y})+\delta(\mathbf{x}-\mathbf{y}')\right], \tag{6.175}$$

and the inverse kernel $\mathcal{F}^{-1}_{\omega\omega}$ in (6.173) is constructed, as usual, as a series of the type (6.74), the first term in which is generated by the graph  .

Let us first analyze the block $h_{\alpha\omega}$. Differentiation with respect to $\omega(\mathbf{y},\mathbf{y}')$ is equivalent to removing one line ω from the graph in all possible ways. The graphs of the derivative $\mathcal{F}_\omega$ will contain two labeled vertices, the ones connected by the removed line. Then, acting on $\mathcal{F}_\omega$ with the operator $\alpha(\mathbf{x})\delta/\delta\alpha(\mathbf{x})$, we successively label each of the vertices of the graph of $\mathcal{F}_\omega$ with the index $\mathbf{x}$. Among the resulting graphs will be ones in which the operator $\alpha\delta/\delta\alpha$ labels the same two vertices that were already labeled by removing the line ω. Clearly, the second term on the right-hand side of (6.175) exactly cancels all such graphs, so that $h_{\alpha\omega}$ is the part of the graphs of $\alpha\mathcal{F}_{\alpha\omega}$ in which the operator $\alpha\delta/\delta\alpha$ labels a new, not yet labeled, vertex. All such graphs must have at least three vertices. Two-vertex graphs—melons—do not contribute at all to $h_{\alpha\omega}$; the first graph which contributes is the triangle given in (6.166).

It follows from the above discussion about the structure of the graphs of $h_{\alpha\omega}$ that the last term on the right-hand side of (6.173) cannot contain either simple chains of the type  or melons, so it cannot generate either polygons or melons in iterations.

We shall now show that all the graphs of $\bar{\bar{\mathcal{F}}}$ except for those in (6.172) are 2-irreducible in the vertices. We shall call a vertex 2-section *complex* if at least one of the cut blocks is neither a simple chain nor a melon. There are no complex 2-sections in the graphs of (6.172), but any other 2-reducible graph must contain such a section. Therefore, our problem is to prove the absence of graphs with complex 2-sections. As always, we shall prove this by induction: we assume that there are none in the lowest-order graphs, and show that then there also cannot be any in graphs of the next highest order obtained by iterating (6.173). In fact, let us assume that in $\mathcal{F}$ a graph has appeared with complex 2-section of type (6.176*a*):

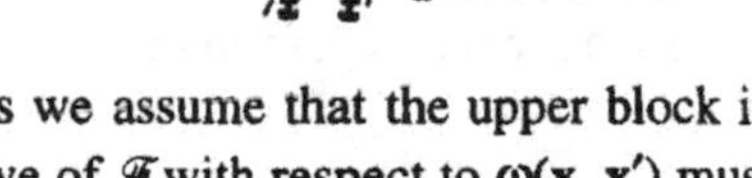

(6.176)

For definiteness we assume that the upper block in (6.176*a*) is complex. Then the derivative of $\mathcal{F}$ with respect to $\omega(\mathbf{x},\mathbf{x}')$ must contain graphs of the type (6.176*b*), obtained from the graphs (6.176*a*) by removing one of the lines of the lower block which was connected to one of the vertices of the 2-section.

Now let us show that under the conditions of the induction assumption, the right-hand side of (6.173) cannot contain graphs of the type (6.176*b*).

Actually, the graphs of the derivatives $\alpha\mathcal{F}_{\alpha\alpha}\alpha$ and $\omega\mathcal{F}_{\omega}$ differ from the differentiated graphs of $\mathcal{F}$ by only the labels of the vertices, so that the induction assumption forbids the presence of complex 2-sections in them. Let us now turn to the last term on the right-hand side of (6.173). Expanding the quantity $\mathcal{F}^{-1}_{\omega\omega}$ in it in a series (see above), we obtain a progression; the structure of one of its terms is shown in (6.176*c*). By analyzing the variants, it can be verified fairly quickly that the presence of complex 2-sections of the type (6.176*b*) in the graph (6.176*c*) always contradicts the induction assumption, according to which the blocks  and  do not have complex 2-sections. The desired statement is thereby proved.

Therefore, the functional $\mathcal{F}$ for the gas is the sum of 2-reducible graphs (6.172) and all graphs of Γ which are 2-irreducible in the vertices. There are very few such graphs in the lowest orders. In addition to the graphs of (6.166) with trivial 2-sections, there are only two of them in the first eight orders: the tetrahedron ("envelope") in sixth order and the pyramid in eighth order. Of the forty-two star Mayer graphs of ninth order, only three are 2-irreducible in the vertices (see Appendix 2).

The nonideal classical gas is the only theory for which the structure of the 2-reducible graphs of the second transform has been fully described. As far as we know, for typical field interactions with power-law or polynomial generating vertex, there have been no studies at all of the Legendre transforms (neither the first nor the second) of the logarithm of the S matrix. So far these objects have been popular only in classical statistics.

6.3.6 *The stationarity equations and the self-consistent field approximation*

Knowing the graphs of the Legendre transforms of Γ and $\mathcal{F}$, we can express all the correlation functions in terms of the variables of these functionals; these correlation functions are (see the definitions in Secs. 5.2.1 and 5.3.1) multiple derivatives of the logarithm of the partition function $W(\varphi, \Delta)$ with respect to the variable φ. For this it is sufficient to write the raising operator $\mathcal{D} \equiv \delta/\delta\varphi$ [the analog of the operators (6.34) and (6.56)] in terms of the needed variables. This is very easily done for the first transform using (6.131): $\mathcal{D} = \delta\alpha/\delta\varphi\cdot\delta/\delta\alpha = W_{\varphi\varphi}\cdot\delta/\delta\alpha = -\Gamma^{-1}_{\alpha\alpha}\cdot\delta/\delta\alpha$. After rather lengthy but uncomplicated algebra, for the second transform (6.162) we obtain

$$\mathcal{D} \equiv \frac{\delta}{\delta\varphi} = [\alpha\omega + 1]\,\{\alpha\frac{\delta}{\delta\alpha} - h_{\alpha\omega}\mathcal{F}^{-1}_{\omega\omega}\frac{\delta}{\delta\omega}\}\,. \tag{6.177}$$

All quantities have been defined in the preceding section; the product of the expression in the square brackets and the expression in the curly brackets should be understood as a contraction.

Acting with the raising operator $\mathcal{D}$ on the first correlation function $W_\varphi = \alpha$, we express all the higher-order correlation functions in terms of the variables of the corresponding Legendre transform: α, Δ for Γ, and α, ω for $\mathcal{F}$. We recall that ω is related to the second correlation function $W_{\varphi\varphi}$ via (6.164).

According to the general rules of Sec. 6.2.1, the equilibrium values of the dressed variables (α for Γ and α, ω for $\mathcal{F}$) are found from the given values of the bare variables φ, Δ by solving the stationarity equations. Below we shall study only the first transform, for which the stationarity equation has the form $\Gamma_\alpha = -\varphi$.

Let us begin with the Ising model, in which $\varphi = \beta h$ is the reduced external field, and the conjugate variable α is the magnetization. We assume that only the zeroth-order approximation (6.150) and the first of the graphs in (6.138) are kept in Γ. The stationarity equation in this approximation takes the form $-\varphi = F_\alpha + \Delta\alpha = -\mathrm{arctanh}\,\alpha + \Delta\alpha$, which up to the notation coincides with the well known Weiss equation for the magnetization [41]. In the language of the ordinary graphs of the logarithm of the partition function, it is obtained by constructing the self-consistent equation for the magnetization by summing all the tree graphs [41], and in the language of the functional integral (5.49) it corresponds to the stationary-phase approximation. In the language of the ordinary graphs of W, the inclusion of the second graph of (6.138) (the simple loop) in Γ solves the problem of finding the correction to the Weiss approximation taking into account not only simple trees, but also all trees with any number of insertions of this loop (in general, with insertions of all the graphs which are kept in Γ). This problem was solved in [80] in the language of ordinary graphs.

We also note that the well known Ornstein–Zernike approximation for the correlation function $W_{\varphi\varphi}$ (Eq. (2.77) in [41]) is obtained from the same approximation for Γ as the Weiss equation. Actually, as shown at the beginning of this section, $W_{\varphi\varphi} = -\Gamma^{-1}_{\alpha\alpha}$. If only the contribution of the zeroth- order approximation and the first-order graph are kept in Γ, we obtain

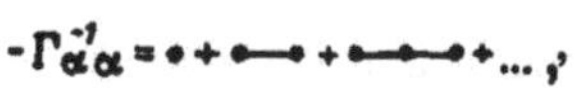

where Δ corresponds to a line and the "fluctuation-suppressing" factor $1-\alpha^2$ corresponds to a point. For a uniform system the progression is easily summed after going to momentum space, and the answer up to notation coincides with Eq. (2.77) in [41].

By comparing these calculations with the corresponding graphical constructions in [41] or [80], one can judge how much simpler the entire proof is when the Legendre transform is used. All the difficult combinatorial problems are essentially solved when the structure of the graphs of Γ is determined, and after this is done it is elementary to obtain the usual approximations and corrections to them. However, we do not obtain the correct critical behavior in any finite order of Γ, because (6.156) in finite order is analytic in α and β and therefore leads to the same values of the critical exponents as the classical Landau theory of phase transitions. In the language of ordinary perturbation theory, the use of a finite number of graphs of Γ corresponds to the summation of infinite classes of graphs, and to obtain the correct critical behavior it is not sufficient. The technique of Legendre transforms makes it possible to find an anomalous solution only in some simple approximation which is valid far from the critical point. Then this approximation can be systematically improved by including more graphs of Γ, so that the citical point is approached more and more closely. However, the behavior of the solution at the critical point itself apparently cannot be determined without summing the infinite number of graphs of Γ (see footnote in Sec. 6.1.6).

Let us now turn to the classical gas. As already mentioned in Sec. 6.3.2, for a spatially uniform system in zero external field, α is the density, $a = \exp\varphi$ is the activity, and the specific value $\gamma = \Gamma/V$ is given by the series [see (6.141) and (6.148)]

$$\gamma(\alpha, \beta) = \alpha - \alpha \ln\alpha + \sum_{n=2}^{\infty} \alpha^n S_n, \tag{6.178}$$

the general term of which is the sum of the contributions of all star graphs with n vertices. The self-consistent field approximation, i.e., the van der Waals equation for the pressure [7,8], is much more complicated to obtain here than in all the other cases and does not reduce to selecting the first few graphs of the Legendre transform. This again shows that in some respects the statistics of the nonideal classical gas is more complicated than quantum field theory.

The phenomenon of condensation described by the van der Waals equation occurs when the two-body interaction potential $\mathcal{V}(\mathbf{x}, \mathbf{x}')$ corresponds to attraction at large distances and fairly strong repulsion at small ones. These two regions are treated differently in constructing the self-consistent field approximation. Repulsion is often described by the hard-sphere model, in which the potential $\mathcal{V}$ is assumed to be $+\infty$ for $|\mathbf{x} - \mathbf{x}'| \leq d$, where d is the sphere diameter. For $|\mathbf{x} - \mathbf{x}'| > d$ the potential is negative, corresponding to attraction.

According to the definition in Sec. 5.3.1, the line Δ is related to the potential $\mathcal{V}$ as $\Delta = -\beta\mathcal{V}$, and in the hard-sphere model it is not only possible but necessary to change over to the Mayer line $g = -1 + \exp\Delta$, taking the finite value -1 at $\mathcal{V} = +\infty$.

To obtain the self-consistent field approximation ("mean-field theory"), it is sufficient to take into account the attractive potential, as in the Ising model, only in the lowest order for $\bar{\Gamma}$, i.e., in the graph $\bullet\!\!-\!\!\!-\!\!\bullet = \alpha\Delta\alpha$. This gives the term $c\beta\alpha^2$ on the right-hand side of (6.178), where c is some positive (owing to the positivity of Δ for $\mathcal{V} < 0$) constant.

The situation with repulsion is more complicated. It would be best to calculate the function (6.178) exactly for the model of simple hard spheres without attraction, but it has not been possible to do this for real three- dimensional space. In the language of the Legendre transform, the following expression for γ is a rough estimate of the partition function for a gas of hard spheres leading ultimately to the van der Waals equation (see, for example, [8], p. 79):

$$\gamma^{\mathrm{rep}}_{\mathrm{vdW}} = \alpha - \alpha\ln\alpha + \alpha\ln(1 - b\alpha)\,, \tag{6.179}$$

where b is four times the volume of a hard sphere.

Equation (6.179) can be understood as the result of a very rough summation of the contributions of the infinite number of graphs of $\bar{\Gamma}$. The specific form of the additional term $\alpha\ln(1 - b\alpha)$ is actually not very important: any term whose derivative with respect to α becomes infinite at some finite value of α leads to an equation of the van der Waals type. Physically, this corresponds to the existence of a limiting density which cannot be exceeded, in other words, the finiteness of the volume occupied by the gas molecules is taken into account. It is this physical idea which forms the basis of the van der Waals equation.

Adding the contribution of the attractive forces to (6.179), we obtain

$$\gamma_{\mathrm{vdW}} = \alpha - \alpha\ln\alpha + \alpha\ln(1 - b\alpha) + c\beta\alpha^2. \tag{6.180}$$

The corresponding stationarity equation $\partial\gamma/\partial\alpha = -\varphi = -\ln a$ determines the dependence of the density α on the activity a. The pressure p is given by (6.147); after substituting $\Gamma = V\gamma$ from (6.180) into it we obtain $\beta p = \gamma - \alpha\gamma_\alpha = \alpha(1 - b\alpha)^{-1} - c\beta\alpha^2$, which up to notation coincides with the usual van der Waals equation [7,8]. We note that the well known Maxwell equal-area rule eliminating the unphysical (in the equilibrium theory) supercooling and superheating segments is obtained automatically when the function (6.180) is replaced by its convex envelope.

Appendix 1

NONSTATIONARY PERTURBATION THEORY FOR A DISCRETE LEVEL

Here we derive the asymptotic expressions of nonstationary perturbation theory for an isolated discrete level which we need to prove (1.74), expressing the energy shift of a level in terms of the S matrix, and (1.71), used to go from the Heisenberg picture to the interaction picture in the Green functions. The unperturbed level in (1.71) and (1.74) is assumed to be nondegenerate, but for generality we shall also consider the case of a degenerate level, which is important for the theory of the atom.

The expressions of nonstationary perturbation theory for a discrete level are usually obtained using the Gell-Mann–Low adiabatic formalism [81]; a detailed derivation can be found in [82,83]. In the adiabatic formulation the cutoff factor $\exp(-\alpha|t|)$ is introduced into the interaction operator **V** and then the limit $\alpha \to 0$ is studied.

The introduction of an adiabatic cutoff is inconvenient in several respects. First, for understandable reasons it complicates the transformation to momentum space in the graphs of perturbation theory. Second, the appearance of an explicit time dependence in the interaction Hamiltonian makes it impossible to use the first of the representations in (1.55) for the evolution operator. Only the second representation in (1.55), the Volterra T-exponential, remains valid. Finally, the choice of cutoff factor in the form $\exp(-\alpha|t|)$ is completely arbitrary. The same results can be obtained using, for example, a Gaussian cutoff $\exp(-\alpha t^2)$ or any other. This arbitrariness would be irrelevant if the final expressions were independent of the choice of cutoff, but actually this is not so, at least for a degenerate level. The Gell-Mann–Low adiabatic theory is characterized by the appearance of the operator $g\,d/dg$ in the final expressions, where g is the coupling constant in the interaction. This operator introduces an additional combinatorial factor of n for the nth-order contribution of perturbation theory. If the calculations for Gaussian cutoff could be completed, the results would certainly contain some combinatorial operator different from $g\,d/dg$.

We shall use a "natural cutoff," taking the evolution operator on a finite time interval and then letting this interval go to infinity. For this cutoff the final expressions, as we shall see below, will not contain any artificial combinatorial operators.

It should also be noted that the final result of the adiabatic theory for a degenerate level is expressions for the secular and wave operators constructed in terms of the evolution operator [82,83], but there are no explicit asymptotic representations for the evolution operator itself in the theory. Such expressions can be obtained using the natural cutoff. This has been done in [84,85], which we shall follow with slight changes. For the case of a nondegenerate level, the asymptotic expressions in the theory with natural cutoff were known long ago, but they were usually obtained on the basis of the Gell-Mann--Low adiabatic formulas (see, for example, the appendix in the paper by Hubbard [86]).

Now for the derivation itself. Let $\mathbf{H} = \mathbf{H}_0 + \mathbf{V}$ be the full Hamiltonian acting in some Hilbert space X, $\mathbf{P}$ be a projector onto the eigensubspace of the free Hamiltonian $\mathbf{H}_0$ corresponding to some isolated, discrete level of energy E_0, $X_0 = \mathbf{P}X$ be the unperturbed eigensubspace of $\mathbf{H}_0$, and $\mathbf{P}' = 1 - \mathbf{P}$ be the projector onto the orthogonal complement to X_0. For a nondegenerate level the subspace X_0 is one-dimensional.

To impart rigor to the expressions we derive, we shall introduce an intermediate regularization into the theory by means of the substitution $\mathbf{H}_0 \to \mathbf{H}_{0\varepsilon} \equiv \mathbf{H}_0 - i\varepsilon\mathbf{P}'$, $\varepsilon > 0$. This substitution does not introduce an explicit time dependence into the Hamiltonian, so that it does not come into conflict with the use of the evolution operator in the form of an ordinary product of exponentials (1.55). All quantities in the final expressions will be regular in ε near $\varepsilon = 0$, so that we can simply set $\varepsilon = 0$ in them. This regularization should not be confused with the cutoff of the time integrals, which we shall accomplish simply by using a finite evolution time.

Let $\mathbf{U}(\tau_1, \tau_2)$ be the evolution operator in the interaction picture (1.55). Owing to time-translational invariance, which is not spoiled by the introduction of the regularization, the projection $\mathbf{PU}(\tau_1, \tau_2)\mathbf{P}$ depends only on the time difference $T \equiv \tau_1 - \tau_2$, and we denote it as $\mathbf{U}_{PP}(T)$. We shall show that for $T \to \infty$

$$\mathbf{U}_{PP}(T) \cong \exp[-i\mathbf{Q}T]\,\mathbf{Z} = \mathbf{Z}\exp[-i\mathbf{Q}^+T] = \mathbf{Z}^{1/2}\exp[-i\overline{\mathbf{Q}}T]\,\mathbf{Z}^{1/2}. \tag{A.1}$$

This is an asymptotic expression, valid in each order of perturbation theory up to exponentially small corrections $\sim\exp(-\varepsilon T)$; $\mathbf{Q}$, $\mathbf{Z}$, and $\overline{\mathbf{Q}} \equiv \mathbf{Z}^{-1/2}\mathbf{Q}\mathbf{Z}^{1/2}$ are operators in X_0 which are independent of T and regular in ε in the neighborhood of $\varepsilon = 0$, and we shall construct their explicit perturbation series.

For $\varepsilon = 0$ the operators $\mathbf{Z}$ and $\overline{\mathbf{Q}}$ are Hermitian and $\mathbf{Q}$ is not, and

$$\mathbf{QZ} = \mathbf{ZQ}^{+}. \tag{A.2}$$

The Hermiticity of $\overline{\mathbf{Q}}$ is a consequence of this equation.

For the "half S-matrices" we obtain the following asymptotic representations valid with the same accuracy as (A.1):

$$\mathbf{U}(0, -T)\,\mathbf{P} \cong \Omega \exp\,[-i\mathbf{Q}T]\,\mathbf{Z};\;\; \mathbf{PU}(T, 0) \cong \mathbf{Z}\exp\,[-i\mathbf{Q}^{+}T]\,\Omega'. \tag{A.3}$$

Here $\mathbf{Q}$, $\mathbf{Q}^{+}$, and $\mathbf{Z}$ are the same operators as in (A.1), Ω is the operator from X_0 to X, and Ω' is the operator from X to X_0. Comparing (A.1), (A.2), and (A.3), we see that

$$\mathbf{P}\Omega = \mathbf{P},\;\; \Omega'\mathbf{P} = \mathbf{P},\;\; \Omega'\Omega = \mathbf{Z}^{-1}. \tag{A.4}$$

Before proving (A.1)–(A.3), let us explain the physical meaning of the operators entering into these equations. We recall that a *secular operator* or *secular matrix* is, in a broad sense, any linear operator in the eigensubspace X_0 whose eigenvalues are the desired shifts of the energy levels. In the narrow sense [83] a secular operator is an operator whose eigenvectors are the projections of the exact wave functions Ψ_α onto the subspace X_0. The term *wave operator* refers to an operator from X_0 to X taking projections of the exact wave functions $\mathbf{P}\Psi_\alpha$ into these functions themselves.

We assume that Eqs. (A.1)–(A.3) are valid, and show that then Ω is a wave operator and $\mathbf{Q}$ is a secular operator in the narrow sense.

Let Φ be an arbitrary vector from X_0, Φ_α be a set of eigenfunctions of the operator $\mathbf{Q}^{+}$ in X_0, and ω_α be the corresponding eigenvalues. We expand Φ in the basis Φ_α: $\Phi = \Sigma c_\alpha \Phi_\alpha$. Using the first of the representations in (A.3) and Eq. (A.2), for the vector $\mathbf{U}(0, -T)\Phi \equiv \Psi(T)$ we obtain

$$\psi(T) = \sum_\alpha c_\alpha \Omega \mathbf{Z}\Phi_\alpha \exp\,(-i\omega_\alpha T). \tag{A.5}$$

This is an asymptotic (for $T \to \infty$) representation valid up to exponentially small corrections in each order of perturbation theory (regularization is assumed to have been done). On the other hand, according to the first equation in (1.55), $\mathbf{U}(0, -T) = \exp(-i\mathbf{H}T)\exp(i\mathbf{H}_0T)$ and $\mathbf{U}(0, -T)\mathbf{P} = \exp i(E_0 - \mathbf{H})T\cdot\mathbf{P}$ owing to the equation $\mathbf{H}_0\mathbf{P} = E_0\mathbf{P}$. From this it is clear that each vector $\Psi(T) = \mathbf{U}(0, -T)\Phi$, $\Phi \in X_0$ satisfies the Schrödinger equation $i\partial\Psi/\partial T = (\mathbf{H} - E_0)\Psi$. Substituting $\Psi(T)$ in the form (A.5) into this and taking into account the fact that c_α is arbitrary, we conclude that $(\mathbf{H} - E_0)\Omega\mathbf{Z}\Phi_\alpha = \omega_\alpha\Omega\mathbf{Z}\Phi_\alpha$ for any α. This proves that $\Psi_\alpha \equiv \Omega\mathbf{Z}\Phi_\alpha$ are eigenvectors of the full Hamiltonian $\mathbf{H}$, i.e., the exact wave functions, and the ω_α are the desired energy shifts. The latter implies that $\mathbf{Q}^+$ is a secular operator.

It follows from the equation $\mathbf{P}\Omega = \mathbf{P}$ in (A.4) that the vectors $\mathbf{Z}\Phi_\alpha$ are the projections of the exact wave functions Ψ_α in X_0: $\mathbf{Z}\Phi_\alpha = \mathbf{P}\Omega\mathbf{Z}\Phi_\alpha = \mathbf{P}\Psi_\alpha$. Therefore, Ω is a wave operator. Finally, from the definition of the Φ_α as eigenvectors of $\mathbf{Q}^+$ and (A.2) we find that the projections $\mathbf{Z}\Phi_\alpha$ are eigenvectors of $\mathbf{Q}$: $\mathbf{Q}\mathbf{Z}\Phi_\alpha = \mathbf{Z}\mathbf{Q}^+\Phi_\alpha = \omega_\alpha\mathbf{Z}\Phi_\alpha$. Therefore, $\mathbf{Q}$ is a secular operator in the narrow sense, as we wanted to prove.

We note that the expression $i\partial\mathbf{U}(0, -T)\mathbf{P}/\partial T = (\mathbf{H} - E_0)\mathbf{U}(0, -T)\mathbf{P}$ and the representation (A.3) give the equations

$$\Omega\mathbf{Q} = (\mathbf{H} - E_0)\,\Omega, \quad \mathbf{Q} = \mathbf{P}\mathbf{V}\Omega. \tag{A.6}$$

The second equation follows from the first when the expressions $\mathbf{P}\Omega = \mathbf{P}$ and $\mathbf{P}(\mathbf{H} - E_0) = \mathbf{P}\mathbf{V}$ are taken into account.

Using the asymptotic representations (A.1) and (A.3), we can express the secular and wave operators in terms of the evolution operator:

$$\mathbf{Q} = \lim_{T\to\infty} \frac{i\partial\mathbf{U}_{PP}(T)}{\partial T}\mathbf{U}_{PP}^{-1}(T)\,; \quad \Omega = \lim_{T\to\infty} \mathbf{U}(0, -T)\,\mathbf{U}_{PP}^{-1}(T)\,. \tag{A.7}$$

For a nondegenerate level the subspace X_0 is one-dimensional and all the operators in it are ordinary numbers. The number $\mathbf{Q} = \mathbf{Q}^+ = \Delta E$ is the shift of the energy level, and after taking the logarithm of (A.1) we obtain (1.74). Equation (1.74) can be made considerably more precise. From (A.1) we see that for a nondegenerate level the leading term in the asymptotic representation of the logarithm of the expectation value of the evolution operator is of order T with coefficient $-i\Delta E$, and the next term is of order $T^0 = 1$ with coefficient $\ln \mathbf{Z}$. The corrections to these two terms are already exponentially small, i.e., terms of order T^{-1}, T^{-2}, etc., which might have been expected to appear, are in fact absent.

We obtain (1.71) from (A.3) for the case of a nondegenerate level. If Φ is a normalized unperturbed eigenstate, then $\Psi = \Omega\mathbf{Z}\Phi$ is the exact normalized eigenstate, because isometry of $\Omega\mathbf{Z}$ obviously follows from the unitarity of the evolution operator.

Let us now go directly to the proof of (A.1)–(A.3). The starting point is the usual representation (1.55) of the evolution operator:

$$\mathbf{U}(\tau_1, \tau_2) = \sum_{n=0}^{\infty} (-i)^n \int_{\tau_2}^{\tau_1} \ldots \int_{\tau_2}^{\tau_1} dt_1 \ldots dt_n \theta(t_1 \ldots t_n) \mathbf{V}(t_1) \ldots \mathbf{V}(t_n) . \quad \text{(A.8)}$$

The general term in this series can be represented graphically by the chain , where the operator factor $-i\mathbf{V}(t_k)$ is associated with each point, and the function $\theta(t_1 - t_2)$ is associated with a directed line . Each time t_k is integrated from τ_2 to τ_1.

Inserting the factor $1 = \mathbf{P} + \mathbf{P}'$ between all the interaction operators in (A.8), we can represent the series (A.8) by the progression

(A.9)

with the operator blocks

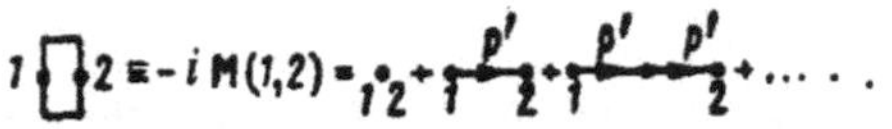

(A.10)

The lines θ in (A.9) contain the projector **P**, and the lines in (A.10) contain the projector **P′**. Each of the blocks **M** is a function of two arguments, the times of the endpoints of the chains (A.10). The times of the other points in (A.10) are integrated over (between finite limits, because they are all enclosed between the times of the endpoints). The series (A.10) can be written analytically as

$$-i\mathbf{M}(t, t') = \sum_{n=1}^{\infty} (-i)^n \int \ldots \int dt_1 \ldots$$

$$\ldots dt_n \delta(t - t_1) \delta(t' - t_n) \theta(t_1 \ldots t_n) \mathbf{V}(t_1) \mathbf{P}'\mathbf{V}(t_2) \ldots \mathbf{P}'\mathbf{V}(t_n) . \quad \text{(A.11)}$$

We see from this definition that the operator $\mathbf{M}(t, t')$ is retarded, i.e., it is nonzero only for $t \geq t'$.

The general term of the series (A.9) with n blocks contains $2n$ time arguments, two in each block. All these arguments are integrated over from τ_2 to τ_1.

We are interested in the projections **PUP**, **P′UP**, and **PUP′**, which for brevity we denote as $\mathbf{U}_{PP}$, $\mathbf{U}_{P'P}$, and $\mathbf{U}_{PP'}$. We see from (A.9) that

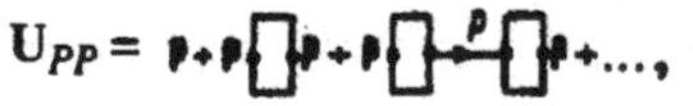

(A.12)

 (A.13)

and similarly for $\mathbf{U}_{PP'}$.

Let us first consider the operator $\mathbf{M}(t, t')$. Replacing each of the operators $\mathbf{V}(t)$ in (A.11) by $\exp(i\mathbf{H}_0 t)\mathbf{V}\exp(-i\mathbf{H}_0 t)$ and noting that the exponentials appearing between adjacent operators $\mathbf{V}$ depend only on time differences, we conclude that

$$\mathbf{M}(t, t') = \exp(i\mathbf{H}_0 t)\,\mathbf{G}(t-t')\exp(-i\mathbf{H}_0 t'), \tag{A.14}$$

where $\mathbf{G}$ is a retarded operator which is a function of the difference $t - t'$. Transforming to energy representation

$$\mathbf{G}(\tau) = \frac{1}{2\pi}\int dE\; \mathbf{g}(E)\exp(-iE\tau), \tag{A.15}$$

it is easy to use (A.11), (A.14), and (A.15) to calculate the operator $\mathbf{g}$:

$$\mathbf{g}(E) = \mathbf{V} + \mathbf{V}\frac{\mathbf{P}'}{E-\mathbf{H}_0+i0}\mathbf{V} + \mathbf{V}\frac{\mathbf{P}'}{E-\mathbf{H}_0+i0}\mathbf{V}\frac{\mathbf{P}'}{E-\mathbf{H}_0+i0}\mathbf{V} + \dots\,. \tag{A.16}$$

The infinitesimal imaginary additions $+\,i0$ in the denominators are also present in the unregularized theory. They arise from the standard spectral representation of the θ function and reflect the fact that $\mathbf{G}$ is retarded. The introduction of regularization leads to the replacement $\mathbf{H}_0 \to \mathbf{H}_0 - i\varepsilon\mathbf{P}'$ in (A.16), which is equivalent to the replacement $i0 \to i\varepsilon$. This shows that we are dealing with the natural regularization of a retarded function: the contour of the E integration in (A.15) is shifted from the upper lip of the real axis into the complex plane.

The replacement $i0 \to i\varepsilon$ in the denominators of (A.16) is equivalent to the replacement $E \to E - i\varepsilon$ in the argument of the exponential in (A.15), i.e., in the regularization

$$\mathbf{G}(\tau) \to \mathbf{G}_\varepsilon(\tau) = \mathbf{G}(\tau)\exp(-\varepsilon\tau). \tag{A.17}$$

Therefore, the regularized operator $\mathbf{G}(\tau)$ falls off exponentially for $\tau \to \infty$, so that the integral of $\mathbf{G}$ over τ converges well (we recall that $\mathbf{G} = 0$ for $\tau < 0$).

Let us now see how regularization affects the various projections of the operator (A.14). According to (A.17), upon regularization the operator $\mathbf{G}(t-t')$ acquires the cutoff $\exp \varepsilon (t'-t)$. The exponential operators in (A.14) acquire additional factors upon regularization only when they are multiplied by the projector $\mathbf{P}'$. From this we see that the projection $\mathbf{M}_{PP}(t, t')$ acquires the same factor from the regularization as $\mathbf{G}(t-t')$, the projection $\mathbf{M}_{P'P}(t, t')$ acquires altogether the factor $\exp \varepsilon t'$, and the projection $\mathbf{M}_{PP'}(t, t')$ acquires the factor $\exp(-\varepsilon t)$. Therefore, the factor from regularization in the projection $\mathbf{M}_{PP}$ is always a cutoff, in $\mathbf{M}_{P'P}$ it will be a cutoff only for $t' < 0$, and in $\mathbf{M}_{PP'}$ it will be one for $t > 0$. This is the reason that the projection $\mathbf{U}_{PP}$ of the evolution operator, expressed, as seen from (A.9), only in terms of $\mathbf{M}_{PP}$, can be considered in any time interval, whereas the projections $\mathbf{U}_{P'P}$ and $\mathbf{U}_{PP'}$ must respectively be considered on the negative and positive time semi-axes [which explains (A.1) and (A.3)].

For what follows, we note that the operator $\mathbf{g}(E)$ is regular in E near the point $E = E_0$, because we have assumed that the level E_0 is isolated (of course, only in the orders of perturbation theory). From this it follows that $\mathbf{g}(E)$ itself and all its derivatives are Hermitian operators at the point $E = E_0$.

Let us consider the projection $\mathbf{U}_{PP}(\tau_1, \tau_2)$ which, as already mentioned, depends only on the difference $T \equiv \tau_1 - \tau_2$. We shall therefore take $\tau_1 = T$, $\tau_2 = 0$ without loss of generality.

The general term of the series (A.9) is obviously a polynomial in T, the highest degree of which is equal to the number of blocks $\mathbf{M}_{PP}$ in the chain (our parameter T is the analog of $1/\alpha$ in the adiabatic formalism). To obtain the asymptotic expression (A.1) we need to explicitly isolate the entire T dependence and show that the resulting power series in T sums to give an exponential operator.

Let us start with the solution of the first problem. The projection $\mathbf{M}_{PP}(t, t') = \mathbf{G}(t-t')\exp iE_0(t-t')$ depends only on the difference $t - t'$. In the energy representation

$$\mathbf{M}_{PP}(t,t') = \frac{1}{2\pi}\int dE \; \mathbf{m}(E) \exp iE(t'-t) \tag{A.18}$$

we have $\mathbf{m}(E) = \mathbf{g}_{PP}(E + E_0)$. If for each of the blocks $\mathbf{M}_{PP}(t, t')$ in (A.9) we introduce the new variables $x \equiv (t + t')/2$ and $\tau \equiv t - t'$, the dependence on the variables x enters only into the arguments of the θ functions in (A.9), so that we can attempt to integrate over these variables explicitly. For the variables t, t' the integration in (A.9) runs over the region $T \geq t \geq t' \geq 0$, so that for fixed $\tau = t - t' > 0$ the integration over x runs from $\tau/2$ to $T - \tau/2$. In the general term of the series (A.9) with n blocks $\mathbf{M}_{PP}$ the integral over the variables $x_1 \ldots x_n$ from the product of θ functions looks like

$$\int_{\tau_1/2}^{T-\tau_1/2} dx_1 \ldots \int_{\tau_n/2}^{T-\tau_n/2} dx_n \theta\left(x_1 - x_2 - \frac{\tau_1+\tau_2}{2}\right) \ldots \theta\left(x_{n-1} - x_n - \frac{\tau_{n-1}+\tau_n}{2}\right).$$

It can be done explicitly and is equal to (all the τ_i are nonnegative)

$$\frac{1}{n!}\theta\left(T - \sum_{k=1}^{n} \tau_k\right)\left(T - \sum_{k=1}^{n} \tau_k\right)^n.$$

Therefore, for the projection $\mathbf{U}_{PP}(T)$ we obtain

$$\mathbf{U}_{PP}(T) = \sum_{n=0}^{\infty} \frac{(-i)^n}{n!} \int_0^T \ldots \int d\tau_1 \ldots d\tau_n \mathbf{M}_{PP}(\tau_1, 0) \ldots$$

$$\ldots \mathbf{M}_{PP}(\tau_n, 0)\, \theta\left(T - \sum_{k=1}^{n} \tau_k\right)\left(T - \sum_{k=1}^{n} \tau_k\right)^n. \qquad \text{(A.19)}$$

This is exact, because we have not yet made any approximations.

The regularized operator $\mathbf{M}_{PP}(\tau, 0)$ falls off exponentially for $\tau \to \infty$, so all the integrals over τ in (A.19) converge well. We therefore see that after discarding the θ function in (A.19) and extending the integration over all the τ_k to infinity, we introduce only an exponentially small error. This leads to the asymptotic expression

$$\mathbf{U}_{PP}(T) \cong \sum_{n=0}^{\infty} \frac{(-i)^n}{n!} \int_0^\infty \ldots \int d\tau_1 \ldots d\tau_n \mathbf{M}_{PP}(\tau_1, 0) \ldots$$

$$\ldots \mathbf{M}_{PP}(\tau_n, 0)\left(T - \sum_{k=1}^{n} \tau_k\right)^n. \qquad \text{(A.20)}$$

The entire T dependence has been isolated explicitly, and we now need only to sum the power series in T to an exponential. Using the identity

$$\left(T-\sum_{k=1}^{n}\tau_k\right)^n = \left(T+i\frac{d}{dE}\right)^n \exp\left(iE\sum_{k=1}^{n}\tau_k\right)\Bigg|_{E=0}$$

and Eq. (A.18), we write the right-hand side of (A.20) as a double series:

$$\mathbf{U}_{PP}(T) \cong \sum_{n=0}^{\infty}\sum_{k=0}^{n}\frac{(-i)^n}{k!\,(n-k)!}T^k(i\mathscr{D})^{n-k}\mathbf{m}^n. \tag{A.21}$$

Here and below $\mathbf{m} \equiv \mathbf{m}(E)$, $\mathscr{D} \equiv d/dE$, and it is understood that after differentiating it is necessary to set $E = 0$.

Changing the order of the summation in (A.21), we obtain the series

$$\mathbf{U}_{PP}(T) \cong \sum_{k=0}^{\infty}\frac{1}{k!}\mathbf{Z}_k(-iT)^k, \tag{A.22}$$

in which

$$\mathbf{Z}_k \equiv \sum_{n=0}^{\infty}\frac{1}{n!}\mathscr{D}^n\mathbf{m}^{n+k}. \tag{A.23}$$

We immediately note that for $\varepsilon = 0$ all the operators $\mathbf{Z}_k$ are Hermitian owing to the Hermiticity of the operator $\mathbf{m}(E) = \mathbf{g}_{PP}(E + E_0)$ and all its derivatives near the point $E = 0$.

Now we show that

$$\mathbf{Z}_k = \mathbf{Q}^k\mathbf{Z},\quad k = 0, 1, 2, \ldots, \tag{A.24}$$

where $\mathbf{Z} \equiv \mathbf{Z}_0$, and $\mathbf{Q}$ is an operator given by

$$\mathbf{Q} = \sum_{n=0}^{\infty}\frac{1}{n!}\mathscr{D}^n\mathbf{m}\cdot\mathbf{Q}^n. \tag{A.25}$$

For the proof we use the equation

$$\mathscr{D}^n\mathbf{m}^{1+n+k} = \sum_{s=0}^{n}\frac{n!}{s!\,(n-s)!}\mathscr{D}^{n-s}\mathbf{m}\cdot\mathscr{D}^s\mathbf{m}^{n+k},$$

substitution of which into (A.23) gives a system of equations for $\mathbf{Z}_k$:

$$\mathbf{Z}_{k+1} = \sum_{n=0}^{\infty} \frac{1}{n!} \mathcal{D}^n \mathbf{m} \cdot \mathbf{Z}_{n+k} = \mathbf{m}\mathbf{Z}_k + \mathcal{D}\mathbf{m} \cdot \mathbf{Z}_{k+1} + \ldots. \tag{A.26}$$

Let us define the raising operator $\mathbf{Q}_k$ by the expression $\mathbf{Z}_{k+1} = \mathbf{Q}_k\mathbf{Z}_k$. This operator can be constructed as a series in $\mathbf{m}$ using (A.26). From (A.23) we see that the minimum order in $\mathbf{m}$ in $\mathbf{Z}_k$ is k, so the minimum order in $\mathbf{m}$ for $\mathbf{Q}_k$ is the first rather than zeroth order. Therefore, to find $\mathbf{Q}_k$ in lowest order in $\mathbf{m}$ only the first term on the right-hand side of (A.26) need be kept, so in this approximation $\mathbf{Z}_{k+1} = \mathbf{m}\mathbf{Z}_k$ and $\mathbf{Q}_k = \mathbf{m}$. In the next order it is necessary to also include on the right-hand side of (A.26) the contribution of $\mathbf{Z}_{k+1}$ in the approximation found earlier, but the terms with $\mathbf{Z}_{k+2}$ and higher can still be dropped. Therefore, in the first two orders $\mathbf{Z}_{k+1} = \mathbf{m}\mathbf{Z}_k + \mathcal{D}\mathbf{m}\cdot\mathbf{m}\mathbf{Z}_k$, i.e., $\mathbf{Q}_k = \mathbf{m} + \mathcal{D}\mathbf{m}\cdot\mathbf{m}$, and so on. It is important that Eqs. (A.26) are completely sufficient for the iterative construction of all the $\mathbf{Q}_k$. Finally, we should note that the iteration process is exactly the same for all k, so that the resulting operator $\mathbf{Q}_k$ actually is independent of k, from which (A.24) follows.

If we simply substitute the operators $\mathbf{Z}_k$ in the form (A.24) into (A.26), the factor $\mathbf{Q}^k\mathbf{Z}$ cancels and the entire system (A.26) reduces to the single equation (A.25) for $\mathbf{Q}$. We have thereby proved that $\mathbf{Q}_k = \mathbf{Q}$ for all k is the solution of the system (A.26). However, this is actually not a simplification, because we still must verify that the system (A.26) determines all the $\mathbf{Q}_k$ uniquely.

In proving (A.24), we also proved the first of the equations in (A.1), because when (A.24) is substituted into (A.22) the exponential series is obtained. The other equations in (A.1) can be obtained using the known Hermiticity of all the operators $\mathbf{Z}_k$ in (A.22) and the representation (A.24). From this $\mathbf{Z}_k = \mathbf{Q}^k\mathbf{Z} = \mathbf{Z}(\mathbf{Q}^+)^k = \mathbf{Z}_k^+$, which proves (A.2) along with the other expressions in (A.1).

Equation (A.25) is easily solved by iteration. In the first four orders in $\mathbf{m}$ we obtain

$$\mathbf{Q} = \mathbf{m} + \mathbf{m}'\mathbf{m} + \frac{1}{2}\mathbf{m}''\mathbf{m}\mathbf{m} + \mathbf{m}'\mathbf{m}'\mathbf{m} + \mathbf{m}'\mathbf{m}'\mathbf{m}'\mathbf{m} + \frac{1}{2}\mathbf{m}'\mathbf{m}''\mathbf{m}\mathbf{m} +$$

$$+ \frac{1}{2}\mathbf{m}''\mathbf{m}'\mathbf{m}\mathbf{m} + \frac{1}{2}\mathbf{m}''\mathbf{m}\mathbf{m}'\mathbf{m} + \frac{1}{6}\mathbf{m}'''\mathbf{m}\mathbf{m}\mathbf{m} + \ldots, \tag{A.27}$$

where $\mathbf{m}$, $\mathbf{m}'$, $\mathbf{m}''$, and $\mathbf{m}'''$ respectively denote the operator $\mathbf{m}(E)$ and its first three derivatives at the point $E = 0$. Using the definition $\mathbf{m}(E) = \mathbf{g}_{PP}(E + E_0)$

and the explicit expression (A.16) for $\mathbf{g}$, from (A.27) we obtain the perturbation series in powers of $\mathbf{V}$. In the first four orders

$$\mathbf{Q} = \mathbf{PVP} + \mathbf{PVAVP} + \mathbf{PVAVAVP} - \mathbf{PVA^2VPVP} + \mathbf{PVAVAVAVP} - \\ - \mathbf{PVAVA^2VPVP} - \mathbf{PVA^2VAVPVP} - \mathbf{PVA^2VPVAVP} + \\ + \mathbf{PVA^3VPVPVP} + \ldots, \quad \text{(A.28)}$$

where we have written $\mathbf{A} \equiv \mathbf{P}'(E_0 - \mathbf{H}_0)^{-1}$. We note that only the first three terms of (A.27) contribute to (A.28), because for any of the derivatives of $\mathbf{m}$ the expansion begins at second order in $\mathbf{V}$, while for $\mathbf{m}$ it begins at first order. We also give the operator $\mathbf{Z}$ through terms of order $\mathbf{V}^3$:

$$\mathbf{Z} = \mathbf{P} - \mathbf{PVA^2VP} - \mathbf{PVA^2VAVP} - \mathbf{PVAVA^2VP} + \\ + \mathbf{PVA^3VPVP} + \mathbf{PVPVA^3VP} + \ldots.$$

It is useful to note that in the case of a nondegenerate level, when $\mathbf{Q} = \Delta E$ is a simple number, (A.25) can be rewritten as $\Delta E = \mathbf{m}(\Delta E) = \mathbf{g}_{PP}(E_0 + \Delta E)$, since the right-hand side of (A.25) is then an ordinary Taylor series.

Let us now prove the asymptotic expressions (A.3). From (A.10) we see that

$$\mathbf{U}_{P'P}(0, -T) = -i \int_{-T}^{0} dt \int_{-T}^{t} dt' \mathbf{M}_{P'P}(t, t')\, \mathbf{U}_{PP}(t', -T) . \quad \text{(A.29)}$$

In the regularized theory the integrals over t and t' are well convergent, so that to obtain the asymptotic expression we can extend the integration over t and t' to $-\infty$, and replace the operator $\mathbf{U}_{PP}$ by its asymptotic expression $\exp[-i\mathbf{Q}(t' + T)]\cdot\mathbf{Z}$. We obtain

$$\mathbf{U}_{P'P}(0, -T) \cong \mathbf{\Omega}_{P'P} \exp[-i\mathbf{Q}T] \cdot \mathbf{Z}, \quad \text{(A.30)}$$

where

$$\mathbf{\Omega}_{P'P} \equiv -i \int_{-\infty}^{0} dt \int_{-\infty}^{t} dt' \exp(i\mathbf{H}_0 t)\, \mathbf{G}_{P'P}(t - t') \exp[-i(E_0 + \mathbf{Q})t'] . \quad \text{(A.31)}$$

Equation (A.30) together with (A.1) proves the first of the asymptotic expressions in (A.3), because $\mathbf{U}(0, -T)\mathbf{P} = \mathbf{U}_{PP}(0, -T) + \mathbf{U}_{P'P}(0, -T) = [\mathbf{P} + \boldsymbol{\Omega}_{P'P}]\exp[-i\mathbf{Q}T]\cdot\mathbf{Z}$. Knowing the expansions (A.16) and (A.28) for the operators $\mathbf{g}$ and $\mathbf{Q}$, we can use (A.31) to calculate any number of terms of the perturbation series for the wave operator $\boldsymbol{\Omega} = \mathbf{P} + \boldsymbol{\Omega}_{P'P}$. In the first three orders in $\mathbf{V}$ we obtain

$$\begin{aligned}\boldsymbol{\Omega} = {} & \mathbf{P}+\mathbf{AVP}+\mathbf{AVAVP}-\mathbf{A}^2\mathbf{VPVP}+\mathbf{AVAVAVP}-\mathbf{AVA}^2\mathbf{VPVP} \\ & -\mathbf{A}^2\mathbf{VAVPVP}-\mathbf{A}^2\mathbf{VPVAVP}+\mathbf{A}^3\mathbf{VPVPVP}+\dots\end{aligned} \tag{A.32}$$

In agreement with the second equation in (A.6), this series can be obtained by simply discarding the left factor $\mathbf{PV}$ on the right-hand side of (A.28). The expansions (A.28) and (A.32) are, of course, the same as those obtained in the adiabatic theory [83] and in stationary perturbation theory.

The asymptotic representation (A.3) for the second half of the S matrix is obtained in a completely analogous fashion.

If the perturbation expansions are not used, the integration over t and t' in (A.31) can be performed explicitly only when at least one of the two exponential operators in (A.31) is actually a number rather than an operator. This will happen, in particular, for a nondegenerate level: the operator $\mathbf{Q}$ is then a simple number. For a degenerate level the operator $\mathbf{Q}$ can be replaced by its eigenvalue $\omega_\alpha = \Delta E_\alpha$ if we are interested in the action of the wave operator on the eigenvector $\mathbf{P}\Psi_\alpha = \mathbf{Z}\Phi_\alpha$ of the operator $\mathbf{Q}$ (we recall that the Ψ_α are the exact wave functions, $\mathbf{P}\Psi_\alpha = \mathbf{Z}\Phi_\alpha$ are their projections in X_0, and the Φ_α are the eigenvectors of $\mathbf{Q}^+$ in X_0). Replacing in (A.31) $\mathbf{Q}$ by ΔE_α and $\mathbf{G}$ by its spectral representation (A.15), we can perform the integration over t and t' and then integrate over E in (A.15) (residues in the upper E half-plane). It is necessary to retain the regularizing parameter ε in the intermediate expressions, as it ensures the convergence of the time integrals and determines the location of the poles in the E plane; we can take $\varepsilon = 0$ in the final expression. We give only the result:

$$\psi_\alpha = \boldsymbol{\Omega}\mathbf{P}\psi_\alpha = \boldsymbol{\Omega}\mathbf{Z}\Phi_\alpha = \left[\mathbf{P}+\frac{1}{\mathscr{E}_\alpha-\mathbf{H}_0}\mathbf{g}_{P'P}(\mathscr{E}_\alpha)\right]\mathbf{Z}\Phi_\alpha .$$

Here $\mathscr{E}_\alpha \equiv E_0 + \Delta E_\alpha$ are the exact eigenvalues of $\mathbf{H}$ and $\mathbf{g}_{P'P}(\mathscr{E}_\alpha)$ is the projection of the operator $\mathbf{g}(E)$ for $E = \mathscr{E}_\alpha$.

Appendix 2

GRAPHS AND SYMMETRY COEFFICIENTS

Here we give the various types of graphs in the lowest orders (the order corresponds to the number of lines). All the graphs are accompanied by their symmetry coefficients.

A. Mayer graphs which are 1-irreducible in the vertices (stars). There are 10 of these through sixth order:

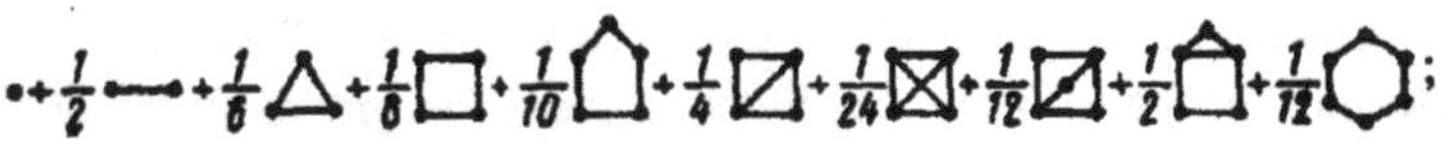

7 in seventh order:

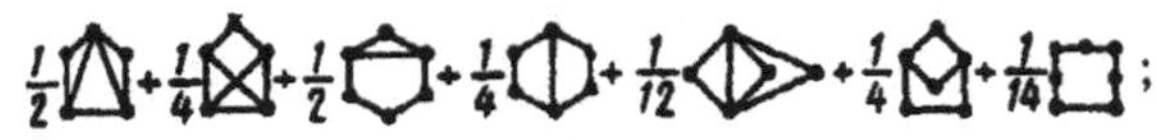

and 16 in eighth order:

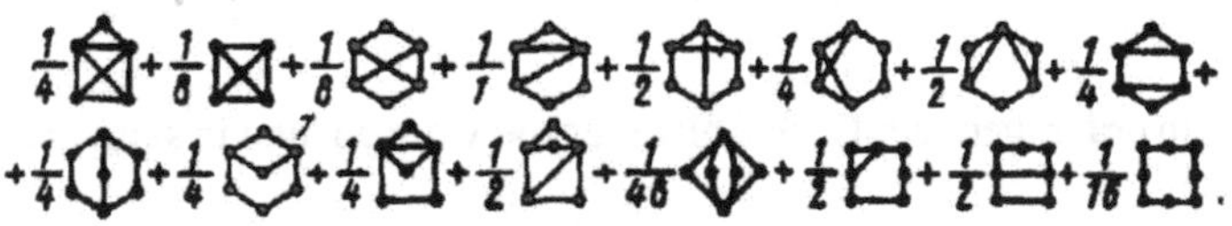

There are 42 such graphs in ninth order; we do not give them here.

These graphs enter into the first Legendre transform for the classical gas (the virial expansion) and the Ising model.

B. All graphs which are 2-irreducible in the vertices. Through ninth order they are:

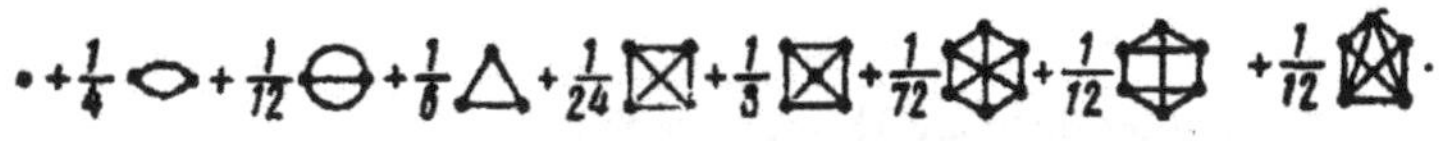

They enter into the second Legendre transform for the gas.

C. All connected Mayer graphs, i.e., graphs without self-contracted, double, triple, etc. lines. They are obtained from the graphs of group A by constructing "star trees" [8]:

in which each of the blocks represents the sum of all the graphs of group A.

Through fourth order there are 11 such graphs:

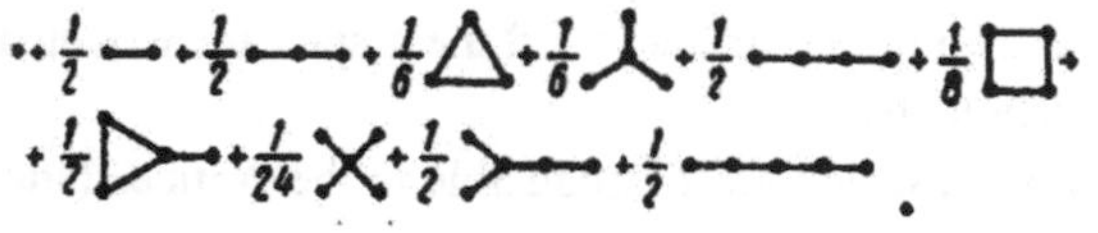

In fifth order there are 12, and in sixth order there are 30.

D. Connected graphs with ordinary (not Mayer) lines, but without self-contracted lines. They are obtained from the graphs of group C by replacing each line by

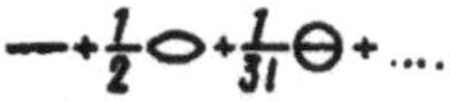

Through third order there are 9 of them:

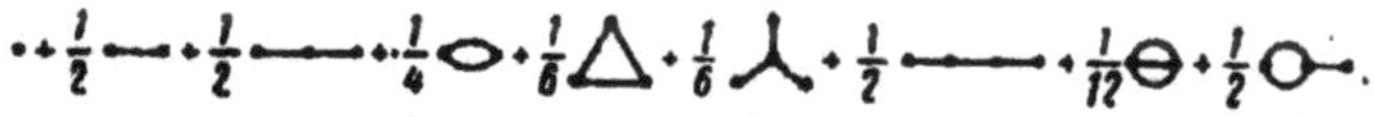

In fourth order there are 12, in fifth order there are 33, and in sixth order there are 102.

E. All connected graphs (the logarithm of the *S* matrix). These are obtained from the graphs of group D by replacing the reduced vertices (see Sec. 1.4.1) by ordinary ones, i.e., by adding to each vertex self-contracted lines according to the rule

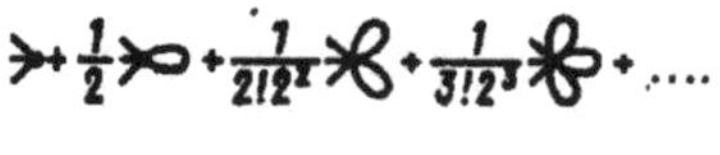

Through third order there are 18 such graphs:

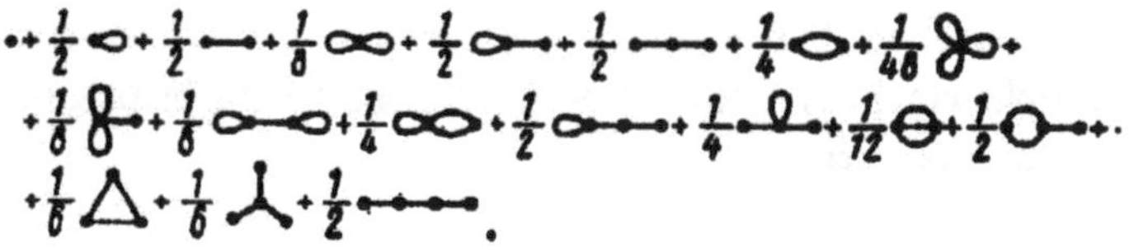

In fourth order there are 30, in fifth there are 95, and in sixth there are several hundred.

F. The graphs of the generating functional of connected Green functions $W(A)$ for a theory with triple and quadruple bare vertices. By definition, $W(A) = \ln G(A)$, where $A \equiv \{A_n,\ n = 1, 2, 3, 4\}$ and $G(A) = \text{const} \int D\varphi \exp(\Sigma A_n \varphi^n / n!)$. It is known (see Sec. 1.7.4) that $W(A) = \text{const} + 1/2 \cdot \text{tr} \ln \Delta$ plus the sum of all connected graphs with line $\Delta = -A_2^{-1}$ and vertex factors A_n with $n = 1, 3, 4$. They are obtained by selecting the graphs of group E which have only vertices with $n = 1, 3$, and 4, where n is the number of lines converging at a vertex. Through fourth order there are 15 such graphs:

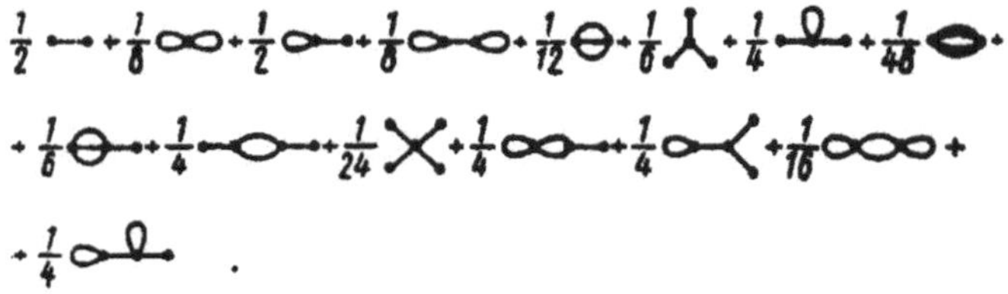

G. The graphs of the first Legendre transform of the functional $W(A)$. By definition, $\Gamma(\alpha; A_2, A_3, A_4) = W(A) - \alpha A_1$, where $\alpha = \delta W / \delta A_1$. It is known (see Sec. 1.8.3) that $\Gamma = \text{const} + 1/2 \cdot \text{tr} \ln \Delta - 1/2 \cdot \alpha \Delta^{-1} \alpha$ plus the sum of all 1-irreducible graphs of the functional $W(A_1 \to \Delta^{-1}\alpha, A_2, A_3, A_4)$ without the first of the graphs in group F. The graphs of $\Gamma(\alpha; A_2, A_3, A_4)$ are obtained from the graphs of group F by replacing the "rays" ΔA_1 connected to vertices with $n = 3, 4$ by legs α connected directly to the vertices and discarding the resulting graphs which turn out to be 1-reducible. Through third order there are 17 such graphs:

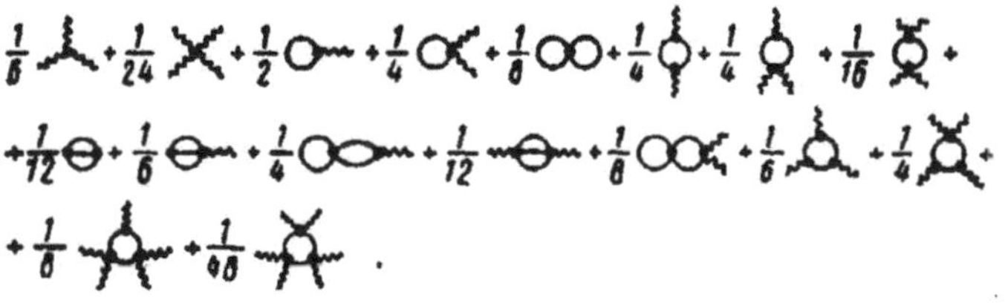

In fourth order there are another 20 of them.

For the second, third, and fourth Legendre transforms we discard those graphs of group G which do not possess the needed irreducibility properties. This greatly reduces the number of graphs. We shall not give the graphs of the higher transforms here, because they can be found in the relevant sections of Chap. 6 (for the second transform they are given through fifth order, and for the third and fourth transforms they are given through eighth order).

References

1. Bogolyubov, N. N. and Shirkov, D. V. (1973) *Introduction to the Theory of Quantized Fields*, 3rd ed. Wiley, New York.
2. Berezin, F. A. (1966) *Method of Second Quantiation*. Academic Press, New York.
3. Hori, S. (1952) On the well ordered *S* matrix. *Prog. Theor. Phys.* **7**, 578.
4. Konopleva, N. P. and Popov, V. N. (1972) *Gauge Fields*. Harwood, Chur.
5. Harary, F. (1969) *Graph Theory*. Addison-Wesley, Reading, Mass.
6. Vasil'ev, A. N. and Radzhabov, R. A. (1974) Legendre transformations in the Ising model. *Theor. Math. Phys. (USSR)* **21**, 963.
7. Mayer, J. E. and Mayer, M. G. (1977) *Statistical Mechanics*, 2nd ed. Wiley, New York.
8. Uhlenbeck, G. E. and Ford, G. W. (1963) *Lectures in Statistical Physics*. American Mathematical Society, Providence, R.I.
9. Efimov, G. V. (1974) Proof of unitarity of the *S* matrix. Preprint R2-8340, JINR, Dubna [in Russian].
10. Umezawa, H. and Takahashi, Y. (1953) The general theory of the interaction representation. Part I: The local field. *Prog. Theor. Phys.* **9**, 14; (1954) Part II: General fields and interactions. *Prog. Theor. Phys.* **9**, 501.
11. Medvedev, B. V., Pavlov, V. P., and Sukhanov, A. D. (1971) The Lagrangian and the scattering matrix. *Theor. Math. Phys. (USSR)* **8**, 737.
12. Pavlov, V. P. and Tavluev, G. A. (1971) The interaction Hamiltonian in quantum field theory. *Theor. Math. Phys. (USSR)* **9**, 993.
13. Feynman, R. P. and Hibbs, A. R. (1965) *Quantum Mechanics and Path Integrals*. McGraw-Hill, New York.
14. Berezin, F. A. (1971) Non-Wiener functional integrals. *Theor. Math. Phys. (USSR)* **6**, 141.
15. Alimov, A. L. (1974) *The Feynman Path Integral*. Author's Abstract of Candidate's Dissertation, Leningrad State University, Leningrad [in Russian].
16. Englert, F. and de Dominicis, C. (1968) Lagrangian field theory of Green's functions. Part I. Generalized perturbation theory and inversion theorems. *Nuovo Cimento* **53A**, 1007; Part. II. Theory of self-generating interactions. *Nuovo Cimento* **53A**, 1021.
17. Englert, F. and de Dominicis, C. (1967) Potential-correlation function duality in statistical mechanics. *J. Math. Phys.* **8**, 2143.
18. Schwinger, J. (1951) On the Green's functions of quantized fields. *Proc. Natl. Acad. Sci. U.S.A.* **37**, 452.
19. Kadanoff, L. P. and Baym, G. (1964) *Quantum Statistical Mechanics: Green's Function Methods in Equilibrium and Nonequilibrium Problems*. Benjamin, New York.
20. Goldstone, J. (1961) Field theories with "superconductor" solutions. *Nuovo Cimento* **19**, 154.
21. Abrikosov, A. A., Gor'kov, L. P., and Dzyaloshinskii, I. E. (1962) *Methods of Quantum Field Theory in Statistical Physics*. Prentice-Hall, Englewood Cliffs, N. J.
22. McWeeny, R. and Sutcliffe, B. T. (1969) *Methods of Molecular Quantum Mechanics*. Academic Press, New York.
23. Thouless, D. J. (1972) *The Quantum Mechanics of Many-Body Systems*, 2nd ed. Academic Press, New York.
24. Feynman, R. (1948) Space-time approach to nonrelativistic quantum mechanics. *Rev. Mod. Phys.* **20**, 367.

25. Yang, C. N. and Mills, R. L. (1954) Conservation of isotopic spin and isotopic gauge invariance. *Phys. Rev.* **96**, 191.
26. Faddeev, L. D. (1969) The Feynman integral for singular Lagrangians. *Theor. Math. Phys. (USSR)* **1**, 1.
27. Popov, V. N. and Faddeev, L. D. (1967) Feynman diagrams for the Yang–Mills field. *Phys. Lett.* **25B**, 29.
28. Popov, V. N. and Faddeev, L. D. (1967) Perturbation theory for gauge- invariant fields. Preprint ITF-67-36, Institute of Theoretical Physics, Kiev [in Russian].
29. Vasil'ev, A. N. and Pis'mak, Yu. M. (1975) The *S*-matrix generating functional in gauge field theory. Part I. *Vestnik Leningrad. Univ.* No. 10, 7 [in Russian]; Part II. The massless Yang–Mills field. *Vestnik Leningrad. Univ.* No. 16, 7 [in Russian].
30. Slavnov, A. A. (1972) Ward identities in gauge theories. *Theor. Math. Phys. (USSR)* **10**, 99.
31. 't Hooft, G. (1971) Renormalization of massless Yang–Mills fields. *Nucl. Phys. B* **33**, 173.
32. Osterwalder, K. and Schrader, R. (1973) Axioms for Euclidean Green's functions. *Commun. Math. Phys.* **32**, 83.
33. Gel'fand, I. M. and Vilenkin, N. Ya. (1964) *Generalized Functions*, Vol. 4. Academic Press, New York.
34. Matsubara, T. (1955) A new approach to quantum statistical mechanics. *Prog. Theor. Phys.* **14**, 351.
35. Martin, P. C. and Schwinger, J. (1959) Theory of many-particle systems. Part I. *Phys. Rev.* **115**, 1342.
36. Fradkin, E. S. (1965) The Green function method in the theory of quantized fields and in quantum statistics. *Proc. of the Physics Institute, USSR Academy of Sciences* **29**, 7 [in Russian].
37. Landau, L. D. and Lifshits, E. M. (1980) *Statistical Physics*, Parts 1 and 2, 3rd ed. Pergamon Press, Oxford.
38. Kac, M. (1959) *Probability and Related Topics in Physical Sciences.* Interscience, London.
39. Ginibre, J. (1965) Reduced density matrices of quantum gases. Part I. Limit of infinite volume. *J. Math. Phys.* **6**, 238; Part II. Cluster property. *J. Math. Phys.* **6**, 252.
40. Ruelle, D. (1969) *Statistical Mechanics: Rigorous Results*. Benjamin, New York.
41. Brout, R. H. (1965) *Phase Transitions*. Benjamin, New York.
42. Fisher, M. (1967) The theory of equilibrium critical phenomena. *Rep. Prog. Phys.* **30**, 615.
43. Stanley, H. E. (1971) *Introduction to Phase Transitions and Critical Phenomena*. Clarendon Press, Oxford.
44. Izyumov, Yu. A., Kassan-ogly, F. A., and Skryabin, Yu. N. (1974) *Field Methods in Ferromagnetism*. Nauka, Moscow [in Russian].
45. Stinchcombe, R., Horwitz, G., Englert, F., and Brout, R. (1963) Thermodynamic behavior of the Heisenberg ferromagnet. *Phys. Rev.* **130**, 155.
46. Radzhabov, R. A. (1975) Extension of the Mayer theorem to a gas with many-body forces. *Theor. Math. Phys. (USSR)* **23**, 484.
47. Patashinskii, A. Z. and Pokrovskii, V. L. (1979) *Fluctuation Theory of Phase Transitions*. Pergamon Press, Oxford.
48. Dyson, F., Montroll, E., Kac, M., and Fisher, M. (1973) *Stability and phase transitions* [Russian translation from the English].
49. Vasil'ev, A. N. (1973) Consequences of convexity of Legendre transformations. *Theor. Math. Phys. (USSR)* **15**, 550.
50. Goldstone, J., Salam, A., and Weinberg, S. (1962) Broken symmetries. *Phys. Rev.*

127, 965.

51. Bogolyubov, N. N. (1961) Quasi-averages in problems of statistical mechanics. Preprint R-1451, JINR, Dubna [in Russian].
52. de Dominicis, C. and Martin, P. C. (1964) Stationary enthropy principle and renormalization in normal and superflulid systems. Part I. Algebraic formulation. *J. Math. Phys.* **5**, 14; Part II. Diagrammatic formulation. *J. Math. Phys.* **5**, 31.
53. Belyaev, S. T. (1958) Application of the methods of quantum field theory to a system of Bose particles. *Sov. Phys. JETP* **7**, 289.
54. Lee, T. D. and Yang, C. N. (1959) Many-body problem in quantum statistical mechanics. Part I. General formulation. *Phys. Rev.* **113**, 1165; (1960) Part IV. Formulation in terms of average occupation number in momentum space. *Phys. Rev.* **117**, 22.
55. Luttinger, J. M. and Ward, J. C. (1960) Ground-state energy of a many-fermion system. Part II. *Phys. Rev.* **118**, 1417.
56. de Dominicis, C. (1962) Variational formulations of equilibrium statistical mechanics. *J. Math. Phys.* **3**, 983.
57. Jona-Lasinio, G. (1964) Relativistic field theories with symmetry- breaking solutions. *Nuovo Cimento* **34**, 1790.
58. Dahmen, H. D. and Jona-Lasinio, G. (1967) Variational formulation of quantum field theory. Part I. *Nuovo Cimento* **52A**, 807; (1969) Part II. A study of a functional derivative equation related to the gA^3 theory. *Nuovo Cimento* **62A**, 889.
59. Vassiliev, A. N. (1972) Studies of broken symmetries using functional methods. In *Cargèse Lectures in Physics*, Vol. 5, edited by D. Bessis, Gordon and Breach, New York [in French].
60. Vasil'ev, A. N. and Kazanskii, A. K. (1973) Equations of motion for a Legendre transformation of arbitrary order. *Theor. Math. Phys. (USSR)* **14**, 215.
61. Vasil'ev, A. N., Kazanskii, A. K., and Pis'mak, Yu. M. (1974) Equations for higher Legendre transformations in terms of 1-irreducible vertices. *Theor. Math. Phys. (USSR)* **19**, 443.
62. Vasil'ev, A. N. and Kazanskii, A. K. (1972) Legendre transformations of generating functionals in quantum field theory. *Theor. Math. Phys. (USSR)* **12**, 875.
63. Pis'mak, Yu. M. (1974) Proof of 3-irreducibiliy of the third Legendre transformation. *Theor. Math. Phys. (USSR)* **18**, 211.
64. Vasil'ev, A. N., Kazanskii, A. K., and Pis'mak, Yu. M. (1974) Diagrammatic analysis of the fourth Legendre transformation. *Theor. Math. Phys. (USSR)* **20**, 754.
65. Pis'mak, Yu. M. (1974) Higher Legendre transformations in quantum field theory. *Vestnik Leningr. Univ.* No. 16, 130 [in Russian].
66. Pis'mak, Yu. M. (1975) Combinatorial analysis of the overlap problem for vertices with more than four legs. Part I. Some topological properties of the Feynman graphs. *Theor. Math. Phys. (USSR)* **24**, 649; Part II. Higher Legendre transformations. *Theor. Math. Phys. (USSR)* **24**, 755.
67. Higgs, P. W. (1966) Spontaneous symmetry breakdown without massless bosons. *Phys. Rev.* **145**, 1156.
68. Kibble T. W. B. (1967) Symmetry breaking in non-Abelian gauge theories. *Phys. Rev.* **155**, 1554.
69. Ginzburg, V. L. and Landau, L. D. (1950) On the theory of superconductivity. *Zh. Eksp. Teor. Fiz.* **20**, 1064 [in Russian].
70. Bloch, C. (1965) Diagram expansions in quantum statistical mechanics. In: *Studies in Statistical Mechanics*, Vol. 3, edited by J. De Boer and

G. E. Uhlenbeck. North-Holland, Amsterdam.
71. Larkin, A. I. and Khmel'nitskii, D. E. (1969) Phase transition in uniaxial ferroelectrics. *Sov. Phys. JETP* **29**, 1123.
72. Polyakov, A. M. (1968) Microscopic description of critical phenomena. *Sov. Phys. JETP* **28**, 533.
73. Polyakov, A. M. (1969) Properties of long- and short-range correlations in the critical region. *Sov. Phys. JETP* **30**, 151.
74. Polyakov, A. M. (1970) Conformal symmetry of critical fluctuations. *JETP Lett.* **12**, 381.
75. Vasil'ev, A. N. and Radzhabov, R. A. (1975) Analysis of the non-star graphs of the Legendre transformation in the Ising model. *Theor. Math. Phys. (USSR)* **23**, 575.
76. Bogolyubov, N. M., Vasil'ev, A. N., Korzhenevskii, A. L., and Radzhabov, R. A. (1976) Numerical values and symmetry coefficients of Mayer star graphs up to eighth order inclusive for various lattices. *Vestnik Leningrad. Univ.* No. 4, 7 [in Russian].
77. Bogolyubov, N. M., Brattsev, V. F., Vasil'ev, A. N., Korzhenevskii, A. L., and Radzhabov, R. A. (1976) High-temperature expansions for arbitrary magnetization in the Ising model. *Theor. Math. Phys. (USSR)* **26**, 230.
78. Bloch, C. and Langer, J. S. (1965) Diagram renormalization, variational principles, and the infinite-dimensional Ising model. *J. Math. Phys.* **6**, 554.
79. Morita, T. and Hiroike, K. (1961) A new approach to the theory of classical fluids. Part III. *Prog. Theor. Phys.* **25**, 537.
80. Krivnov, V. Ya., Ol'khov, O. A., Provotorov, B. N., and Sarychev, M. E. (1970) The self-consistent field near the critical point in the antiferromagnetic Ising model. *Theor. Math. Phys. (USSR)* **2**, 177.
81. Gell-Mann, M. and Low, F. (1951) Bound states in quantum field theory. *Phys. Rev.* **84**, 350.
82. Tolmachev, V. V. (1963) *The Field Form of Perturbation Theory For the Many-Electron Problem in Atoms and Molecules*. Tartu University Press, Tartu, Estonia [in Russian].
83. Tolmachev, V. V. (1973) *The Theory of the Fermi Gas*. Moscow State University Press, Moscow [in Russian].
84. Vasil'ev, A. N. and Kitanin, A. L. (1975) Nonstationary perturbation theory for energy shifts of a degenerate level. *Theor. Math. Phys. (USSR)* **24**, 786.
85. Kitanin, A. L. (1975) On nonstationary perturbation theory for a degenerate discrete level. *Theor. Math. Phys. (USSR)* **25**, 1224.
86. Hubbard, J. (1957) The description of collective motions in terms of many-body perturbation theory. *Proc. R. Soc. London A* **240**, 539.
87. Campbell, W. B., Finkler, P., Jones, C. E., and Misheloff, M. N. (1975) Path-integral formulation of scattering theory. *Phys. Rev. D* **12**, 2363.
88. Wilson, K. G., Kogut, J. B. (1974) The renormalization group and ε-expansion. Phys. Rep. 12C, 75.

Subject Index

This index does not include terms contained in the table of contents

Milton Keynes UK
Ingram Content Group UK Ltd.
UKHW021623071024
449327UK00020BA/1170

9 780367 579265